Applications and Engineering of Monoclonal Antibodies

Applications and Engineering of Monoclonal Antibodies

DAVID J. KING

Celltech Therapeutics
Slough, UK

TAYLOR & FRANCIS
ALERE FLAMMAM
1798 – 1998

UK Taylor & Francis Ltd, 1 Gunpowder Square, London EC4A 3DE
USA Taylor & Francis Inc., 325 Chestnut Street, 8th Floor, Philadelphia, PA 19106

British Library Cataloguing-in-Publication Data

A catalogue record for this book is available from the British Library

ISBN 0–7484–0422–8 (hb)
0–7484–0423–6 (pb)

Library of Congress Cataloging Publication Data are available

Cover design by Youngs Design in Production

Typeset in Times 10/12pt by Graphicraft Limited, Hong Kong

Printed by T.J. International Ltd, Padstow, UK

Acknowledgements

I would like to thank my many colleagues at Celltech for their helpful suggestions and advice, and particularly Alastair Lawson and Martyn Robinson for their comments on the manuscript. Thanks also to Tina Jones for assistance with the figures.

And lastly, sincere thanks to my wife, Jane, for her support and patience through the long hours and weekends spent writing this book.

Contents

1 **Preparation, structure and function of monoclonal antibodies** *page* 1
1.1 Introduction 1
1.2 The role of antibodies in the immune response 2
1.3 Structure and function of antibodies 2
1.4 The organisation of antibody genes 11
1.5 Antigen-binding affinity and avidity 13
1.5.1 Affinity 13
1.5.2 Avidity 14
1.6 Generation of monoclonal antibodies 14
1.6.1 Hybridoma technology 14
1.6.2 Human monoclonal antibodies 18
1.6.3 Human MAbs from transgenic mice 20
1.6.4 Isolation of antibodies by phage display 21

2 **Antibody engineering: design for specific applications** 27
2.1 Introduction 27
2.2 Isolation of variable region genes 28
2.3 Overcoming immunogenicity 29
2.3.1 Chimeric and humanised antibodies 29
2.3.2 Antibody fragments to reduce immunogenicity 37
2.3.3 Chemical modification to reduce immunogenicity 38
2.3.4 Immunosuppressive therapy 40
2.4 Antibody fragments 40
2.4.1 Antibody fragments from proteolysis of IgG 40
2.4.2 Recombinant antibody fragments 42
Fab-based fragments 43
Fv-based fragments 45
Multivalent antibody fragments 48
2.5 Antibodies with multiple specificities 50
2.6 Engineering effector functions 54

2.6.1 Engineering natural effector functions 54
2.6.2 Attachment of diagnostic or therapeutic agents 58
Chemical conjugates 58
Site-specific attachment 63
Fusion proteins 65
2.7 Engineering pharmacokinetics and biodistribution 67
2.7.1 Pharmacokinetics of IgG 68
2.7.2 Pharmacokinetics of antibody fragments 71
2.7.3 Clearance 74
2.7.4 Chemical modification 75
2.7.5 Fc region to extend half-life 75

3 Monoclonal antibodies in research and diagnostic applications 77
3.1 Introduction 77
3.2 Immunoassays in diagnostics and research 77
3.2.1 Radioimmunoassay 78
3.2.2 Immunoradiometric assay 79
3.2.3 Non-isotopic immunoassays 83
3.2.4 Improving sensitivity 85
3.2.5 Assay formats 89
3.2.6 Advantages of monoclonal antibodies in immunoassay 91
3.3 Immunosensors 92
3.3.1 Mass-detecting immunosensors 93
3.3.2 Electrochemical immunosensors 94
3.3.3 Optical immunosensors 95
3.4 Immunocytochemistry 97
3.5 Flow cytometry and cell sorting (FACS) 99
3.6 Western blotting (immunoblotting) 100
3.7 Immunopurification 102
3.8 Antibodies in structural biology 106
3.9 *In vivo* diagnostics 106
3.9.1 Radioimmunodetection of human tumours 107
Tumour-associated antigens 107
Form of antibody 109
Radioisotopes 110
Two- and three-step targeting approaches 114
3.9.2 Radioimmunoguided surgery 116
3.9.3 Non-tumour radioimmunodetection 116

4 Monoclonal antibodies in therapeutic applications 119
4.1 Introduction 119
4.2 Cancer 120
4.2.1 Cancer therapy with unmodified (naked) antibodies 121
4.2.2 Anti-idiotype antibodies 123
4.2.3 Bispecific antibody-mediated effector cell targeting 125
4.2.4 Other approaches to recruit the immune system using MAbs 127
4.2.5 Radioimmunotherapy 128
Form of antibody 129
Radioisotopes 132

Clearance mechanisms and two-step targeting 136
Clinical results with RIT 137
4.2.6 Immunotoxins 139
4.2.7 Drug conjugates 142
4.2.8 Antibody-directed enzyme prodrug therapy (ADEPT) 144
4.2.9 Vascular targeting 146
4.3 Infectious disease 147
4.3.1 Antiviral antibodies 147
4.3.2 Bacterial sepsis 148
4.4 Cardiovascular disease 149
4.4.1 Inhibition of platelet aggregation 149
4.4.2 Thrombolysis 150
4.5 Disorders of the immune system/inflammatory diseases 151
4.5.1 The inflammatory response 151
4.5.2 Blocking inflammatory mediators 153
Anti-TNF antibodies in rheumatoid arthritis and inflammatory bowel disease 153
Anti-C5 154
4.5.3 Blocking adhesive interactions 154
4.5.4 Antibodies which directly inhibit T cell activation and proliferation 155
4.5.5 Antibody treatment of allergy 158

5 Production of monoclonal antibodies 161
5.1 Introduction 161
5.2 Expression of antibodies in mammalian cells 161
5.2.1 Transient expression systems 162
5.2.2 Stable expression systems 163
5.2.3 Expression of antibody fragments in mammalian cells 167
5.3 Expression in *Escherichia coli* 167
5.3.1 Intracellular expression of antibody fragments in *E. coli* 168
5.3.2 Secretion of antibody fragments from *E. coli* 170
5.4 Expression in other microbial systems 172
5.5 Expression in plants 173
5.6 Production in transgenic animals 174
5.7 Expression in insect cells 175
5.8 Production of monoclonal antibodies – cell culture 175
5.9 Purification of monoclonal antibodies 176
5.9.1 Purification of IgG 176
5.9.2 Purification of IgM 181
5.9.3 Purification of monoclonal antibody fragments 181
5.9.4 Purification for therapeutic use 185

6 Prospects for engineered antibodies in biotechnology 187
6.1 Gene therapy 187
6.1.1 Intracellular antibodies 187
6.1.2 Other applications of MAbs in gene therapy 188
6.2 Applications of antibodies in plants 189
6.3 Catalytic antibodies 189

6.4 Towards drug design 190
6.5 Improving affinity 191
6.6 Summary and prospects 192

7 References 195

Index 241

1

Preparation, Structure and Function of Monoclonal Antibodies

1.1 Introduction

Antibodies are proteins produced by an individual in response to the presence of a foreign molecule in the body. These foreign molecules are known as antigens, and they usually result from invading organisms such as bacteria, fungi or viruses. Antibodies bind to antigens and elicit a range of effector mechanisms to destroy the invading organism. Therefore, the generation of an antibody response is a key step in the immune system which has evolved to protect individuals from invading pathogenic organisms. However, antibodies are not restricted in specificity to pathogens, but can be formed to a huge variety of antigens including proteins, carbohydrates and organic compounds, including totally novel structures.

In 1975 Kohler and Milstein described a method for the 'production of antibodies of predefined specificity'. This technical breakthrough allowed, for the first time, the production of antibody molecules of a single specificity which could be characterised and defined. Such monoclonal antibodies immediately became valuable research tools, and applications in the diagnosis and therapy of human disease began to be widely investigated. Monoclonal antibodies (MAbs) have become increasingly important as diagnostic agents allowing precise molecular structures to be mapped and analysed. However, initial enthusiasm for their development as therapeutic agents was premature and many problems limited their use in humans (see Chapter 4). A second technical revolution has now arrived in the ability to manipulate antibody genes and to design and produce antibody molecules tailor-made for their application. Such redesigned antibody molecules are now rapidly becoming valuable reagents for therapy of human diseases as well as improved diagnostics and research reagents. The scope of these applications, and how antibody molecules are redesigned, or engineered, in the most suitable form for a particular application is the subject of this book. To understand the nature of this process requires an understanding of antibody structure and function and of how antibody specificities are generated.

1.2 The role of antibodies in the immune response

The immune system can be considered in two parts. Innate immunity is mediated by a variety of physical and biochemical barriers and by cells responding non-specifically to the foreign organism or molecule. Innate immunity is characterised by a similar reponse on re-exposure to the same foreign agent. Adaptive immunity, the second part of the immune system, is mediated by cells termed lymphocytes and is characterised by improved efficiency on re-exposure to the same foreign agent. It is in this system that antibodies play a major role.

Two major types of lymphocytes are involved in the adaptive immune response, T lymphocytes and B lymphocytes. T lymphocytes can be subdivided into cytotoxic T lymphocytes and 'helper' T lymphocytes. Cytotoxic T lymphocytes bind to foreign or infected cells through a surface antigen receptor, the T cell receptor, and lyse them. Helper T lymphocytes play a regulatory role in controlling the response of both T and B lymphocytes. B lymphocytes exert their effect through producing antibody molecules which bind to the foreign agent and invoke specific mechanisms for its elimination.

Antigen is recognised by B lymphocytes through the use of antibody molecules on the surface of the cell. Each B lymphocyte carries antibody molecules on its surface with a single specificity as a consequence of the rearrangement of immunoglobulin genes by the individual cell during its development in a random, antigen-independent process. Each individual will have many millions of B cells at one time thus comprising a 'polyclonal' population. When an antigen enters the body it will be recognised by any B cell which has antibody molecules able to bind to antigenic determinants (epitopes) present on the antigen. This recognition leads to activation of the cell leading to proliferation and differentiation (Figure 1.1).

In the case of protein antigens, bound antigen molecules are internalised into the B cell and degraded into peptides. Some of these antigen peptides are then bound to major histocompatibility complex (MHC) class II molecules to form a complex which is transported to the cell surface and 'presented' by the B cell. Although B cells are relatively weak at antigen presentation compared to other cell types such as dendritic cells, these complexes can be recognised by receptors present on T lymphocytes, the T cell receptors. If appropriate recognition takes place then the T cell may deliver 'help' in the form of signals back to the B cell to stimulate proliferation and differentiation. This process therefore gives rise to a 'T-cell-dependent B-cell response'.

Proliferation results in a clone of identical cells which can differentiate to form either plasma cells capable of secreting large amounts of soluble antibody of the same specificity as the original activated B cell, or memory cells which mount an accelerated immune response on re-exposure to the original antigen. The memory cell-mediated 'secondary' response results in the production of higher affinity antibodies due to hypermutation in the immunoglobulin gene loci followed by antigen driven selection, a process known as affinity maturation. This process of B cell clonal selection and the generation of antibody responses is covered in more detail in an excellent review (Rajewsky, 1996).

1.3 Structure and function of antibodies

Antibody molecules have two principal functions, firstly to bind to antigen and secondly to trigger its elimination from the body. Antibodies are therefore 'adaptor' molecules which have both the ability to bind to the antigen molecule and the ability to bind and

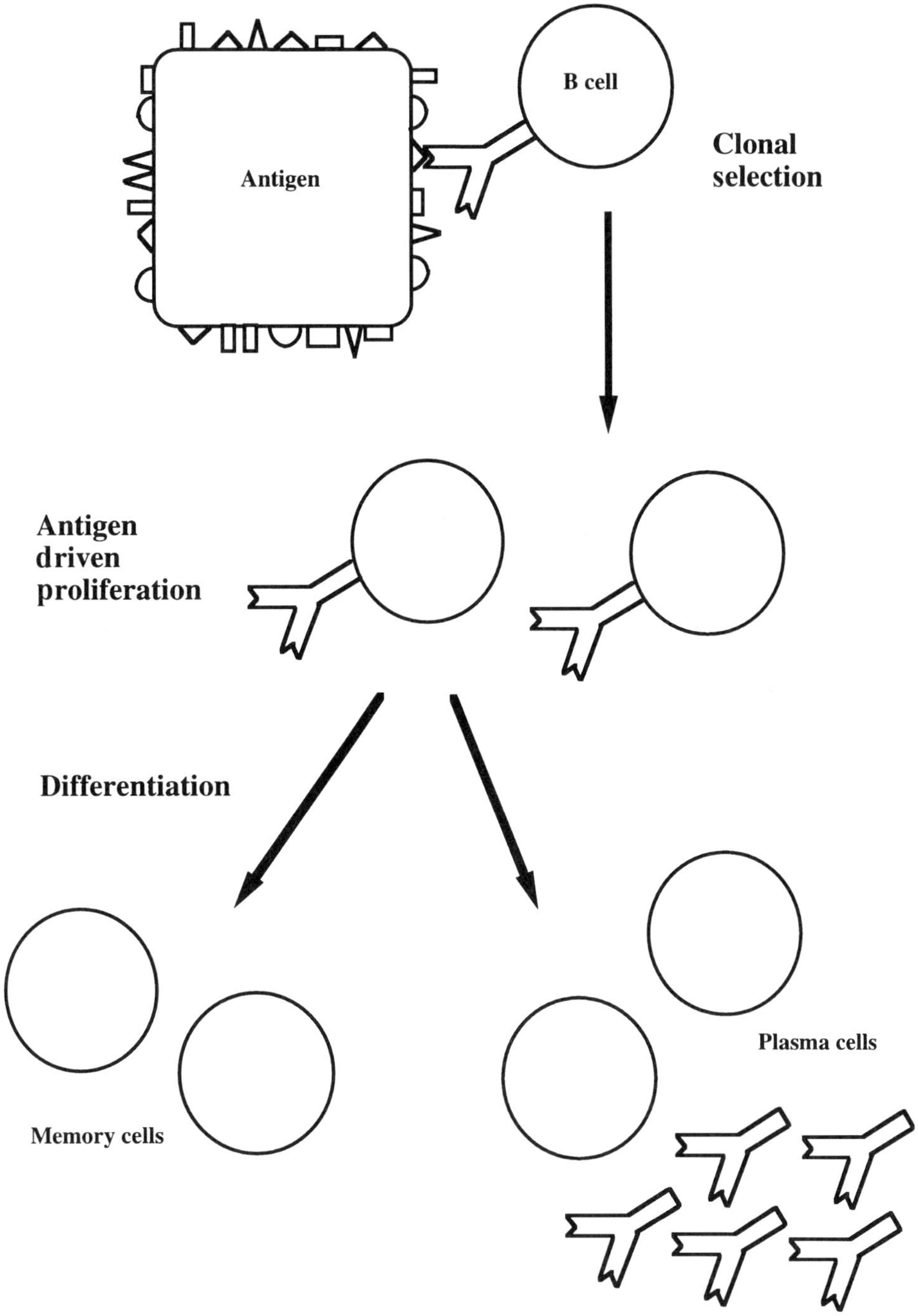

Figure 1.1 Differentiation of B cells to antibody-secreting plasma cells

bring into action molecules of the effector system which can then remove the foreign material. For effective defence systems to operate it is essential that an individual is able to recognise a wide variety of foreign material and thus antibody molecules of many different binding specificities are required. However, in each case the antibody must also

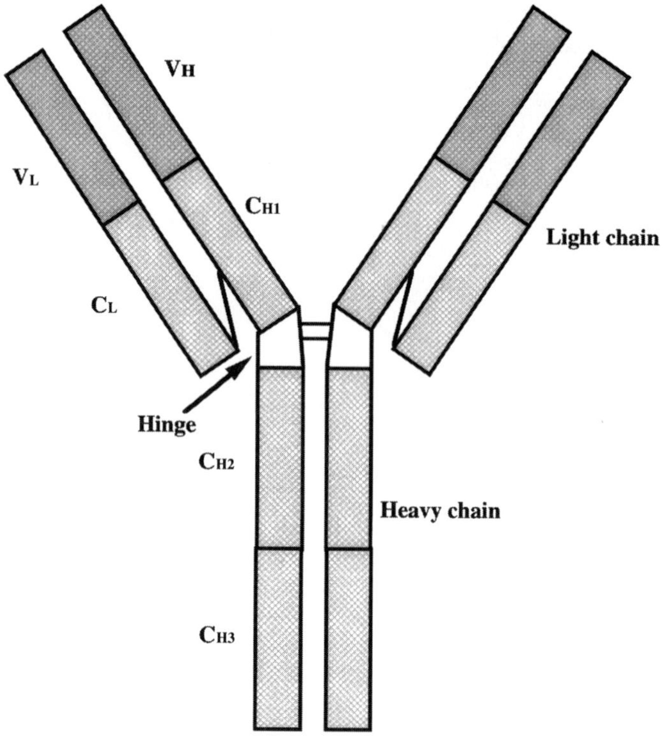

Figure 1.2 Representation of the organisation of protein chains of an IgG molecule

be able to trigger the same effector systems. To achieve this the antibody molecule has evolved variable regions which vary in protein sequence and structure to accommodate the development of binding specificities to a wide range of different antigens, and constant regions which are largely the same in each antibody. The constant regions maintain a common structure of the molecule and allow interaction with the effector systems such as complement binding and binding to Fc receptors on macrophages to activate phagocytosis.

Antibody molecules are also known as immunoglobulins. The term 'immunoglobulin' applies to the antibody protein whether or not the binding specificity of the molecule is characterised, whereas an antibody is an antigen-specific immunoglobulin. In practice the two terms are usually used interchangeably.

Higher mammals have five classes (isotypes) of immunoglobulin, termed IgG, IgM, IgA, IgE and IgD. The most abundant of these, which is used in most applications of antibodies, is the IgG class, the main class of antibody generated by the secondary immune response. The IgG molecule consists of four polypeptide chains, two heavy chains of approximately 50 kDa and two light chains of approximately 25 kDa (Figure 1.2). Each of these is divided into discretely folded structural domains of approximately 110 amino acids stabilised by an internal disulphide bond. These are linked together by short regions of comparatively flexible protein chain which allow movement of the domains relative to one another. Each light chain comprises a variable domain (VL) and a constant domain (CL), and each heavy chain a variable domain (VH) and three constant domains (CH1, CH2 and CH3). The heavy and light chain variable domains associate to form the antigen-binding site. The IgG molecule thus has two antigen-binding sites and is capable of

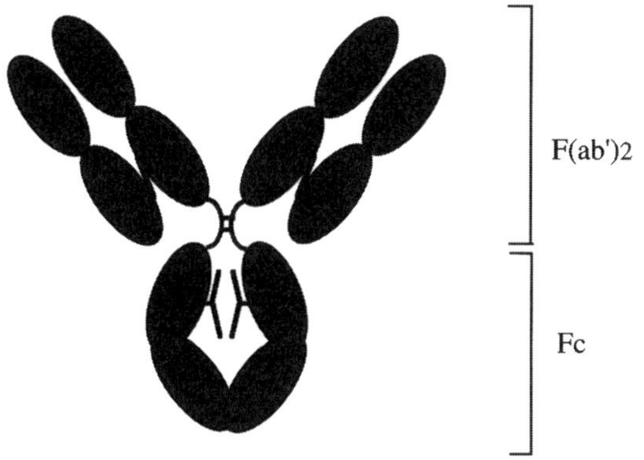

Figure 1.3 Domain structure of IgG, demonstrating the antigen-binding F(ab′)$_2$ region and the Fc region with carbohydrate attached at the CH2 domain

binding antigens divalently to allow high binding avidity (see Section 1.5). The heavy and light chains are linked by a disulphide bond between the CL and CH1 domains. A flexible region, known as the hinge, links the CH1 and CH2 domains of each heavy chain and it is at this point that the two heavy chains are linked by disulphide bonds. The CH2 domain is normally glycosylated and the CH3 domains of each heavy chain associate with each other by non-covalent interactions.

Much early work on the structure and function of antibodies made extensive use of proteolysis of the antibody protein and this has led to commonly used terminology for the different regions of the antibody (Figure 1.3). The area around the antibody hinge is more susceptible to proteolysis than the tightly folded domains and thus this is the point at which proteolytic cleavage usually takes place. Proteolysis above the disulphide bonds in the hinge region results in monovalent Fab (fragment antigen-binding) fragments which comprise light chain together with the N-terminal two domains of the heavy chain. The Fab fragment thus contains one functional antigen binding site. Proteolysis immediately below one or more of the hinge disulphide bonds results in the divalent F(ab′)$_2$ fragment. The CH2 and CH3 domains together make up the Fc fragment (fragment crystalline) which contains the sites for binding effector molecules. These fragments were the starting points for crystallographic determinations of antibody structure, as the flexibility of the intact IgG molecule prevented crystallisation until fairly recently. Thus while there are more than 50 structures for Fab fragments in the literature and several for hinge deleted immunoglobulins (reviewed by Padlan, 1994) there is only one fully defined structure of an intact IgG (Harris *et al.*, 1997).

Crystallographic structure determinations of many Fab fragments and several Fc fragments of antibodies have revealed that the folded domains have similar overall structures. This structure comprises two stacked β-sheets twisted into a characteristic fold, termed the immunoglobulin fold, and stabilised by a disulphide bond (Poljak *et al.*, 1973). In the constant domains one sheet has three and the other four antiparallel beta strands, while in the variable domains there are nine strands.

The primary sequences of several thousand antibody molecules are known, representing the largest number of known sequences in one protein family. Analysis of these sequences has revealed that the variability of antibody variable domains is largely restricted to three 'hypervariable' regions in each of the heavy and light chain variable domains, with comparatively little variation in the intervening 'framework' regions (Kabat *et al.*, 1991). These hypervariable regions, also known as complementarity determining regions, or CDRs, form loops at the tip of the Fab structure making up an antigen-binding surface of approximately 2800 Å^2. Structures of several Fab:antigen complexes have confirmed that antigen binding takes place at this surface; the hypervariability of these loops is therefore responsible for the different binding specificities of different antibody molecules. The CDR loops vary in both amino acid sequence and length between antibodies. This results in the variety of antigen-binding surfaces, some of which contain grooves or clefts in the surface to accommodate antigen binding whereas others are comparatively smooth. In some cases three-dimensional structures have been determined for antibody fragments with and without bound antigen which show that the antigen-binding site is not always completely rigid and may move to allow tight antigen binding in an 'induced fit' mechanism (Bhat *et al.*, 1990; Rini *et al.*, 1992). In some cases this may also lead to some conformational changes at sites in the antibody further away from the antigen-binding region (Guddat *et al.*, 1994).

There are four different subclasses (or isotypes) of IgG, called IgG1, IgG2, IgG3 and IgG4 in humans and IgG1, IgG2a, IgG2b and IgG3 in mice. Although all within the IgG class, these subclasses vary to some extent in their structure and function. The subclasses have different heavy chains termed γ1, γ2, γ3 and γ4 in humans and each of these can use light chains from one of the two light chain isotypes, κ or λ. The structures of the subclasses vary in their pattern of disulphide bonding (Figure 1.4). The number of disulphide bonds between the heavy chains varies from two in human IgG1 and IgG4 to 15 in human IgG3. Also, the position of the disulphide bond between the heavy and light chains varies from between the CL and CH1 region to between the CL and VH/CH1 interdomain region.

The antibody effector functions are mediated through the constant regions. The effector functions can be mediated through complement activation or by cellular interactions through specific Fc receptors expressed on a range of cell types, which can result in the process of antibody-dependent cellular cytotoxicity (ADCC). Complement activation via the classical complement cascade is initiated through binding of the IgG to the complement component C1q. This is subject to conformational restraints which are only partially understood, and requires two molecules of IgG bound to an antigenic surface to bind C1q efficiently. The binding site for C1q on IgG has been known for some time to be localised to the CH2 domain, with evidence that the N-terminal region of CH2 is important for binding C1q in the case of human IgG1 (Morgan *et al.*, 1995) and also that the C-terminal region of CH2 is required for efficient complement lysis (Tao *et al.*, 1993; Greenwood *et al.*, 1993). Mutagenesis studies have suggested that three amino acid residues on the surface of the CH2 domain of murine IgG2b, numbers 318, 320 and 322, are involved in the binding interaction (Duncan and Winter, 1988), although these residues are also present on antibodies which cannot activate complement, suggesting that these residues are necessary but not sufficient to activate complement. Mutagenesis studies with human IgG1 have also suggested that C1q binding alone is not sufficient for complement activation but that interactions of the IgG molecule with other steps in the complement pathway are also required (Tao *et al.*, 1993).

Fc receptors have been identified for all the classes of immunoglobulin, IgG, IgM, IgD, IgE and IgA. Of these the best characterised are those for IgG and IgE. Three types of Fc receptors have been identified for IgG in humans; FcγRI (also known as CD64), FcγRII (also known as CDw32), and FcγRIII (also known as CD16). These receptors are structurally related and distributed on various blood cell types. FcγRI is a high affinity receptor, capable of binding monomeric IgG, which plays a key role in ADCC and is

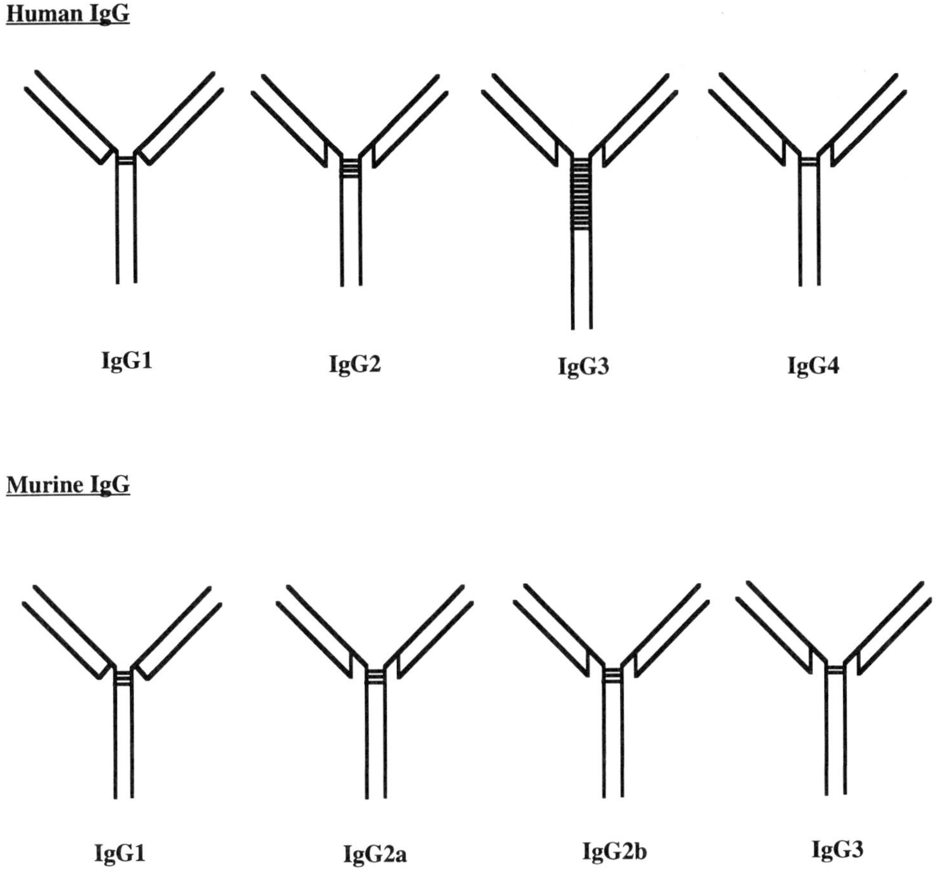

Figure 1.4 Variation in the disulphide bonds between heavy chains in IgG subclasses

found on macrophages, monocytes and neutrophils. FcγRII and FcγRIII are low affinity receptors which bind aggregated IgG. FcγRII is found on most leukocytes including monocytes, macrophages and neutrophils and FcγRIII is found on neutrophils, macrophages and NK cells. These low affinity receptors are also capable of eliciting ADCC and phagocytosis. The binding of a single antibody to Fc receptor is reversible and does not elicit a response. If several antibodies are clustered together Fc receptor clustering will take place and a cellular response will be elicited (Figure 1.5). All three Fcγ receptors recognise sites on the lower hinge/CH2 domain of IgG but not in an identical fashion. In particular, the sequence of heavy chain residues at positions 234–237 is involved in the FcγR1 receptor binding site (Duncan *et al.*, 1988; Sarmay *et al.*, 1992).

The ability to elicit effector functions varies between isotypes of IgG (Table 1.1), and this may reflect important differences in the functions of the individual isotypes. For example, human IgG1 and IgG3 are highly active isotypes with respect to complement activation and elicitation of ADCC responses whereas IgG2 and IgG4 are relatively inactive, being only poorly able to activate complement through C1q binding with little binding to Fc receptors.

The presence of carbohydrate attached to the CH2 domain has been shown to be required for both Fc receptor binding and complement activation (Tao and Morrison,

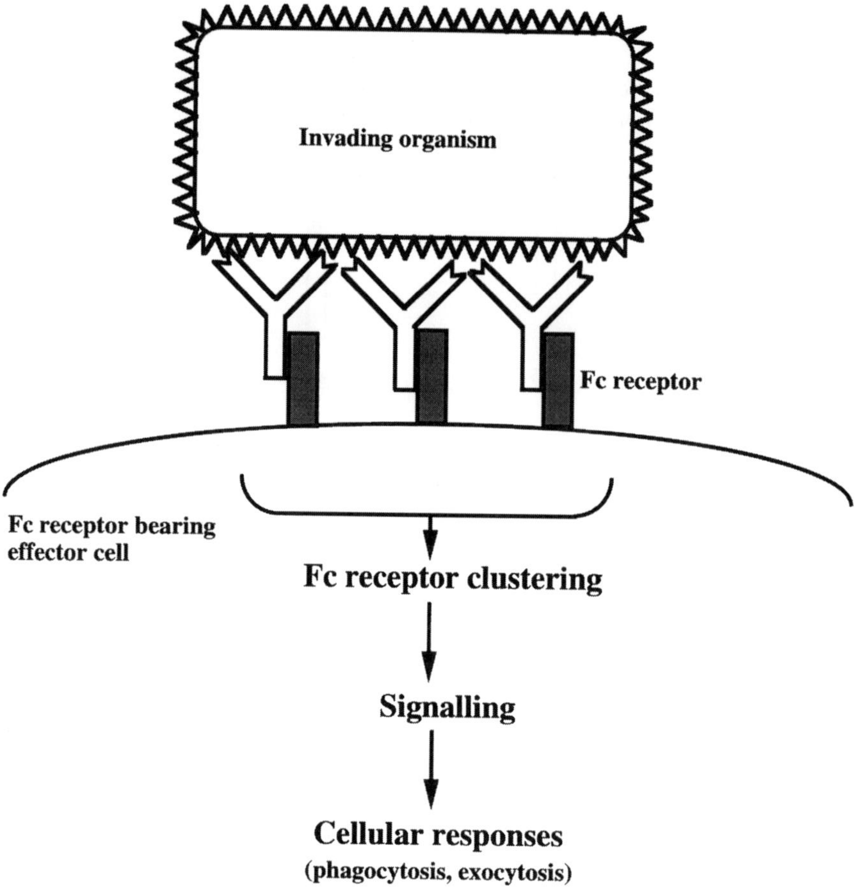

Figure 1.5 Fc receptor clustering due to multiple antibody molecules binding to antigen leading to cellular responses

Table 1.1 Effector functions of human and mouse IgG subclasses

	Human IgG				Mouse IgG			
	IgG1	IgG2	IgG3	IgG4	IgG1	IgG2a	IgG2b	IgG3
Complement activation	+++	+	++	—	+	++	+++	++
Human FcγRI (CD64)	+++	—	+++	+	—	++	—	++
Human FcγRII (CDw32)	++	+	++	—	+	+	+	—
Human FcγRIII (CD16)	++	—	++	—	—	+	+	++

Table 1.2 Human immunoglobulins

Antibody	Approximate molecular weight (kDa)	Heavy chain	Approximate concentration in normal human serum (mg/ml)
IgG1	146	γ1	9
IgG2	146	γ2	3
IgG3	170	γ3	1
IgG4	146	γ4	0.5
IgM	970	μ	1.5
IgA1	160	α1	3
IgA2	160	α2	0.5
IgD	200	δ	0.03
IgE	200	ε	0.0001

1989). The role of the carbohydrate is not fully understood but it is believed to be important in maintaining the tertiary structure and disposition of the CH2 domains. IgG without carbohydrate has been shown to have an altered conformation in the lower hinge region and may result in disruption of the interaction sites (Lund *et al.*, 1990).

IgG represents approximately 70–75% of the total immunoglobulin in human serum, the remainder being molecules of the other classes mentioned above, IgM, IgA, IgD and IgE. Although the IgG structure is a good general model for antibody structure, these immunoglobulin classes differ in their structure, reflecting differences in their normal functions. The different classes of immunoglobulin use light chains of the same type (κ or λ) and thus differ predominantly in their heavy chains (Table 1.2).

IgM is the predominant antibody raised in primary responses to many antigens and represents approximately 10% of the immunoglobulin in human serum. IgM antibodies have a pentameric structure of the basic four chain unit such that a total of ten antigen-binding sites per molecule are present (Figure 1.6). The μ heavy chain does not have a hinge region but has an extra constant region domain, which is inserted between the CH1 and CH2 domains in the analogous IgG structure, such that the CH4 domain of IgM is analogous to the CH3 domain of IgG. There is also an extra tail of 19 amino acids on the heavy chain involved in polymer assembly. The heavy chains are held together by disulphide bonds between CH3 domains and there is also an extra polypeptide chain, the J chain, whose role is not fully understood although it may assist the process of assembly of the pentameric IgM molecule. The μ heavy chain is also more heavily glycosylated than the γ heavy chain, with sugar attached at five glycosylation sites. Although it is presented as a pentameric structure in Figure 1.6, this is not always the case and the occurrence of IgM molecules with different numbers of binding units is now well established. The production of hexamers and tetramers has been reported (Eskeland and Christensen, 1975) and in some cases hexamers may be a predominant structure (Cattaneo and Neuberger, 1987; Davis *et al.*, 1988). This multivalent structure provides the IgM molecule with a high binding avidity, to help compensate for the relatively low affinity of each binding site. Another consequence of the multivalent structure is the ability of one molecule of IgM to activate complement, which is thought to occur via binding to the IgM CH3 domain. Therefore IgM molecules are well adapted for their role in the primary immune response.

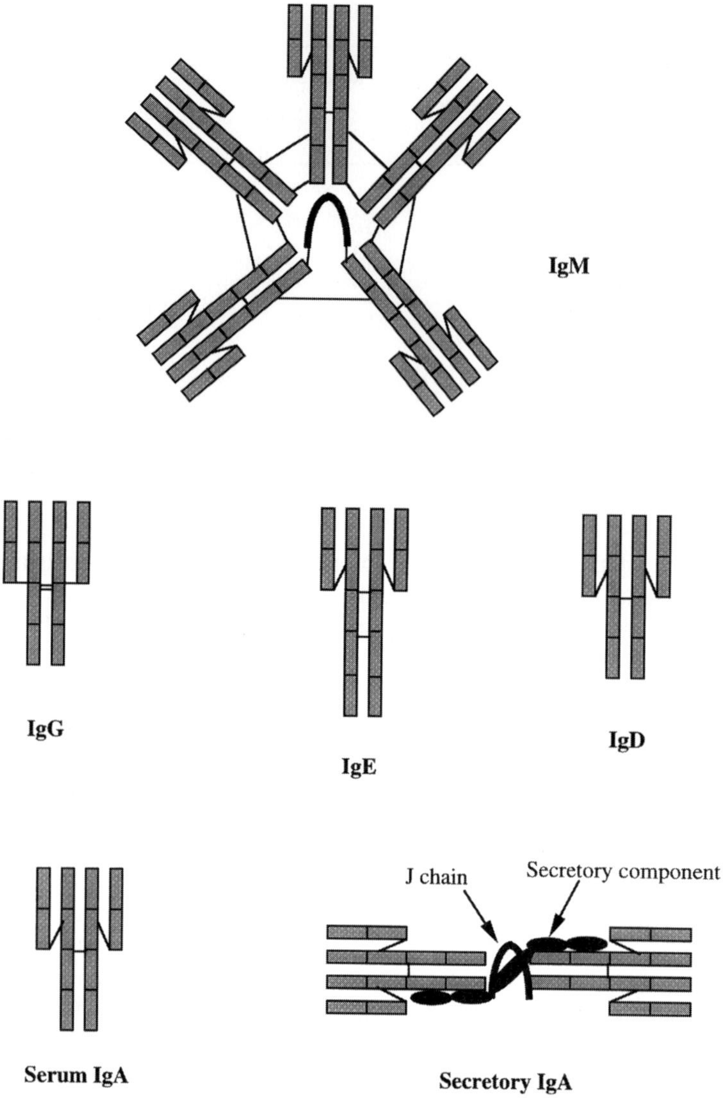

Figure 1.6 Organisation of protein chains in the immunoglobulin classes

There are two isotypes of IgA present in humans (IgA1 and IgA2) which together make up 15–20% of the total serum immunoglobulin. In addition, a form of IgA, termed secretory IgA or sIgA, is the predominant immunoglobulin present in external secretions such as saliva, milk, tears, genitourinary secretions and tracheobronchial secretions (Tomasi, 1992). The mucosal surfaces bathed in these secretions are a major site of exposure to the environment and thus secretory IgA is an important defence mechanism against invading organisms. In terms of quantity, sIgA is the major type of Ig produced, with estimates of 2g per day being produced in humans, more than other Ig forms. The IgA present in serum is largely monomeric and consists of two heavy chains and two light chains assembled in a similar manner to IgG. IgA1 is the predominant isotype, making up approximately 90% of the serum IgA. IgA heavy chains have three constant domains with IgA1 having a longer hinge region than IgA2. In addition, IgA heavy chains also have a similar tail to IgM which allows polymerisation, and binding of the J chain to the α chain, through a C-terminal cysteine residue in a similar manner to IgM. sIgA exists predominantly in dimeric form with a molecular weight of approximately 385,000. sIgA also contains a secretory component which is an extra polypeptide chain of approximately 70 kDa, synthesised by epithelial cells and not the plasma cells. The secretory component is part of the receptor involved in transport of IgA into mucosal secretions which is cleaved during transport across endothelial surfaces to release the 70 kDa secretory component. This then becomes bound to the IgA by disulphide bonds (Fallgreen-Gebauer *et al.*, 1993). sIgA is more stable to proteolytic attack than the serum form and part of this increased stability may result from the association of the secretory component which protects the immunoglobulin from degradation (Tomasi, 1992).

Human IgD and IgE are relatively poorly understood although the overall structures are known (Figure 1.6). IgD has been known to exist in a membrane-bound form on B cells for some time, where along with membrane-bound IgM it plays an important role in initiating the antibody response as described above. In the circulation, IgD is present at low levels, less than 1% of the circulating immunoglobulin, and may also be involved in regulation of the immune response (Roes and Rajewsky, 1993). IgD has a structure similar to IgG except that the heavy chain is more heavily glycosylated and there is only one disulphide bond between the two heavy chains even though the hinge region is extended and similar in length to that of IgG3 which has 11 inter-heavy chain disulphide bonds. IgE is present at very low levels in normal human serum although levels are raised significantly in allergic diseases and parasitic infections. IgE is believed to have evolved for protection against parasitic diseases, though in industrialised countries its role in allergy and asthma is now more important (Sutton and Gould, 1993). IgE also contains two heavy chains and two light chains in an assembled IgG-like structure. The ε heavy chain, however, contains four constant domains and multiple glycosylation sites (Figure 1.6).

1.4 The organisation of antibody genes

Antibody heavy and light chains are encoded in the genome at separate genetic loci; the two light chain loci, κ and λ, and the heavy chain locus, IgH. In humans and mice these loci are found on different chromosomes. Within each locus there are exons which encode the constant region domains and clusters of exons which are used to form the unique variable domains by each B cell. There are three families of repeated gene segments which are used to form the variable region at the heavy chain locus, variable (V), diversity (D) and joining (J), while the light chains use only V and J gene segments. During the development of the B cell a series of recombination events takes place to form the variable domain exon. In the mouse, the light chain variable region is recombined from one V out of a choice of approximately 200, and one J out of a choice of 4. The heavy chain chooses one V out of a choice of approximately 100, one D out of approximately 50 and one of 4 J region gene segments. The variable region along with the signal sequence exon (to direct secretion of the antibody) and the promoter/enhancer (involved in transcription) is then recombined with the constant region exons for subsequent expression (Figure 1.7). In this way each B cell forms a unique antibody from a total of approximately 10^{10} possible combinations. This is a random process and requires antigen itself to select out B cells producing antibodies of useful specificity (Section 1.2). Each individual mouse probably has B cells expressing about 10^8 specificities at one time. Further diversity in

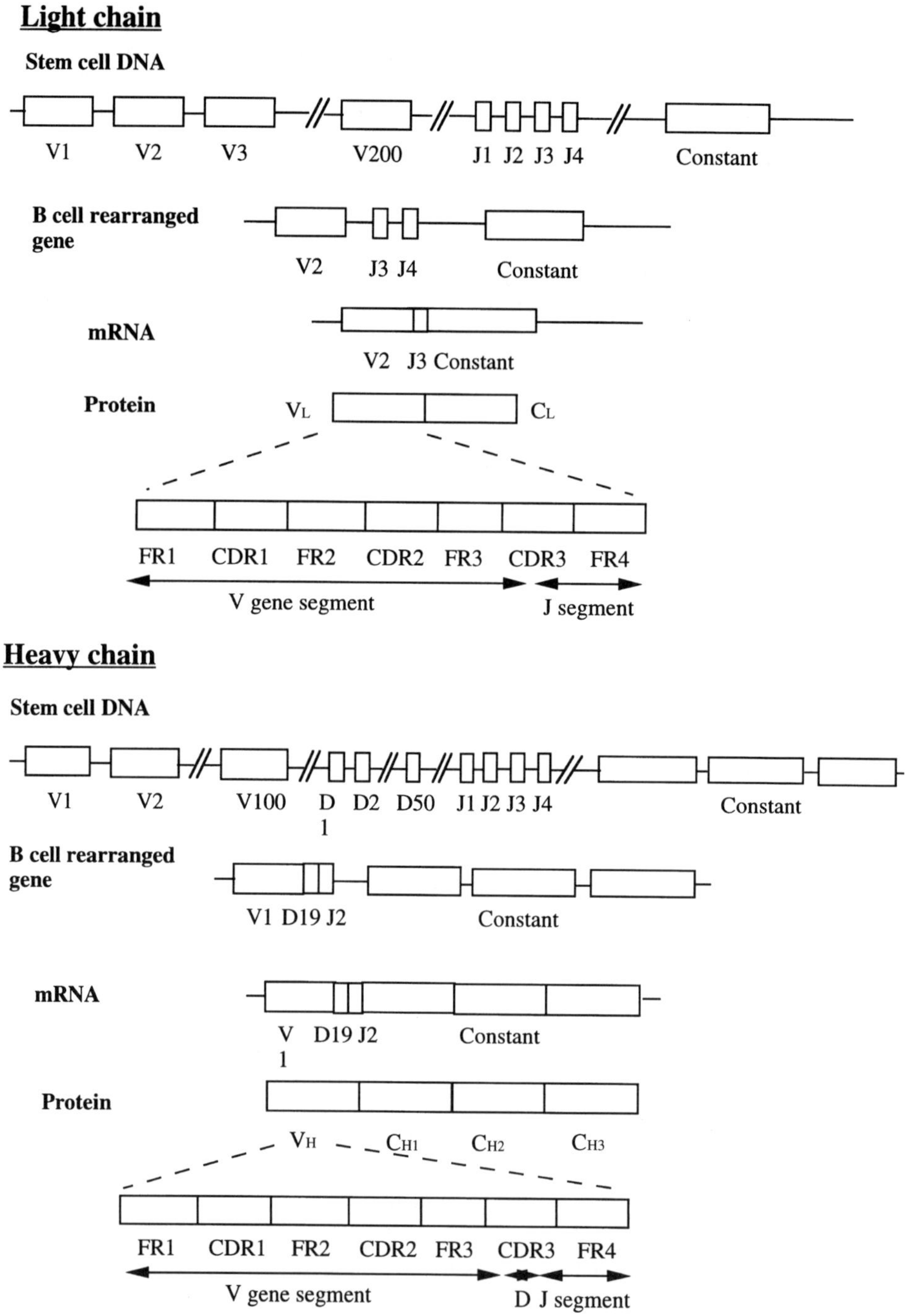

Figure 1.7 Gene organisation of IgG1, demonstrating an example of gene rearrangement to produce light and heavy chains

the antibody genes is generated because of imprecision in joining the V, D and J regions, as well as occasional incorporation of extra nucleotides at the junctions (Lieber, 1996).

The constant region exons next to the variable regions determine the isotype of the antibody produced. For the heavy chain this is initially μ, leading to the production of

IgM. As the immune response progresses, B cells rearrange the genes such that other classes of immunoglobulin are produced: for example, switching the μ exons for γ1 exons to produce an IgG1. This phenomenon is known as class switching. Also antigen selects for B cells producing antibody of high binding affinity. This is achieved through hypermutation of the variable region sequences and selection of the cells producing antibody of the highest affinity. These cells then follow the same pathway as before, some differentiating into plasma cells and some into memory cells. This is the genetic basis of the process known as affinity maturation.

There is also some genetic variation in the constant region determined isotypes between individuals. These result from small variations in the constant region sequences of a particular isotype between populations, usually between different racial groups, and are known as allotypes. In humans, allotypes have been identified for IgG1, IgG2 and IgG3 as well as IgA, IgE and κ light chains.

1.5 Antigen-binding affinity and avidity

1.5.1 *Affinity*

A measure of the strength of the antibody:antigen interaction is essential in comparing MAbs to the same or different antigens. The strength of the antibody:antigen interaction is measured through the binding affinity. This can be considered as the sum of all the non-covalent interactions between antibody and antigen involved in the binding reaction. The antibody affinity is quantified through the association constant K_a.

For the binding reaction:

$$\text{Ab} + \text{Ag} \rightarrow \text{AbAg complex}$$

K_a is the equilibrium constant of this reaction, and can be determined from the equilibrium concentrations of the reactants and complex using the law of mass action:

$$K_a = \frac{[\text{AbAg complex}]}{[\text{Ab}]\,[\text{Ag}]}$$

The units of K_a are therefore M^{-1}.

Frequently the dissociation constant K_d of the reaction is used in preference to the association constant. The dissociation constant is the affinity constant of the reverse reaction, and is therefore given by:

$$K_d = \frac{[\text{Ab}]\,[\text{Ag}]}{[\text{AbAg complex}]}$$

K_d is also, of course, the reciprocal of the association constant: $K_d = 1/K_a$ and the units of K_d are M.

The popularity of K_d as a measure of antigen-binding strength stems from the determination of affinities through measuring the concentration of antibody required to complex with half of the antigen. At the 50% bound point the concentration of Ag = AbAg complex by definition. These therefore cancel each other out in the equation above, and K_d = concentration of antibody required to complex 50% of the antigen. High affinity antibodies will require a low concentration of antibody to achieve this, e.g. 10^{-8}–10^{-12} M.

In some circumstances the kinetic rate constants of the binding reaction, the binding on-rate k_{on} and the dissociation rate or off-rate k_{off}, may also be important. These can be considered measures of the rate of binding and dissociation whereas the overall strength of the reaction is given by K_a or K_d. The rate of the binding reaction is dependent on the concentrations of Ag and Ab and the rate constant k_{on}. Units for k_{on} are M^{-1} s^{-1}. The rate of dissociation is independent of the concentration of AbAg complex and depends only on the dissociation rate constant k_{off}. Units for k_{off} are s^{-1}. At equilibrium, by definition, the rate of binding is equal to the rate of dissociation, $k_{on} = k_{off}$.

$$k_{off}\,[\text{AbAg complex}] = k_{on}\,[\text{Ab}]\,[\text{Ag}]$$

Rearranging this equation gives:

$$\frac{k_{off}}{k_{on}} = \frac{[\text{Ab}]\,[\text{Ag}]}{[\text{AbAg complex}]}$$

Therefore the rate constants can be used to give K_a or K_d:

$$K_d = \frac{[\text{Ab}]\,[\text{Ag}]}{[\text{AbAg complex}]} = \frac{k_{off}}{k_{on}}$$

The values of k_{on} and k_{off} can be measured directly as initial rates using instrumentation such as the optical biosensor BIAcore device (see Chapter 3; Malmqvist, 1993). This allows rapid determination of K_d values without waiting for the reaction to come to equilibrium. Several methods for determination of affinity and kinetic parameters, either equilibrium-based or initial rate-based, are available and have been reviewed (Goldberg and Djavadi-Ohaniance, 1993; Malmqvist, 1993).

1.5.2 *Avidity*

The antibody affinity is a measure of the strength of binding of an individual antibody binding site to a single antigenic site. However, antibody molecules usually have more than one binding site (e.g. 2 for IgG, 10 for IgM), and many antigens contain more than one antigenic site (e.g. a cell surface or antigen immobilised on a solid support), and therefore multivalent binding may be possible. The strength with which a multivalent antibody binds antigen is termed avidity. The avidity of the antibody depends on the affinities of the individual antibody binding sites, but is greater than the binding affinity as all the antibody:antigen interactions must be broken simultaneously for the antibody to dissociate completely. Thus, although the affinities of the individual binding sites are the same, multivalency leads to a large effect on the dissociation rate resulting in high binding avidity. Avidity is also referred to as functional affinity.

1.6 Generation of monoclonal antibodies

1.6.1 *Hybridoma technology*

The techniques conventionally used for the production of MAbs are based on the original report by Kohler and Milstein (1975). This technology, known as hybridoma technology, has proved to be capable of producing rodent antibodies of predetermined specificity to a wide variety of different antigens, for example proteins, nucleic acids, carbohydrates and haptens. Since each individual B cell produces antibody of a single specificity the production of antibodies of a single specificity requires isolation of individual B cells. However, B cells are not normally capable of growth in culture, and thus the production of MAbs directly from B cells is not possible. Hybridoma technology allows the production of hybrid cell lines from B cells which secrete a single, monoclonal, antibody with one binding specificity which can potentially be produced in unlimited quantitities.

The general scheme for the production of MAbs by hybridoma technology is shown in Figure 1.8. Mice are immunised by injection of the antigen to which a MAb is required. Rats are also commonly used to generate MAbs, and in some cases (particularly when antibodies to mouse antigens are required) other species have been used such as hamsters (Houlden *et al.*, 1991), chickens (Nishinaka *et al.*, 1991) and rabbits (Speiker-Polet *et al.*, 1995). When an immune response has been raised, B cells are harvested from the rodent, usually from the spleen, and these are fused with myeloma cells by promoting fusion of the cell membranes with, for example, polyethylene glycol. Myeloma cells are derived from a mutant cell line of a B cell tumour, and cells are used which do not secrete any immunoglobulin, and which are essentially immortal and can be easily grown in culture. Fusion results in the production of immortalised hybrid cells or hybridomas. The remainder of the process is then one of isolation and propagation of hybridoma cells which have retained the ability of the B cell to produce antibody and the good growth characteristics of the myeloma cell. The process of hybridoma production can therefore be considered in three parts: immunisation, fusion and selection of the required hybridoma. Many protocols for immunisation and generation of hybridomas are available (e.g. Harlow and Lane, 1988; Donohoe *et al.*, 1995), and should be referred to for practical guidance.

Immunisation of the animal to raise an immune response is of key importance. Immunogens of different molecular types can be used and they do not need to be pure materials, as a hybridoma secreting a MAb of the required specificity can be selected out later. Different molecules vary greatly in their immunogenicity and thus require different immunisation protocols for an optimal response. Factors such as the form of the antigen, number and route of immunisations, carrier, adjuvant and the species and strain of animal used need to be considered. It is important to consider the conformation of the molecule which the MAb is desired to bind to. For example, will a protein be in its native conformation or denatured? MAbs of exquisite specificity can be generated, and therefore a desired isomer or molecular form of a particular antigen might be required.

There are two important exceptions to the range of immunogens which can be used. Antibodies will not normally be raised to 'self' antigens, and small antigens (less than approximately 1000 in molecular weight) will not raise a response unless conjugated to a high molecular weight carrier protein. For this purpose proteins are used, such as bovine serum albumin or keyhole limpet haemocyanin, which are very immunogenic in rodents, easily available and readily conjugated to small molecules. The generation of high affinity antibodies normally requires a T-cell-dependent B-cell response as described in Section 1.2, and thus both T and B cell epitopes are needed to allow the memory response and class switching to take place. T cells respond to proteolytically produced peptides of the antigen presented by MHC class II molecules and thus in general non-self proteins and glycoproteins elicit high affinity responses after multiple immunisations. Without protein, carbohydrates and other non-proteinaceous materials elicit only a T cell-independent response with little or no memory or class switching.

The type of antibody produced is also dependent on the number of immunisations. If IgM is required, only one immunisation is carried out as only a primary response is required, whereas IgG antibodies require multiple injections (usually three or four) at intervals of approximately 3–4 weeks to allow an effective secondary response. Adjuvants,

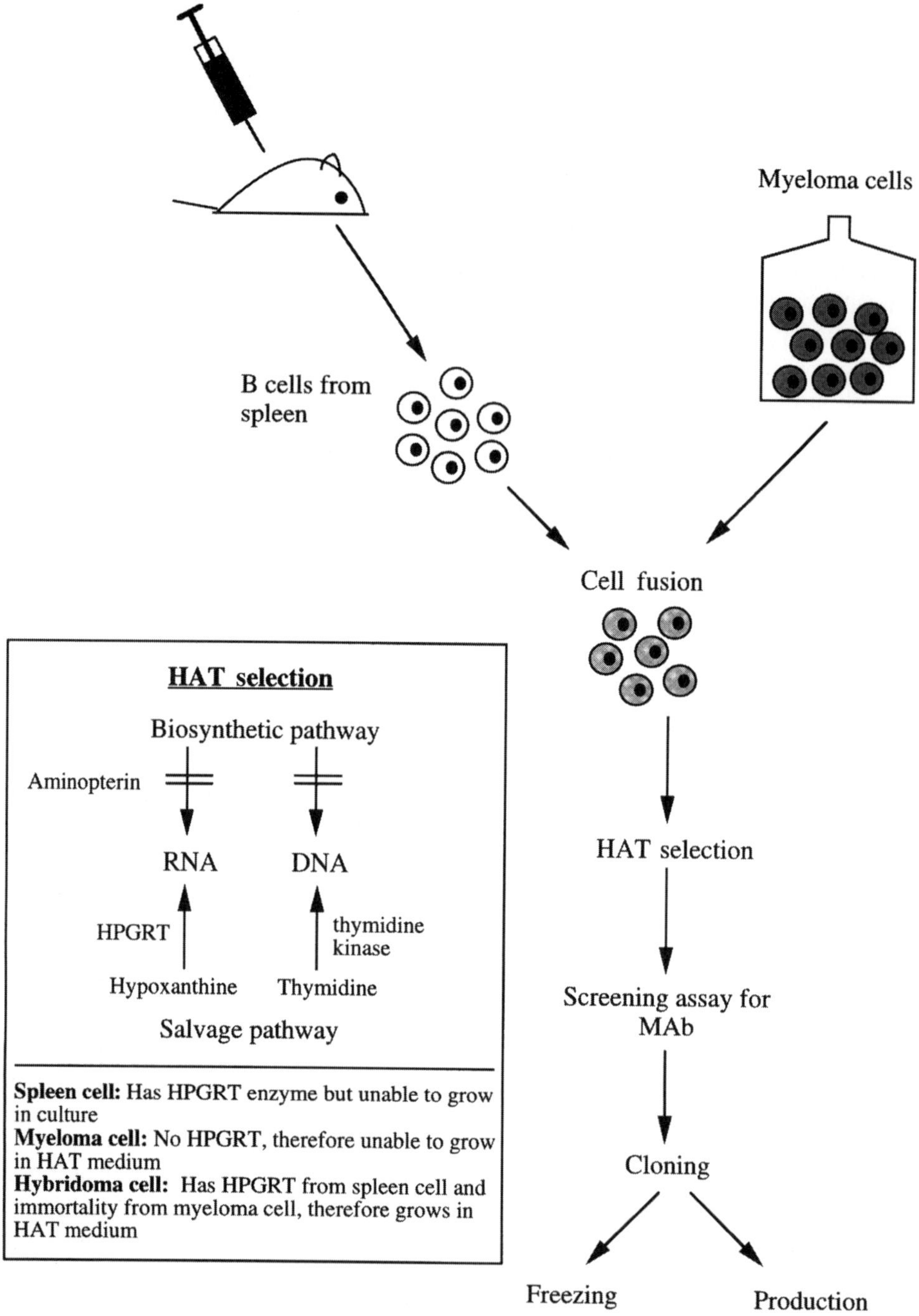

Figure 1.8 Production of monoclonal antibodies by hybridoma technology – see text for details

such as Freund's adjuvant, are widely used to aid the generation of an immune response to soluble antigens injected via the intraperitoneal, intradermal, intramuscular or subcutaneous routes which are usually used for primary immunisations (Harlow and Lane, 1988). Adjuvants are usually colloidal or oils, which are mixed with the soluble antigen to produce an emulsion which is injected into the animal and provides a long-lasting local 'depot' of antigen which is slowly released and thus continually boosts the response. In addition, other components are sometimes included in the adjuvant mixture to activate the immune system non-specifically. For example, complete Freund's adjuvant contains killed mycobacteria for activation as well as oil and an emulsifying agent. Complete Freund's adjuvant is used for the first immunisation and subsequent booster immunisations are carried out in incomplete Freund's adjuvant (without the mycobacteria component) to maximise response to the antigen and avoid excessive inflammatory responses to mycobacteria. Insoluble antigens can often be used without adjuvant, as these form insoluble 'depots' on their own.

Genetic immunisation has recently been developed to allow immunisation with DNA encoding a protein of interest. The DNA is injected directly into mouse tissue, in a form which allows expression of protein from the DNA by the mouse tissue. An immune response is then generated against the protein which allows MAbs to be produced (Paddock *et al.*, 1994). This approach may be particularly useful when the protein is difficult to produce or purify sufficiently to allow direct immunisation. In addition, it may be possible to boost the immune response to poorly immunogenic proteins by producing fusion proteins with a highly immunogenic sequence.

To ensure that an immune response has been made to the antigen before harvesting B cells, blood samples can be obtained from the immunised animals and tested for the presence of specific antibodies. If positive, final boosts of soluble antigen are often made intravenously to increase the number of immune B cells present in the spleen prior to harvest. The spleen is the most common source of immune B cells, although cells can also be harvested from lymph nodes, particularly if soluble antigen is not available and a final intravenous boost cannot be given.

Several myeloma cell lines are available for fusion purposes from mice and rats. As these lines are derived from tumours of antibody-producing B cells, it is important to try to choose a line which does not produce any antibody of its own. If the resulting hybridoma produces two types of light and heavy chains these will probabably associate randomly resulting in a mixture of antibody molecules produced comprising the original myeloma antibody, the desired MAb and various mixed combinations of the two types of heavy and light chains. This mixing can be advantageous for the production of bispecific antibodies (see Chapter 2), but will reduce the efficiency of the initial isolation of a MAb and make purification of the desired MAb difficult. Another requirement of the myeloma cells is that they are deficient in the enzyme hypoxanthinephosphoribosyltransferase and will allow selection of the fused hybridoma cell over the parent myeloma cell (see below).

Fusion of the myeloma cells and the harvested B cells is usually accomplished using polyethylene glycol (PEG) as the agent to induce membrane fusion. Alternative methods are also available, for example the use of Sendai virus as used in early research or electrofusion. However, PEG fusion is achieved simply, cheaply and reliably and thus has become the most common method. Initially only the cell membranes fuse and two nuclei are present in the cell. Nuclear fusion takes place during mitosis and subsequent generations of the hybridoma cell often result in loss of some chromosomes. After fusion a mixture of hybridoma cells, spleen cells and myeloma cells are present and the next stage of the process is the selection of the hybridoma cells over the other cell types. Spleen cells will not grow in culture and thus growth of the cell mixture for a few days will readily remove residual spleen cells. Myeloma cells, on the other hand, will grow rapidly and would make isolation of any hybridoma cells very difficult. For this reason myeloma cells are used for the fusion which are deficient in the enzyme hypoxanthinephosphoribosyltransferase (HPRT) and consequently are not able to use the salvage pathway for RNA synthesis. Selection is then achieved by using HAT medium (Figure 1.8). HAT medium contains hypoxanthine (H), aminopterin (A) and thymidine (T). Aminopterin is an effective inhibitor of RNA and DNA synthesis and thus will block growth of the myeloma cells. However, hybrid cells which have the HPRT enzyme from the spleen cells will be able to use the added hypoxanthine and thymidine to produce RNA via the salvage pathway and survive.

After selection in HAT medium, hybrid cells are grown on and tested to determine which colonies of cells produce antibodies of interest. Not all of the spleen cells which form hybrids are B cells, and thus there will be some hybrids which do not produce antibody at all, and also many cells will be present which produce antibodies which are not of interest. An important step is therefore the screening of clones of cells to identify those which produce the MAbs required. A large number of clones may need to be screened quickly to allow selection of those required for further culture. A suitable screening assay should therefore be rapid, able to screen many samples and capable of giving a result with the small amount of antibody produced by early clones of cells. There are many types of suitable assays to measure antigen-binding properties including ELISA (enzyme linked immunoadsorbent assay), radioimmunoassay, dot blots and western blots, immunohistochemistry and immunofluorescence. In addition, it may be desirable to screen for other properties such as isotype or specific functional properties such as complement activation or the ability to elicit ADCC. If possible, the assay which uses the antibody in the closest way to that required for the intended end use of the antibody is the most suitable, as this will ensure that the optimal properties of the antibody are selected.

Once suitable hybridomas have been identified, these must be cloned several times to ensure a stable, homogeneous colony of cells and to ensure that the antibody is indeed monoclonal. To do this requires the growth of a colony from a single cell. This is achieved by subculturing the cells by limiting dilution so that each well of the culture plate contains an average of less than one cell. Alternatively cloning can be carried out by plating out cells in semisolid agar or by single cell manipulation techniques. Individual clones of cells can than be screened for antibody production as before. Once cloned, the hybridoma cell is ready to be used for the production of large amounts of antibody, but it is also important to ensure that samples of cells are preserved by freezing in liquid nitrogen at each stage so that valuable hybridomas are not lost!

1.6.2 ***Human monoclonal antibodies***

The generation of human MAbs is often desirable for clinical applications. This is because murine antibodies are recognised as foreign when administered to humans and therefore elicit an immune response directed against the administered MAb: a human anti-mouse antibody response or HAMA. This prevents repeat administration of MAb as the HAMA results in formation of immune complexes which are rapidly cleared, rendering the antibody ineffective. In addition, HAMA responses may result in adverse reactions by the patient such as allergic reactions, hepatic dysfunction and in some cases anaphylactic shock. Attempts to overcome this problem have driven much research both in reconstructing rodent antibodies by genetic engineering (see Chapter 2) and in attempts to generate human MAbs. Human antibodies may also be advantageous in that they are compatible with human effector mechanisms such as complement and ADCC which are often not fully activated by rodent antibodies.

Table 1.3 Problems in human MAb production using conventional techniques

Immunisation *in vivo* or *in vitro*
Source of human B lymphocytes
Immortalisation: EBV/choice of fusion partner cell line
Stability of human hybridoma/heterohybridoma cell lines
Selection of isotypes and high affinity MAbs
Scale up to produce large quantities of human MAbs

The production of human MAbs has not been straightforward and attempts have driven many technical innovations and improvements to antibody technology. Production of human MAbs by hybridoma techniques is usually difficult for several reasons (Table 1.3). For many antigens of interest, such as human disease markers, it is not appropriate to immunise humans directly with the antigen for ethical and moral reasons. Therefore it can be difficult to obtain suitably immune lymphocytes. The source of lymphocytes is also a major problem. It is usually not possible to harvest these from the spleen or lymph node and the most available and commonly used source is peripheral blood. B cells are comparatively poorly represented in peripheral blood and the majority of those present express surface IgM. Fusion therefore often results in cells which largely produce low affinity IgM antibodies. The choice of cell line to be used as a fusion partner is also problematic. Human myeloma lines have proved to be comparatively poor with respect to their growth characteristics and fusion efficiencies and usually secrete immunoglobulin of their own which complicates production and purification. Attempts to use murine cell lines, which do not produce any immunoglobulin of their own, as fusion partners have met with some success, although the hybrid cells produced can be unstable and may preferentially lose human chromosomes and thus the ability to secrete human antibody. More successful has been the development of mouse–human heteromyeloma cell lines as fusion partners. These are formed by fusing a human lymphoid tumour line with a mouse myeloma cell line and using the resulting hybrid cell line to fuse with the required B cells. The resulting fused cells are termed heterohybridomas or triomas and these often produce more antibody than cell lines produced with murine fusion partners alone and are relatively stable.

An alternative approach to immortalise human B cells has been the transformation of these cells with Epstein–Barr virus (EBV). EBV is a virus which preferentially infects B cells and activates them to divide and produce antibody. Some of the infected cells may be immortalised but this technique is relatively inefficient and cell lines are often unstable and lose the ability to produce antibody. Combinations of this approach with the use of heteromyeloma fusion are more efficient. For example, stimulating B cells with EBV followed by fusion with a mouse–human heteromyeloma has been used to produce several human MAbs (e.g. Kozbor *et al.*, 1982; Gustafsson and Hinkula, 1994).

To overcome the problem of being often unable to immunise humans *in vivo*, approaches to the *in vitro* immunisation of human B cells in culture have been developed (Borrebaeck, 1989; Koda and Glassy, 1990). Although developed primarily for use in human antibody production, *in vitro* immunisation was first used for the production of mouse antibody and may offer advantages for the production of rodent antibodies in some circumstances, for example when trying to raise antibodies to weakly antigenic epitopes which might be masked by dominant epitopes *in vivo*, or when trying to raise antibodies to hazardous antigens which it would not be possible to administer *in vivo*. *In vitro* immunisation is an attempt to achieve primary activation of B cells in culture and as such requires a complex *in vitro* protocol including co-cultivation with a variety of cytokines and mitogenic factors to stimulate B cells (Danielsson *et al.*, 1987; Darveau *et al.*, 1993). Suppressor cells also need to be eliminated from the population by cell separation or by treatment with agents such as leucyl leucine methyl ester which is cytotoxic to CD8+ T cells and natural killer cells (Borrebaeck *et al.*, 1988). The major drawback of *in vitro* immunisation approaches developed to date is that the antibodies produced are characteristic of a limited primary response and are therefore often of the IgM subclass with low to intermediate affinities for antigen.

Mice with a mutation resulting in severe combined immune deficiency which lack functional B and T cells (SCID mice) have been used in attempts to produce higher affinity human antibodies. Because of their immune deficiency, SCID mice can accept grafts of human lymphocytes isolated from peripheral blood which can be used to re-create a partial human immune system in the mouse (Mosier *et al.*, 1988). Cells can be transplanted from immunised donors or after *in vitro* immunisation and re-immunised in the mouse to generate secondary responses (Duchosal *et al.*, 1992; Carlsson *et al.*, 1992). Immune human B cells can then be harvested from the spleen or lymph nodes to allow antibody production.

In addition, it is possible to use SCID mice with human immune cells to generate primary responses using cells from donors not previously immunised (Sandhu *et al.*, 1994). This raises the prospect of producing a wide range of human antibodies in mice, and research towards developing suitable methods is under way (Walker and Gallagher, 1994). So far only a limited number of antigens have been investigated using this system and the limits to the ability of this system to produce high affinity antibodies to a range of different antigens will be a subject of intensive research over the next few years.

One approach which can successfully overcome the instability of human antibody-producing cell lines once they have been generated and the difficulties involved in their large-scale culture is gene rescue. In this approach the antibody genes are cloned from the cells and expressed in a stable alternative host cell such as a murine myeloma. For example, genes coding for human antibodies to tetanus toxoid and pseudomonal exotoxin A have been rescued from a mouse–human heterohybridoma and an EBV transformed B cell line respectively and re-expressed in a murine myeloma cell line (Gillies *et al.*, 1989b; Nakatani *et al.*, 1989). In both cases stable murine cell lines were obtained which secreted high levels of human antibody, similar to the yields of antibody produced by murine hybridoma cells. This technique also allows engineering of the human antibody, such as changing the isotype of the antibody if required. Engineering of antibodies is discussed in detail in Chapter 2.

1.6.3 *Human MAbs from transgenic mice*

An alternative approach to the generation of human antibodies is the genetic manipulation of mice to disable the production of mouse immunoglobulin and to introduce human immunoglobulin loci. The resulting transgenic mice are capable of producing human antibodies in reponse to challenge with antigen which can be isolated by conventional mouse hybridoma technology (Lonberg *et al.*, 1994; Green *et al.*, 1994). The human immunoglobulin loci are large and contain the V, D and J genes of the variable domains as well as the constant regions as described in Section 1.4. The transfer of the entire locus is technically extremely difficult and thus miniloci have been constructed containing a limited number of the heavy and light chain genes. These genes are introduced into mouse embryonic stem cells and implanted into a surrogate mother to develop. The resulting mouse strain therefore contains both human and mouse antibody genes. A separate mouse strain is produced from embryonic stem cells in which the mouse immunoglobulin heavy and κ light chain genes are disrupted and this is then bred with the strain containing mouse and human genes. Progeny containing only functional human immunoglobulin genes can then be isolated. These mice are capable of immune responses to administered antigen and although the entire diversity of V, D and J regions has not been reproduced, those present are utilised to produce diverse antibody molecules. In some instances somatic mutation has also been demonstrated, suggesting that an extensive diversity can be achieved, with both IgM and IgG antibodies being produced (Lonberg *et al.*, 1994). High avidity antibodies have been isolated; with a panel of human antibodies produced in this way to the human T cell marker CD4, K_a values of 10^9–10^{10} M^{-1} were produced (Fishwild *et al.*, 1996). Progress is also being made on the introduction of larger segments of the human Ig gene loci into mice, with transfer of large DNA segments or chromosome fragments containing the Ig loci now possible (Mendez *et al.*, 1997; Tomizuka *et al.*, 1997). This technology is very much in its infancy and, in the next few years, will no doubt develop to the point where a strain of mice can be produced which is capable of generating a range of useful human antibodies.

Attempts have been made to overcome many of the problems summarised in Table 1.3, and many human MAbs have been generated, but the technology is far from routine and relatively few high affinity antibodies have been produced. Recently an alternative technology for the production of antibodies which can bypass hybridoma technology has been developed, known as phage display. This can be used to produce human antibodies and is probably the method which will dominate human antibody production in the future.

1.6.4 *Isolation of antibodies by phage display*

An alternative approach to isolation of MAbs is to isolate the genes for the antibody directly and thus bypass hybridoma technology altogether. As the difference in antigen-binding specificities between antibodies lies entirely within the variable domains, it is only necessary to isolate genes for the variable regions which can then be joined to constant region genes by recombinant DNA techniques to allow expression of the required antibody molecule (see Chapter 2). Strategies for isolating antibody specificities using this approach have been developed over the past few years (Winter *et al.*, 1994; Burton and Barbas, 1994). These strategies attempt to mimic the key features of antibody generation by the immune system. To achieve this, repertoires of large numbers of antibody genes are isolated and expressed as antibody fragments displayed on the surface of bacteriophage. Ideally each phage displays a single antibody fragment. The phage with the appropriate antibody fragment displayed are then selected by binding to antigen attached to a solid support. Usually several rounds of selection are carried out to isolate the highest affinity antibody fragments. The selected phage can then be used to express soluble antibody fragment and to recover the antibody genes which are used to construct the desired form of antibody (Figure 1.9).

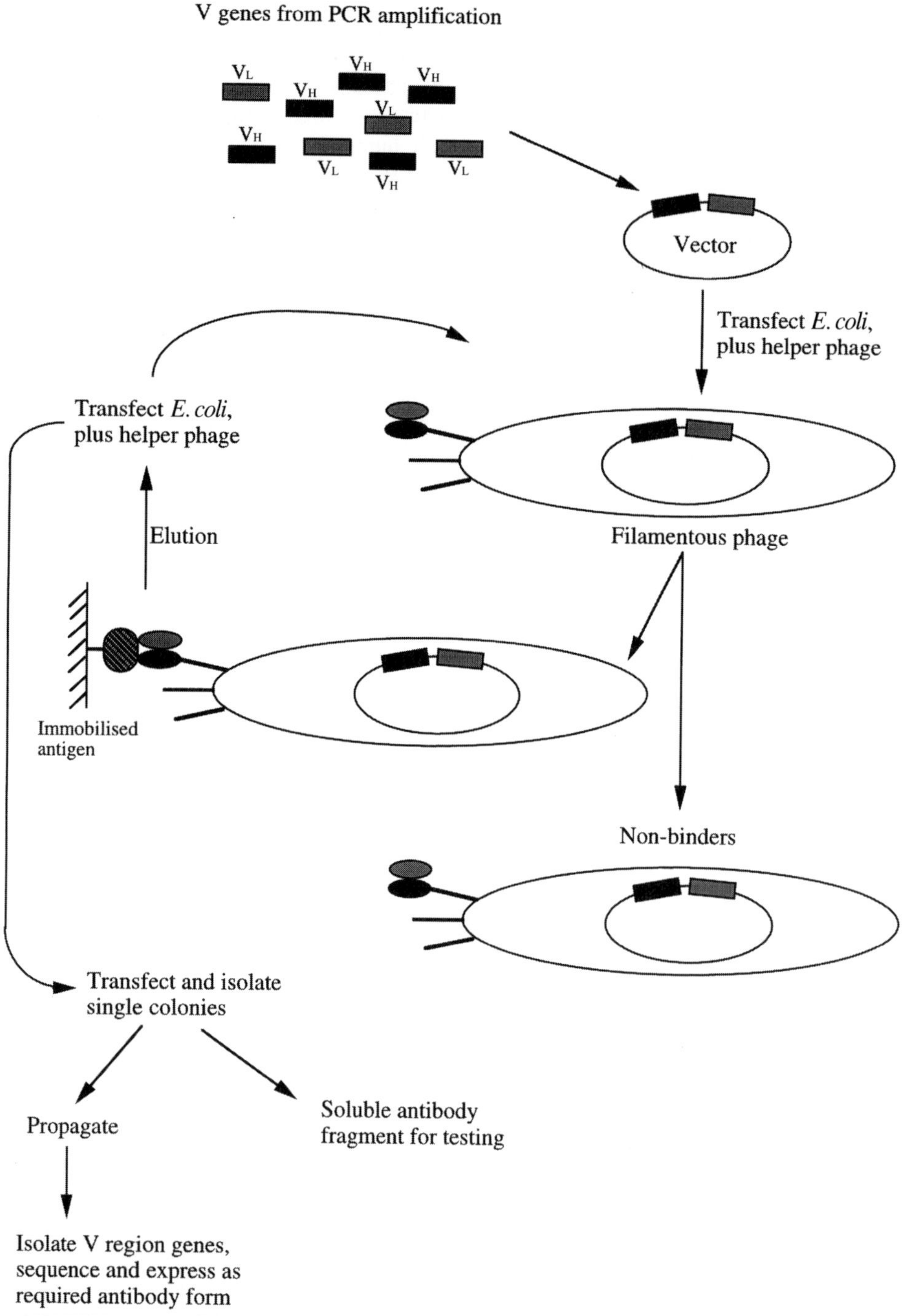

Figure 1.9 Production of monoclonal antibodies by phage display – see text for details

The first requirement is thus a repertoire of antibody genes. Several sources have been used. Genes can be isolated directly from B cells recovered from the spleen, bone marrow or peripheral blood of either immunised or non-immunised animals. Immunised animals contain larger numbers of genes encoding antibodies to the antigen of interest, and may also contain higher affinity antibody genes as a result of secondary immune responses. Therefore it is easier to produce high affinity antibodies from immunised animals. The technique is also applicable to non-immunised donors and has therefore attracted great interest as a technique for the production of human antibodies. As hybridoma technology is bypassed completely, phage display can also circumvent many of the other difficulties in producing human antibodies, such as immortalisation and instability of cell lines, discussed above (Table 1.3). The most common source of variable region genes is cDNA derived from mRNA and amplified by polymerase chain reaction (PCR) technology, although genes can also be recovered by PCR directly from genomic DNA. PCR technology amplifies DNA, producing many copies of the required sequences. Oligonucleotide primers are prepared for binding at the 5′ and 3′ ends of the target DNA on opposite strands which are then extended with a thermostable DNA polymerase. Several cycles of (a) DNA denaturation to single strands, (b) primer annealing and (c) extension by DNA polymerase are carried out in a thermal cycler that cycles between the three temperatures required for these processes. The amount of DNA increases exponentially with each cycle as each new DNA molecule can act as a template to produce new molecules. The flanking sequences of antibody variable region genes have been well characterised and sets of primers have been designed to allow both murine and human antibody genes to be amplified (Orlandi *et al.*, 1989; Marks *et al.*, 1991).

To allow selection of high affinity antibodies a large library of antibody genes is required in order to represent as much of the total repertoire of antibody genes as possible. The larger the library, the greater is the gene diversity and therefore the higher the chance of isolating high affinity specific antibodies. After immunisation of animals, or with humans who have been exposed to the required antigen, for example as a result of infection, the library constructed will have a large proportion of rearranged genes from immune B lymphocytes, and thus smaller numbers of antibody genes need to be screened to identify specific antibodies. Libraries from non-immunised individuals require screening of larger numbers but have the advantage that antibodies to different antigens can be isolated from a single library. Initial libraries were relatively small (10^7–10^9) and resulted in the isolation of weakly binding antibodies. Over the past few years rapid technological advances in library isolation have taken place, including the introduction of synthetic libraries in which random CDR sequences are introduced (Barbas, 1995). Library sizes of 10^{10} have now been achieved and allowed selection of high affinity, specific antibodies (Griffiths *et al.*, 1994; Vaughan *et al.*, 1996).

Two different types of antibody fragment have been widely used for display of antibody specificities on phage, Fab and single-chain Fv (scFv, see Chapter 2). Antibody fragments are used as these are efficiently expressed in bacterial systems (see Chapter 5) and they can be well presented on the surface of the phage. Phage display was first achieved using the phage λ (Huse *et al.*, 1989), but subsequently filamentous phage such as M13, fd or f1 have been found to be more suitable systems (McCafferty *et al.*, 1990; Clackson *et al.*, 1991). These phages are single-stranded DNA phages which consist of a long single strand of DNA coated with approx. 2700 copies of the major coat protein pVIII and a low number of copies of four minor protein components, including the pIII protein which is normally present at 3–5 copies. The pIII protein is located at one end of the phage and is normally involved in infection of a bacterial cell to allow replication. Display at the phage surface is achieved by fusion of the antibody genes to one of the phage surface proteins. It was demonstrated by Smith (1985) that peptides could be presented on the surface of filamentous phage by fusion to the N-terminus of the pIII protein. This was extended to the fusion of antibody fragments resulting in their surface presentation in a form which could bind specifically to antigen (McCafferty *et al.*, 1990; Clackson *et al.*, 1991). Fusions to the pIII protein can be made by direct cloning into the phage genome, although this results in phage presenting an antibody fragment on each of its pIII proteins and results in phage of low infectivity due to preventing the normal function of the pIII protein. Also the presentation of multiple copies of the antibody fragment on each phage is not ideal as multivalent binding may then take place, where the low affinity binding of several copies may result in relatively high avidity interactions which may mask the binding of a single copy of a high affinity antibody fragment (Barbas *et al.*, 1991). A preferred system is therefore the use of phagemid vectors, plasmid vectors containing a viral origin of replication. Phagemids are co-introduced into the bacterial cell with helper phage which provide the genes for native phage proteins. Under appropriate conditions this results in a large number of phage particles containing a single antibody fragment (Garrard *et al.*, 1991). Several different vector systems have now been developed using either the whole or part of the pIII protein fused to either Fab or scFv fragments (Winter *et al.*, 1994; Burton and Barbas, 1994). The pVIII protein has also been used for phage display, although the large number of copies of this protein in the viral coat means that display of antibody fragments results in many antibody molecules on each phage, and thus this system has been relatively little used.

After assembly of the phage display library, selection must be carried out. This is done by binding to antigen which is attached to a solid phase. Antigen is often coated onto microtitre plates for this purpose or can be used in many other formats such as coated to beads or column matrices. Cells containing surface antigen can also be used or antigen in a form which can be readily captured such as biotinylated antigen, which can be captured on streptavidin-coated beads via the high affinity biotin–streptavidin interaction. Phage bound to the solid phase are then eluted either by competition with soluble antigen or by a change in pH or with a chaotropic agent to disrupt the antibody–antigen interaction in much the same way as used in immunoaffinity chromatography (see Chapter 3). Several rounds of selection are usually carried out to ensure isolation of specific binding phage.

After selection, the antibody genes are recovered and may be sequenced and manipulated as required. In some phage vector systems an amber codon is inserted between the pIII protein gene and that for the antibody fragment (Hoogenboom *et al.*, 1991). When grown in a supE supressor strain of *E. coli* the codon is read as glutamine and the antibody fragment is displayed on the phage surface, but when grown in a non-supressor strain the amber codon is read as a stop codon and soluble antibody fragment is secreted from the bacterial cell. This facilitates rapid analysis of the antigen-binding properties of the isolated antibody sequences without further genetic manipulation.

An interesting variation of phage display has been developed using phage which are only infective when carrying the desired antibody specificity (Spada and Pluckthun, 1997). Such selectively infective phage (SIP) are particularly useful in selecting antibody specificities from large libraries as they improve the specificity of the isolation of desired phage, reducing non-specifically binding phage from the system. Selective infectivity is conferred by using a truncated version of the pIII protein for display of the antibody fragment which is not sufficient to allow infectivity on its own. The pIII protein comprises three domains and the two N-terminal domains are required for phage infectivity. Antibody fragments can be displayed on the surface of phage through fusion to the C-terminal domain alone, resulting in non-infectious phage. Infectivity can then be restored through binding antigen fused to the N-terminal domains of pIII. Antigen–pIII fusions can be prepared either as fusion proteins or chemically, therefore allowing both protein and non-protein antigens to be used. As phage are only infective when antibody fragments capable of binding antigen are present, background phage are reduced such that isolation of antibodies is simplified. Using this approach antibodies to a range of different antigens have been described (Spada and Pluckthun, 1997).

Although high affinity antibodies can be isolated directly from large libraries, it is often necessary to improve the antigen-binding affinity of antibodies isolated by phage display. Several strategies for this have been developed. This can be considered to be analogous to the process of affinity maturation undergone in the immune system to generate high affinity secondary response antibodies (see Section 1.2). One approach is to use mutagenesis strategies where sequences in the CDR regions are mutated either randomly or in a more directed fashion and antibodies with higher binding affinities selected by phage display (Yang *et al*., 1995). This can be achieved either by several rounds of

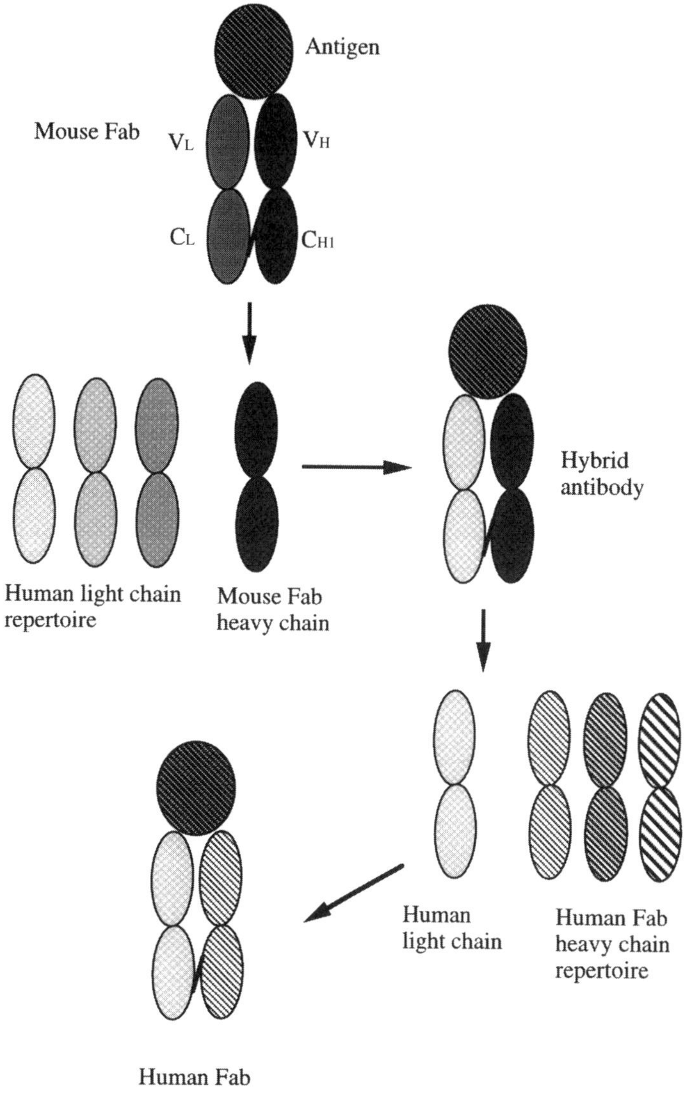

Figure 1.10 Epitope imprinted selection of human Fab from a mouse Fab to the same antigen (adapted from Jespers *et al*., 1994)

sequential mutagenesis of CDR regions or by mutating them in parallel. To date sequential approaches have yielded the best results, with particular improvements coming from mutagenesis of CDR3 regions of the heavy and light chains. Antibodies to both HIV gp120 and the tumour-associated antigen $p185^{HER2}$ have been improved to picomolar affinities by this approach (Yang *et al.*, 1995; Schier *et al.*, 1996). Residues outside the CDR regions may also have an impact on antigen-binding affinity (see Chapter 2) and mutagenesis strategies for these sequences may also need to be taken into account.

An alternative approach is to use chain shuffling for affinity maturation. In this technique one of the two antibody chains (e.g. VH) is used to recombine with a library of light chain genes and the higher affinity binding combination retrieved. The new antibody VL is then used to recombine with a library of heavy chain genes to find a higher affinity heavy chain. This can be repeated, combined with mutagenesis strategies or libraries of synthetic V gene sequences, to introduce further diversity and isolate higher affinity antibodies. In this way antibody affinities have been improved from K_d values of 10^{-7} M to 10^{-9} M (Marks *et al.*, 1992).

An interesting technique for isolating human antibodies to antigens for which mouse antibodies already exist has been described based on chain shuffling (Jespers *et al.*, 1994). This technique, known as epitope imprinted selection, starts with an existing mouse antibody heavy chain and selects light chains from a human light chain library which pair with the mouse heavy chain and bind to antigen. The light chain is then recombined with a library of human heavy chains and a human heavy chain is selected (Figure 1.10). This technique has been used to isolate a human antibody to TNFα with similar binding affinity to the original mouse antibody (Jespers *et al.*, 1994).

The phage display approach offers several potential advantages over hybridoma technology. New antibody specificities can be isolated rapidly, with or without prior immunisation, and high affinity antibodies obtained. Human antibodies can be generated, including to self antigens which is difficult to achieve by other means (Griffiths *et al.*, 1993) and antibodies to weak epitopes can be deliberately selected by masking immunologically dominant epitopes (Ditzel *et al.*, 1995). However, it should not be seen as only a competing technology to hybridoma technology. It is possible to use both techniques to generate antibodies with different properties such that the most appropriate antibody for a specific application can be generated. In addition it is possible to use antibody chains isolated by hybridoma techniques in chain shuffling experiments with phage libraries and isolate novel high affinity antibody chains (Ames *et al.*, 1995). Mutagenesis or chain shuffling techniques combined with phage display can also be used to increase the affinity or even change the epitope recognised by a pre-existing antibody, allowing manipulation of the antibody specificity (Jackson *et al.*, 1995; Ohlin *et al.*, 1996).

2

Antibody Engineering: Design for Specific Applications

2.1 Introduction

Antibody molecules are very amenable to manipulation both chemically and by genetic engineering. As described in Chapter 1, the antibody molecule is made up of a series of domains which are capable of folding into their native structure independently. This has enabled many manipulations of the antibody molecule to be carried out without affecting the conformation of adjacent domains, and allows domains to be moved in position, or substituted with other molecules relatively easily. In addition, the beta sheet structure of the immunoglobulin fold allows some substitutions to be made within domains without loss of activity provided care is taken to ensure the integrity of the domain. In particular, as the antigen binding activity of the V domains is largely conferred by the CDR loops alone, it is possible to make changes in these loops, or switch them from one antibody to another, without disturbing the structure of the framework supporting them. This ease of genetic manipulation has led to the technology of antibody engineering which now allows the specific design of antibody molecules for particular applications.

Several factors must be considered in the design of antibody molecules, and these will be determined by the particular application for which the antibody is to be used. Some major issues to consider are listed in Table 2.1. A large range of modifications to the basic structure are possible, although the choice of one feature may limit the choices in other areas. Firstly, the antibody to be used must be of suitable specificity and affinity for the intended use, and subsequent manipulation should not usually result in alteration of the specificity, as this may lead to undesirable cross-reactivity with other, non-target, antigens. The affinity of the antibody for its antigen should not be drastically reduced, although in some instances this may be an advantage, for example in affinity purification applications. In some cases it has proved possible to improve affinity to a small extent and, in some therapeutic applications particularly, this may be advantageous. The valency of the antibody should also be considered and can be tailored for different applications. The normally divalent antibody can be made into monovalent antibody fragments or multiple binding sites linked together to increase the avidity of antigen binding. In addition, bispecific or even trispecific antibodies can be prepared with binding sites to more than one antigen. Antibody fragments of different sizes can be prepared which may have

Table 2.1 Issues in the design of antibodies for specific applications

Specificity and affinity of the binding site
Valency
Size
Requirement for effector function
Attachment of effector or reporter molecules
In vivo properties: immunogenicity
pharmacokinetics & metabolism
Cost of production

more appropriate properties *in vivo*, for example increased penetration into tissue or altered pharmacokinetics. The native effector or signalling functions of the antibody might be exploited for therapy, or specifically disabled if undesirable. Alternatively, a wide range of potentially therapeutic effector molecules, such as drugs, toxins, radionuclides or cytokines, can be attached to the antibody, as can reporter molecules for diagnostic uses. Specific properties can be designed into the molecule for *in vivo* use, such as appropriate pharmacokinetics and low immunogenicity to allow repeat usage. And of course, for industrial users the costs of producing the molecule required must be economic, and this can also be addressed by engineering. Forms of the antibody can be produced which are suitable for low-cost manufacturing and the use of high-yielding, low-cost expression systems can reduce costs further.

2.2 Isolation of variable region genes

At present it is not possible to design antibody variable domains capable of specific antigen binding *de novo*. Therefore a prerequisite for antibody engineering is the production of a suitable MAb and cloning of the variable region genes. With MAbs isolated by phage display, the gene is isolated directly, and variable region genes can be retrieved from hybridoma cells by several different methods. Early cloning of variable regions was carried out by isolating the rearranged genes from genomic clones (Seidman and Leder, 1978). Libraries of genomic clones constructed in a λ phage vector were size fractionated and screened with gene fragments or synthetic oligonucleotide probes to identify variable region heavy and light chain genes. This approach has the advantage that the Ig promoter and in some cases the Ig enhancer sequences can also be recovered as a single DNA fragment using suitable restriction sites which may be useful in subsequent expression studies (see Chapter 5). Another approach is to recover the genes from mRNA through cDNA cloning (e.g. Bothwell *et al.*, 1981). As there is abundant mRNA for the antibody in most hybridoma lines, only small libraries of cDNA are required (approx. 5000–10 000 clones). The libraries are also screened with gene fragments or oligonucleotides to isolate the required genes. Care must be taken however, as hybridoma cell lines often contain aberrant variable region gene segments, derived from the fusion partner, some of which are transcribed. During both genomic and cDNA cloning these extra antibody sequences can be cloned inadvertently (Cabilly *et al.*, 1984; Carroll *et al.*, 1988; Neumaier *et al.*, 1990), and thus all cloned sequences should be checked to ensure cloning of full length antibody variable regions for both heavy and light chains. In some cases the sequence of the antibody may be known or the N-terminal protein sequence can be determined, although in many cases identification of novel sequences is the goal. The extra light chains most commonly encountered as full length genes from mouse hybridomas are the κ138 or MOPC21 light chain (Cabilly and Riggs, 1985).

These cloning methodologies have now been largely replaced by cloning of variable region genes by PCR using the methods described in Section 1.7. This technique allows rapid isolation of variable region genes through PCR amplification, and can be carried out from very few cells (Liu *et al.*, 1992). An excellent practical guide to isolation of variable region genes by PCR has been published (Adair, 1997). One problem using this technique is the introduction of errors into the sequence due to replication errors or cross-contamination. This is usually overcome by sequencing multiple independent clones. Also aberrant and extra gene sequences originating from the fusion partner may be identified as described above. These can be rapidly screened against with primers that anneal specifically to known sequences in these antibody chains. With this technique, as well as conventional cloning procedures, the best test of accuracy is a functional test of the antigen binding ability of the antibody expressed from the cloned sequence. This can be done rapidly by expressing the antibody either as murine antibody or directly as a chimeric antibody (see Section 2.3.1) in a transient expression system such as COS cells or CHO cells to produce small amounts of recombinant antibody for analysis (Whittle *et al.*, 1987; see Chapter 5).

2.3 Overcoming immunogenicity

The application of MAbs for human therapy and diagnosis *in vivo* has been limited by many factors. One of the major problems encountered with the administration of murine MAbs to humans has been the generation of an immune response to the administered antibody protein known as human anti-mouse antibody response or HAMA. For antibody therapy in many types of diseases, repeat treatment will be required to achieve effective results. However, HAMA prevents repeat administration of antibody due to the formation of immune complexes which are rapidly cleared from the body, rendering repeat therapy ineffective. Also, in some cases, HAMA responses may result in adverse reactions such as allergic reactions or anaphylatic shock. Because of the problems in generating high-affinity human MAbs in many cases (discussed in Section 1.6), alternative solutions to this problem have been sought. These include the use of antibody engineering techniques to produce chimeric or fully humanised antibodies, chemical modification with polymers such as poly(ethylene glycol), the use of antibody fragments and immunosuppressant therapy.

2.3.1 Chimeric and humanised antibodies

One of the most successful approaches to overcoming immunogenicity has been the 'humanisation' of rodent MAbs by genetic engineering. A simple approach to making an antibody more human is the replacement of the constant domains of the antibody with constant domains of a human antibody (Morrison *et al.*, 1984). The resulting chimeric antibody (Figure 2.1) contains only the variable regions of murine origin and would therefore be expected to be less immunogenic in people. Many chimeric antibodies have been prepared and shown to retain the full antigen binding ability of the parent murine antibody as well as taking on the constant region effector functions of the human antibody used.

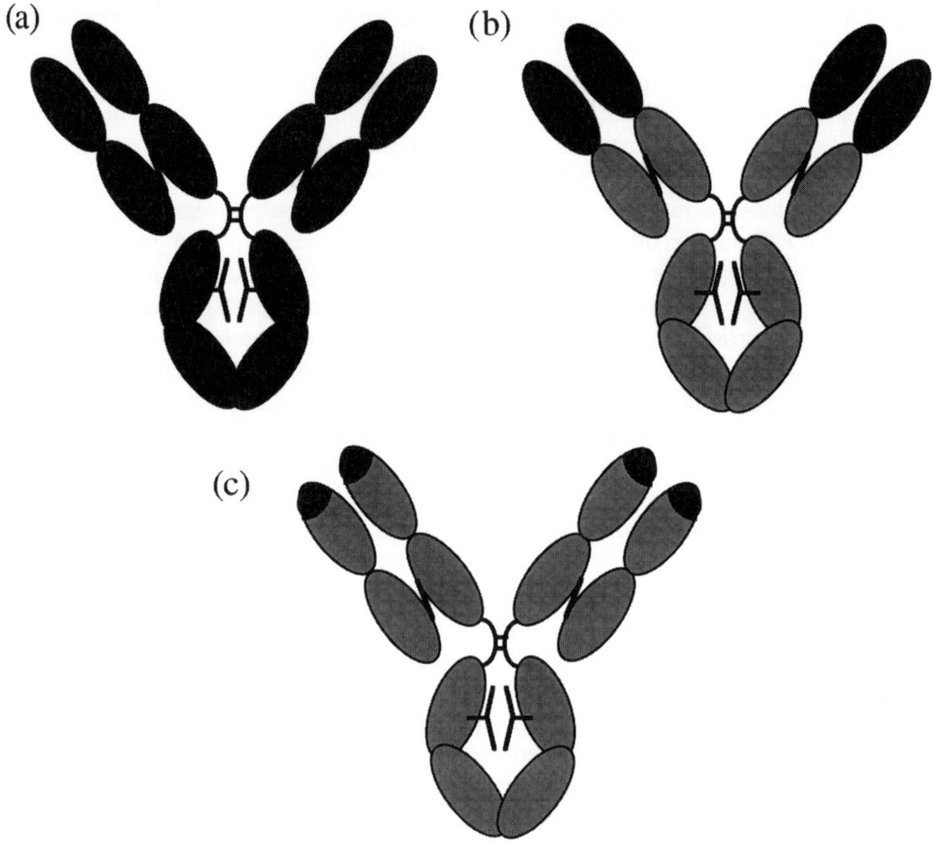

Figure 2.1 Humanisation of IgG: (a) parent murine antibody; (b) mouse:human chimeric antibody with the variable domains retained from the murine antibody and constant regions of human origin; (c) CDR-grafted antibody in which only the antigen-binding loops are retained from the murine antibody, the remainder of the sequence being of human origin

Construction of chimeric antibodies is relatively straightforward and requires simple ligation of the variable region genes 5′ to the constant region DNA and transfer into one of many available expression vectors. The constant region DNA can be in the form of genomic DNA (Whittle *et al.*, 1987) or cDNA (Liu *et al.*, 1987). If cloned by genomic cloning, a restriction site in the intron between the variable domain and constant domain can be used to assemble the required construct. As the intron sequences are removed during RNA splicing, alterations in the intron sequence are of little consequence. Alternatively cDNA clones require site-directed mutagenesis to introduce restriction sites at the V–C junctions to make the chimeric construct, taking care not to alter the required protein sequence. Also oligonucleotide linkers can be used to recreate variable region sequences from a convenient restriction site near the 3′ end of the V regions and introduce a splice donor site immediately after the coding sequence. A convenient restriction site is included to allow assembly of the chimeric gene. For PCR derived genes, primer design can allow incorporation of convenient sites for subsequent manipulation of the cDNA into an appropriate expression construct. PCR approaches using primers at the 5′ end of the coding region will not isolate the signal sequence, and thus this will need to be added to allow secretion of the expressed antibody. Construction of an appropriate vector for expression of a chimeric antibody in mammalian cells requires assembly not just of the antibody heavy and light chain genes themselves but also of a suitable promoter and regulatory elements (see Chapter 5).

Chimeric antibodies have also been produced by homologous recombination in hybridoma cells (Yarnold and Fell, 1994). Homologous recombination is achieved by targeting a plasmid with the human constant region genes of choice to the immunoglobulin locus in the hybridoma cell, such that the murine constant regions are replaced with the targeted human sequences. This approach can be used to chimerise both heavy and light chains but at relatively low efficiency. The other drawbacks of this approach include the limited range of antibody engineering which can be achieved (for example, it is not possible to produce fully humanised antibodies by this technique) and the limitation of the expression level to that seen with the hybridoma, which may be significantly lower than expression levels achieved with many recombinant cell lines.

Transgenic mice have now been developed which can produce chimeric antibodies directly. A bacteriophage derived recombination sysytem known as Cre-*loxP* can be used to replace the mouse antibody constant domain gene segments in embryonic stem cells with human sequences (Zou *et al.*, 1994). Mice producing chimeric light chains have been produced as well as $\gamma 1$ heavy chain recombinants and crossed together to produce offspring capable of producing $\gamma 1$ antibodies with both chimeric heavy and light chains at similar levels to endogenous $\gamma 1$ antibody in wild type mice. These mice are therefore theoretically capable of producing chimeric $\gamma 1$ antibodies to any antigen to which the mice can respond. This technique has the advantage that chimeric constructs do not need to be made for each individual antibody, as it should be possible to recover chimeric antibodies directly by hybridoma or recombinant techniques; however, the disadvantages are similar to the use of recombination in cell lines, in that a relatively restricted amount of engineering is possible.

The HAMA response to murine MAbs is induced in a large proportion (50–100%) of patients receiving antibody (Khazaeli *et al.*, 1994). Table 2.2 summarises some of the data obtained from clinical trials with chimeric antibodies. Although care should be taken in interpreting the figures absolutely, as these are compiled from different investigators using different clinical protocols and different assays to measure the immune response, it is clear that the response to chimeric antibodies is reduced compared with HAMA responses to murine antibodies. The immune responses detected to chimeric antibodies have been directed toward the murine variable domains, and not to the human constant regions. In some cases responses to novel epitopes created at the interface of the murine variable and human constant regions have also been observed (Khazaeli *et al.*, 1991; Baker *et al.*, 1991). The chimeric antibodies vary in their extent of immunogenicity, although the reasons for this are not understood. It is possible that differences seen simply reflect the different immunogenicity of the antibody variable domains, and this has enabled predictions of the immunogenicity of individual chimeric antibodies in man to be made by testing immunogenicity in animals (Cartner *et al.*, 1993). The antibody responses observed towards the chimeric antibody variable domains include responses to both the framework regions and the antigen-binding loops themselves. Antibodies against the antigen-binding loops (the CDRs) are known as anti-idiotype antibodies, and can usually block antigen binding.

An alternative type of chimeric antibody produced from a monkey antibody variable region linked to human constant regions, termed a 'primatised' antibody, has also been produced (Newman *et al.*, 1992). Primate antibody sequences are more homologous to

Table 2.2 Immunogenicity of mouse:human chimeric antibodies in clinical studies

Antibody	Source	Indication	Dosing	Immune response	Reference
c17–1A (γ1)	Centocor	Colorectal carcinoma	1–3 doses, 4–40 mg	1/16, 6%	LoBuglio *et al.*, 1989 Meredith *et al.*, 1991
cB72.3 (γ4)	Celltech	Colorectal carcinoma	1–3 doses, 3–7 mg	16/24, 67%	Khazaeli *et al.*, 1991 Meredith *et al.*, 1992a
cB72.3 (γ4)	Celltech	Colorectal carcinoma	1–4 doses, 10–20 mg	3/6, 50%	Baker *et al.*, 1991
cAnti-CD4 (γ1)	Becton-Dickinson	Mycosis Fungoides	6 doses, 10–80 mg	2/7, 29%	Knox *et al.*, 1991
c14.18 (γ1)	Damon Biotech	Melanoma	1 dose, 5–100 mg	8/13, 62%	Saleh *et al.*, 1992
c14.18 (γ1)	Damon Biotech	Neuroblastoma	1–5 doses, 7–144 mg	3/5, 60%	Uttenreuther-Fischer *et al.*, 1995
cL6 (γ1)	InVitron/ Bristol Myers	Carcinoma (lung, breast or colon)	1–4 doses, 35–700 mg/m^2	4/18, 22%	Goodman *et al.*, 1993
NR-LU-13 (γ1)	NeoRx	Colorectal carcinoma	1 dose, 42 mg	6/8, 75%	Weiden *et al.*, 1993
C2B8	IDEC	B-cell lymphoma	1 dose, 10–500 mg/m^2	0/15, 0%	Maloney *et al.*, 1994
cAnti-CEA (γ4)	Ciba Geigy	Colorectal carcinoma	1 dose, 2–4 mg	2/9, 22%	Buchegger *et al.*, 1995
cMOv18 (γ1)	Centocor	Ovarian carcinoma	1 dose, 3 mg	0/24, 0%	Buist *et al.*, 1995

human than murine sequences and thus it is possible that these may prove less immunogenic than mouse:human chimeric antibodies.

Humanisation of rodent antibodies can be taken further to produce fully humanised antibodies, in the form of reshaped, CDR-grafted or engineered human antibodies, in which much of the variable domain sequences are also replaced by human antibody sequence (Figure 2.1). In these approaches the antigen-binding loops are derived from the rodent antibody and much of the supporting framework is human. The relatively conserved structure of the human and rodent variable domain frameworks allow grafting of the hypervariable loops (CDRs) from one antibody onto another without loss of the overall conformation and structure of the loops and therefore allows antigen-binding ability to be retained. This should reduce any immune response against the variable domain framework. The remaining murine parts of the antibody, the CDRs, are also hypervariable in sequence in human antibodies and therefore more sequence variation may be permitted without generating immune responses.

The simplest form of CDR grafting involves the transfer of the CDR residues from the murine antibody to a human antibody variable domain framework. However, full antigen-binding activity is not achieved by this simple transfer, probably due to small conformational differences in the orientation of the antibody CDRs on the new framework structure. It has therefore proved necessary to introduce a small number of framework residue substitutions from the murine antibody to the humanised antibody, as well as the CDR residues, to restore full binding activity. Successful humanisation therefore requires several key choices to be made, including definition of the CDR residues, identification of human antibody framework sequences, and which if any framework residues to change to murine residues. Several different approaches to humanisation have been developed which largely reflect different approaches to making these choices. The first humanisation in this way was the heavy chain of the hapten binding antibody B1-8 (Jones *et al.*, 1986), and the humanisation of antibodies to complex protein antigens followed soon after (Verhoeyen *et al.*, 1988; Riechmann *et al.*, 1988a). Although difficulty in achieving full antigen binding activity was encountered early in the development of this technology, humanisation of murine antibodies is now relatively routine and many antibodies have been successfully humanised. A review of humanised antibodies, compiled in 1995, revealed over 100 then known examples in which humanisation of the variable domains had been successfully achieved (Adair and Bright, 1995).

The CDR residues are those residues most hypervariable in sequence between antibodies, and have been defined from a comparison of all known antibody sequences (Kabat *et al.*, 1991). All of the six antigen-binding loops are included within the six CDRs, which in several cases extend a short way beyond the structural loops. The choice of human framework is made from the known human antibody variable domain sequences. A human framework can be chosen for which a crystal structure is available, for example the heavy chains from the human antibodies KOL or NEW and the light chain from REI, so that the conformation of the loops can be modelled onto the existing structure and predictions made as to which framework residue substitutions may be required. Alternatively, the heavy and light chains from a single human antibody can be used, such as EU or LAY, or a consensus human antibody sequence. In many cases the human antibody of highest homology to the murine antibody to be humanised is used, as this requires fewer choices to be made with regard to possible framework substitutions. In addition, variable region sequences have been divided into a number of subgroups (Kabat *et al.*, 1991) which are useful in determining the choice of homologous framework.

The remaining choices are which framework residues from the mouse sequence to transfer to the human framework. Some residues not in the CDRs may contribute to antigen binding directly or indirectly by affecting the conformation of the loops, by allowing stable packing of the domain structure or by interacting with the other variable domain at the heavy:light chain variable domain interface (Foote and Winter, 1992). The ideal way to determine which framework residues would be required to maintain an accurate conformation of the CDRs would be by analysis of an X-ray crystallographic structure of the binding site. However, the time taken to generate a crystal structure for each antibody to be humanised is usually prohibitive. Alternatively, these residues can be identified by generating a three-dimensional model of the humanised antibody variable regions and then evaluation of the role of particular residues (e.g. Queen *et al.*, 1989) or by alignment and analysis of the sequences to determine their likely structural location on the basis of known structures and thus enable estimation of the consequences of substitutions (e.g. Riechmann *et al.*, 1988a). In this way some unusual amino acids at particular positions in the individual human framework sequences can also be identified and changed. These approaches to humanisation have been alternatively termed CDR-grafting (Jones *et al.*, 1986), reshaping (Riechmann *et al.*, 1988a), hyperchimerisation (Junghans *et al.*, 1990) and civilisation (Kurrle *et al.*, 1990).

In some cases comparisons of different humanisation methods have been carried out with the same antibody. The antibody 0.5β recognises the gp120 envelope protein of HIV, and has been humanised using the human framework regions from REI for the light chain and from NEW for the heavy chain, both antibody chains for which structural information is available (Maeda *et al.*, 1991). Alternatively the heavy chain was humanised using the consensus sequence of human group I heavy chains with the same light chain. Although several variants were tested, the consensus sequence framework gave significantly more active antibody, probably as a result of its higher degree of homology with the 0.5β murine antibody (Maeda *et al.*, 1991). A similar result was observed with humanisation of the antibody CAMPATH-9 (Gorman *et al.*, 1991). In this case two heavy chain grafted variants were tested based on the framework sequences from NEW and KOL. Again the version using the antibody framework of highest homology was more active, in this case based on KOL. Other examples are the heavy chain of the anti-CD18 antibody 1B4 (Singer *et al.*, 1993), which was humanised using three different frameworks (NEW, GAL and JON), and the anti-interleukin 6 receptor antibody AUK12-20, which was humanised using two different framework sequences, HAX and consensus group I (Sato *et al.*, 1994). In each case the best retention of antigen-binding activity was achieved using the most homologous human antibody framework, be it a consensus sequence or from an individual human antibody.

A further approach to humanisation has been developed, known as veneering or resurfacing (Padlan, 1991). In this approach it is proposed that the immunogenicity of murine framework regions is determined by the surface residues alone and thus identification of the surface residues not involved with antigen binding and their substitution with 'human' residues will reduce immunogenicity. This approach has been successfully applied with retention of antigen-binding activity (Roguska *et al.*, 1994) and also combined with the 'traditional' CDR-grafting technique to humanise an antibody in an approach referred to as framework exchange (Benhar *et al.*, 1994).

Once designed, the humanised antibody variable domain can be assembled by several procedures. Site-directed mutagenesis of a single stranded DNA template (Riechmann *et al.*, 1988a), or by gene synthesis with alternating overlapping oligonucleotides, filling the gaps with DNA polymerase (Queen *et al.*, 1989), or by PCR procedures to gap fill and amplify the DNA (Daugherty *et al.*, 1991). Of these the PCR approach is the most rapid and efficient although errors may be introduced, so several PCR assembled variable region genes may need to be sequenced to identify the correct product. The humanised variable region genes are then joined to constant region genes as described for chimeric antibodies.

For all humanisation approaches it is usually useful to produce several humanised variants of the antibody and produce small amounts of each for analysis of antigen binding or bioactivity. This can be achieved rapidly by transient expression of small amounts of antibody in mammalian cell systems such as COS or CHO cells (Whittle *et al.*, 1987; Baker *et al.*, 1994 – see Section 5.2). In theory it is only necessary to express the variable region genes as an Fv or Fab fragment to produce material for analysis (e.g. Benhar *et al.*, 1994). This can be achieved rapidly in *E. coli* expression systems; however, the analysis of the antigen-binding ability of these monovalent antigen-binding species is complicated by the possibility of formation of small amounts of dimeric or aggregated material which can greatly influence the outcome of the antigen-binding affinity measurement. For example, it has been shown that the presence of only 0.5% of $F(ab')_2$ in a monovalent Fab′ preparation can overestimate the antigen-binding measurement fivefold (Desplancq *et al.*, 1994). For this reason it is preferable to measure the antigen-binding ability of stable dimeric molecules such as IgG.

In some instances unexpected increases in antigen-binding affinity have been reported on humanisation (Co *et al.*, 1992; Carter *et al.*, 1992a, Baker *et al.*, 1994). This may be due to removal of a glycosylation site from the variable region (e.g. Co *et al.*, 1993), although in other cases the reason is unknown and may reflect subtle changes to the conformation of the binding site (e.g. Baker *et al.*, 1994). However, care must be taken to characterise the engineered antibody produced to ensure that properties of the antigen-binding site are not altered in an undesirable fashion, such as leading to a reduced specificity or some unexpected property. Subtle alterations in the recognition of the antigen by the engineered antibody are possible which may only be revealed by a detailed analysis (Kelley *et al.*, 1992).

Several humanised antibodies have now been examined in clinical studies and thus information on the success of humanisation in overcoming immunogenicity is beginning to emerge. Direct comparisons of immunogenicity of humanised antibodies with the parent murine antibody have not been carried out in patients, but several reports have appeared on the immunogenicity of humanised antibodies in monkeys compared to murine antibody (Hakimi *et al.*, 1991; Singer *et al.*, 1993; Stephens *et al.*, 1995). In each case, the immune response was much reduced compared to the parent murine antibodies, with an anti-idiotype response (i.e. directed to the CDRs) developed on repeat dosing. In the case reported in greatest detail (Stephens *et al.*, 1995), the CDRs appear less immunogenic when presented in the human framework since the anti-idiotype response to the humanised antibody was greatly reduced compared to the anti-idiotype component of the immune response to the murine antibody.

The immunogenicity of humanised antibodies in patients is also low, as summarised in Table 2.3. The earliest studies were carried out with the humanised antibody CAMPATH-1H. Two non-Hodgkin lymphoma patients treated with 1–20 mg doses of this antibody for up to 43 days showed no anti-CAMPATH-1H response during the course of treatment (Hale *et al.*, 1988). However, these results should not be over-interpreted, because such patients are somewhat immunocompromised prior to treatment and because the treatment itself is likely to be immunosuppressive. In subsequent studies CAMPATH-1H was administered to eight rheumatoid arthritis patients (Isaacs *et al.*, 1992) and four

Table 2.3 Immunogenicity of humanised antibodies in clinical studies

Antibody	Source	Indication	Dosing	Immune response	Reference
CAMPATH-1H	Univ. Cambridge/ Wellcome	Non-Hodgkin lymphoma		0/2	Hale *et al.*, 1988
CAMPATH-1H	Univ. Cambridge/ Wellcome	Rheumatoid arthritis	5 × 4 mg + 5 × 8 mg	0/8, 3/4 on retreatment	Isaacs *et al.*, 1992
CAMPATH-1H	Univ. Cambridge/ Wellcome	Systemic vasculitis	3 doses, 2–40 mg	1/4	Lockwood *et al.*, 1993
Humanised anti-Tac	PDL/Hoffman La Roche	Graft v. host disease	1–2 doses, 0.5–1.5 mg/kg	0/20	Anasetti *et al.*, 1994
Humanised M195	PDL/MSKCC	Myeloid leukemia	6 or 12 doses, 0.5–10 mg/m^2	0/13 (3 non-specific responses)	Caron *et al.*, 1994
CDP571	Celltech	Healthy volunteers	1 dose, 0.1–10 mg/kg	4/4 at 0.1 mg/kg 0/4 at 2, 5 and 10 mg/kg	Stephens *et al.*, 1995
hMN-14	Immunomedics	CEA producing cancer	1–3 doses, 1–20 mg	0/6 (without exposure to murine ab)	Sharkey *et al.*, 1995
rhuMAb HER2	Genentech	Breast cancer	250 mg + 10 doses 100 mg	0/46	Baselga *et al.*, 1996

patients with systemic vasculitis (Lockwood *et al.*, 1993). In the rheumatoid arthritis study, 4–8 mg doses were given over 10 days. Clinical benefit was seen in seven of the patients, and anti-CAMPATH-1H antibodies were not detectable in any after one course of treatment. Of four patients given a second course of treatment, three showed a detectable anti-CAMPATH-1H response. In the study in systemic vasculitis, one out of four patients developed an anti-idiotype response after treatment. No data are yet available concerning the nature of this response, whether it interferes with efficacy, and if so, whether such interference can be overcome using larger doses. A study with a humanised anti-Tac antibody in patients with graft versus host disease noted no immune responses in any of the patients (Anasetti *et al.*, 1994); however, caution must also be exercised in the interpretation of this study due to the severe immunodeficiency of these patients. The humanised anti-CD33 antibody hM195 has been evaluated in 13 patients with myeloid leukemia at 0.5–10 mg/m^2 with 6 or 12 doses per patient without the onset of any immune response (Caron *et al.*, 1994). The immune response to the humanised anti-tumour necrosis factor antibody CDP571 has been examined after administration of a single dose of 0.1 to 10 mg/kg to fully immunocompetent human volunteers (Stephens *et al.*, 1995). Administration of the lower doses led to the development of a weak anti-idiotype response, predominantly of the IgM isotype, although at higher doses responses were very low or undetectable. The weak IgM anti-idiotype antibodies raised at low doses did not appear to result in rapid clearance of the circulating antibody or to neutralise the ability of the circulating CDP571 to bind TNF, suggesting that this type of response would not be problematic. This type of weak immune response is also seen with the administration of human MAbs (e.g. Steiss *et al.*, 1990) and may be indicative of the humanised antibody behaving in a similar manner to administered human antibody. In a study with a humanised anti-p185^{HER2} antibody, rhuMAb HER2, 46 patients were administered 250 mg of antibody followed by 10 weekly doses of 100 mg each (Baselga *et al.*, 1996). No immune responses were observed. The absence of immune responses after administration of these large doses of antibody is consistent with the results of the other studies. The humanised anti-tumour antibodies hMN-14 and Hu2PLAP have also been studied in humans, but in these cases the analysis of immune response data is difficult due to responses to chemical groups attached to the antibody (Kalofonos *et al.*, 1994) or the administration of murine antibody to the same patients (Sharkey *et al.*, 1995). Nevertheless, in patients receiving hMN-14 alone no immune response was detected (Sharkey *et al.*, 1995). These results taken together demonstrate that fully humanised antibodies have low immunogenicity, and should allow prolonged or repeat therapy in man.

2.3.2 Antibody fragments to reduce immunogenicity

The use of fragments of antibodies has also been advocated as a possible means to reduce the immunogenicity of rodent antibodies in man. Antibody fragments such as Fab, $F(ab')_2$ and Fv can be produced from IgG by proteolytic digestion, or directly by recombinant means (see Section 2.4). Several comparative studies have shown that antibody fragments have reduced incidence of HAMA formation compared to IgG (e.g. Breitz *et al.*, 1992; Buist *et al.*, 1995), presumably as a result of the absence of the immunogenic rodent constant regions. However, there is still a sufficient immune response to murine Fab to prevent repeat dose therapy. To reduce responses, recombinant mouse:human chimeric and humanised antibody fragments have been produced. Similarly to studies with IgG, presentation of the murine variable region in a chimeric Fab fragment leads to less immune reponse than when the murine fragment is used. For example, the Fab fragment of the antibody 7E3 which is directed against the platelet glycoprotein IIb/IIIa receptor has been evaluated in clinical studies as both a murine Fab and a mouse:human chimeric Fab. Immune responses to the chimeric Fab in a large phase III trial of approximately 1400 patients were seen in 5.8% of cases, whereas in similar patients treated with murine Fab, immune responses were detected in approximately 45% of patients (Knight *et al.*, 1995). This was seen even though most of the reponses towards the murine Fab were directed toward the variable region which is identical to that of the chimeric Fab. This result therefore suggests that the presentation of the variable region attached to a human constant region results in a significant advantage over murine antibody fragments.

2.3.3 Chemical modification to reduce immunogenicity

An alternative strategy which has been used to reduce the immunogenicity of many different proteins including MAbs is chemical modification. Addition of polymers such as poly(ethylene glycol) (PEG) and low molecular weight dextran have been shown to be able to reduce the immunogenicity of antibodies in animal model experiments (Kitamura *et al.*, 1991; Fagnani *et al.*, 1990).

PEG can be attached to antibody using a variety of different chemical linkages, usually by attachment to amine groups of lysine residues on the surface of the protein. Some of the methods which have been used for attachment of PEG to antibody are shown in Figure 2.2. Attachment via cyanuric chloride was the first method developed for PEG attachment to protein (Abuchowski *et al.*, 1977), but is relatively harsh and often leads to inactivation of the biological function of the protein. Attachment via *N*-hydroxysuccinimide esters is now preferred, as it is a more gentle procedure (Zalipsky and Lee, 1992). Other lysine-directed attachment methods such as tresyl chloride-activated PEG are now available and have been used for antibody fragments (Delgado *et al.*, 1996). A two-step method for attachment to lysine via thiol groups has also been described (Pedley *et al.*, 1994). In this method stable attachment is achieved under mild conditions by modification of lysine residues with 2-iminothiolane (Traut's reagent) to generate free thiol groups on the antibody, followed by reaction with a thiol-specific, PEG–maleimide reagent. This method has the advantage that the number of thiol groups introduced can be accurately measured and the consequent coupling of PEG to the antibody closely controlled.

PEG conjugation to antibodies results in reduced immunogenicity and also changes in other biological properties such as resistance to proteolysis and increased circulating half-life (see Section 2.7). The magnitude of the effect may be related to the degree of substitution with PEG achieved, i.e. the number of PEG molecules per antibody molecule. Attachment of large numbers of PEG molecules often leads to reduced ability to bind antigen, particularly if harsh attachment methods are used which may also damage other sites in the protein. Therefore, production of suitable conjugates is a compromise between attaching as much PEG as possible and retaining antigen-binding ability. For example, conjugates of a murine MAb prepared by Kitamura *et al.* (1991) showed reduced antigen-binding ability with 10 and 15 PEG molecules per antibody, although a conjugate with 5 PEG molecules attached was relatively unaffected. The conjugate with 10 PEG molecules attached did not elicit an immune response after multiple administrations to rabbits, although the unmodified antibody readily raised an immune response.

An alternative approach has been to use highly substituted IgG–PEG conjugates as tolerogens. A human IgG–PEG conjugate, with 20 PEG molecules per antibody, administered to mice was able to prevent 85–90% of the immune response to subsequently

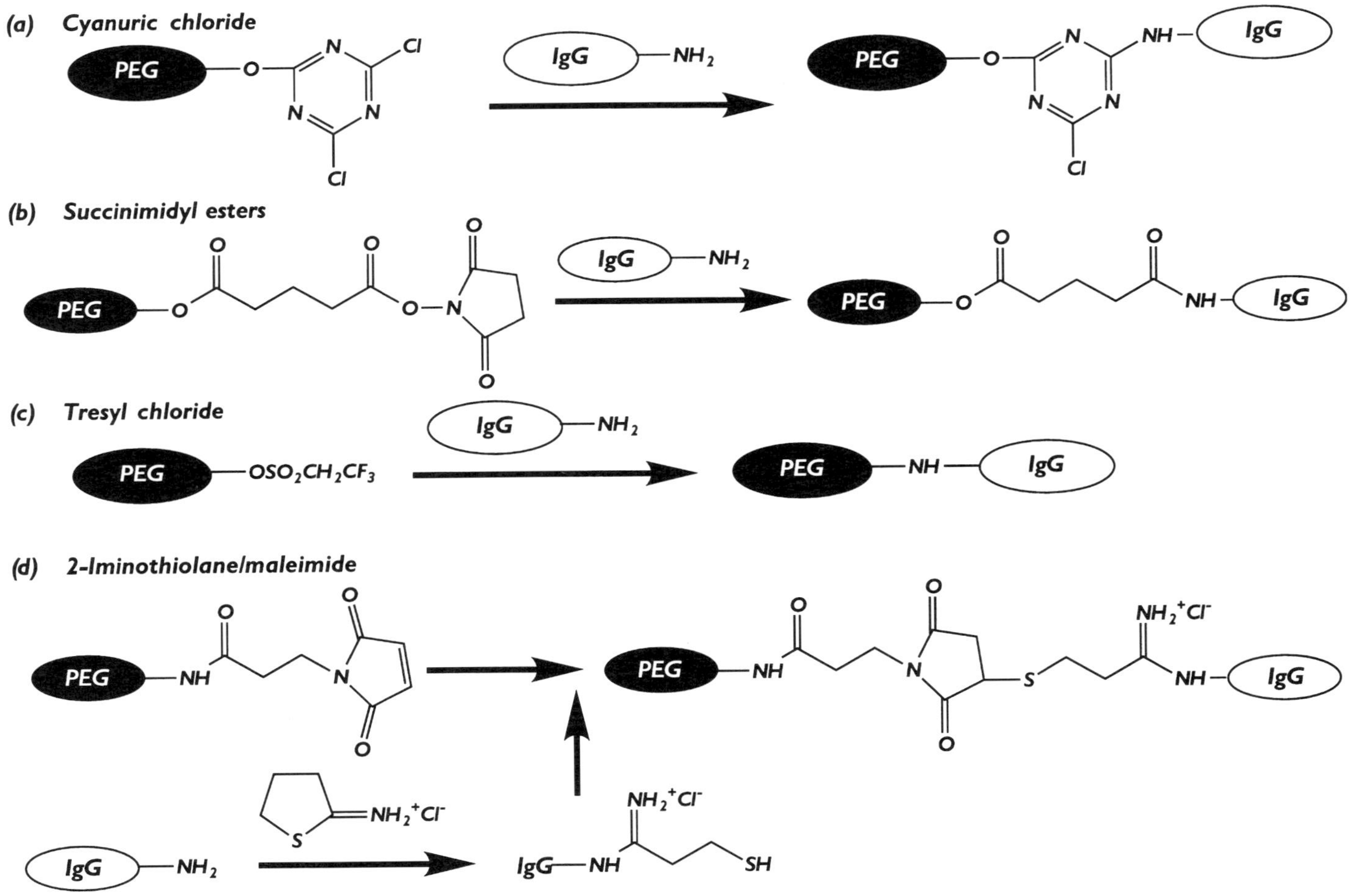

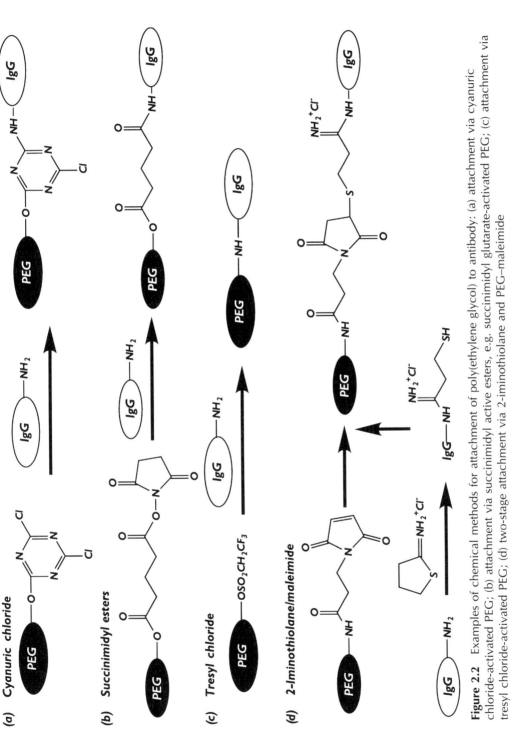

Figure 2.2 Examples of chemical methods for attachment of poly(ethylene glycol) to antibody: (a) attachment via cyanuric chloride-activated PEG; (b) attachment via succinimidyl active esters, e.g. succinimidyl glutarate-activated PEG; (c) attachment via tresyl chloride-activated PEG; (d) two-stage attachment via 2-iminothiolane and PEG–maleimide

administered unmodified human IgG (Wilkinson *et al.*, 1987). Subsequent studies have demonstrated that specific immunosuppression of HAMA can be achieved using this approach, such that reponses to the determinants of individual antibody protein chains are suppressed without affecting responses to other antigens (Bitoh *et al.*, 1993).

The covalent attachment of low molecular weight dextrans has also been shown to reduce immunogenicity of IgG in a similar manner to PEG attachment. Conjugates with up to 17 dextran molecules per antibody have been prepared which retain >50% of their immunoreactivity (Fagnani *et al.*, 1990). Tests of immunogenicity in animals reveal a significant reduction in immune response compared to the unmodified antibody.

2.3.4 *Immunosuppressive therapy*

A further approach to reduce the incidence of HAMA response has been to treat patients with immunosuppressive drugs. For example, cyclosporin A has been used to allow repeat dosing of a murine anti-carcinoembryonic antigen (CEA) antibody to patients with CEA producing tumours (Ledermann *et al.*, 1988). A significant reduction in HAMA was observed which allowed up to four administrations of murine antibody to be given. However, cyclosporin A was effective for only a limited time and secondary responses were observed in some patients. Cyclosporin A and similar drugs are general immunosuppressants, and while their use may be possible in cancer and other severe diseases, their routine use for antibody therapy is unlikely.

2.4 Antibody fragments

Fragments of MAbs have been of interest for many years as small antigen targeting molecules. Antibody fragments are also useful 'building blocks' for the assembly of molecules designed to carry other agents to a desired antigen such as diagnostic reagents or therapeutic conjugates. For *in vivo* applications, fragments of antibodies have been of interest due to their altered pharmacokinetic behaviour which has been investigated for cancer therapy with cytotoxic agents, and for their rapid penetration into body tissues which may offer advantages for imaging and therapy techniques. Also, effector functions can be disabled by removal of the Fc portion, and cell internalisation may be disabled by making a monovalent binding fragment. Initially antibody fragments, notably $F(ab')_2$ and Fab fragments and occasionally functional Fv fragments, were produced by proteolysis of IgG molecules. More recently recombinant techniques have extended the range of antibody fragments available by allowing novel fragments to be produced, and have extended the range of potential applications by opening up the possibilities of economical large-scale manufacturing.

2.4.1 *Antibody fragments from proteolysis of IgG*

The production of Fab and $F(ab')_2$ fragments of antibodies by proteolysis of IgG has been established for many years (Figure 2.3). As described in Chapter 1, the area around the antibody hinge is more susceptible to proteolysis than the tightly folded domains and thus this is the point at which cleavage usually takes place. Proteolysis above the disulphide bonds in the hinge region with, for example, papain results in Fab fragments which are monovalent for antigen binding, whereas proteolysis below the hinge disulphide bonds,

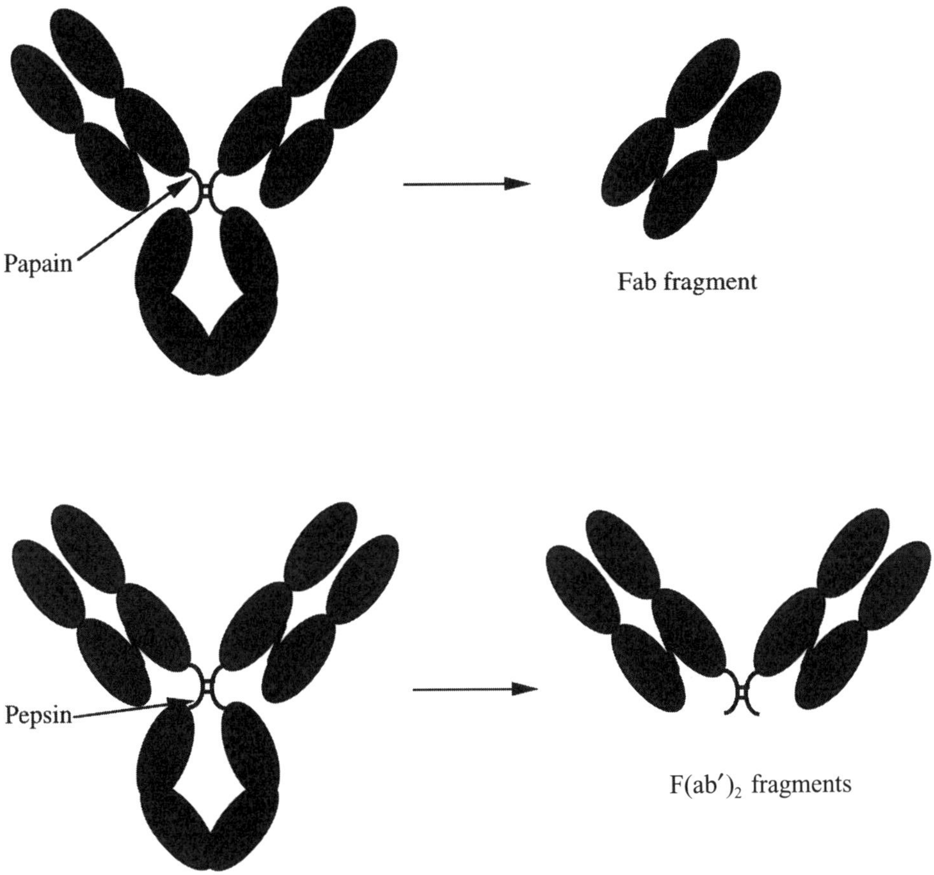

Figure 2.3 Diagram to represent the difference in cleavage point between the proteases pepsin and papain: papain cleaves IgG above the hinge region cysteines leading to the production of monovalent Fab fragments; pepsin cleaves IgG below the hinge region cysteines leading to the divalent $F(ab')_2$ fragment

with enzymes such as pepsin, results in the divalent $F(ab')_2$ fragment. Fc fragment is also produced during the production of Fab fragment by papain digestion, although during the production of $F(ab')_2$ fragments with pepsin the Fc portion is substantially degraded. Typical enzymes used to produce Fab and $F(ab')_2$ fragments from murine IgG are given in Table 2.4. Fab′ fragments, which are Fab fragments with an attached hinge region, can be produced by selective reduction of the hinge disulphide bonds of $F(ab')_2$ fragments. The digestion of IgG is very subclass-dependent and different experimental conditions are required for the production of fragments from different antibody subclasses (Parham, 1986). Individual antibodies within a subclass can also vary in their susceptibility to digestion (e.g. Rea and Ultee, 1993). Therefore, to maximise the yields of active fragments achieved, it is usually necessary to carry out small-scale experiments to optimise the enzyme used, buffer and pH, incubation time and the ratio of enzyme to antibody.

The production of Fab fragments with the non-specific thiol protease papain or other related thiol proteases, such as ficin, is generally successful across a range of antibody types, but the production of active $F(ab')_2$ fragments is more variable. Murine monoclonal IgG1 $F(ab')_2$ can be produced in good yield with bromelain or ficin (Milenic *et al.*, 1989;

Table 2.4 Enzymes commonly used for the production of murine antibody fragments by proteolysis of monoclonal IgG

	Fab	$F(ab')_2$
IgG1	Papain	Bromelain, ficin, pepsin, papain
IgG2a	Papain	Pepsin
IgG2b	Papain, pepsin	Lysyl endopeptidase*
IgG3	Papain	Pepsin

*Based on one report, general applicability currently unknown.

Mariani *et al.*, 1991) and in a few cases with appropriate buffer conditions with pepsin (Parham, 1986; Rea and Ultee, 1993). $F(ab')_2$ fragments can usually be generated from murine IgG2a and IgG3 antibodies with pepsin. However, the production of $F(ab')_2$ from murine IgG2b has not proved possible with pepsin or other enzymes, with rapid degradation to Fab/c and smaller fragments taking place (Parham, 1986), although one report suggests that careful digestion of a murine IgG2b with lysyl endopeptidase can allow $F(ab')_2$ formation (Yamaguchi *et al.*, 1995). Conditions for $F(ab')_2$ production from the rat IgG subclasses have been described (Rousseaux *et al.*, 1983), and human IgG1 and IgG4 $F(ab')_2$ can be produced readily with pepsin and bromelain respectively. It is also possible to produce fragments of IgM antibodies including Fab and $F(ab')_2$ with, for example, pepsin and trypsin (Parham, 1986), although as the individual binding site affinities of IgM antibodies are generally of low affinity (normally compensated for by the high avidity of the pentameric molecule) there has been relatively little use of these for other than experimental investigations. Similarly, murine IgA Fab, IgD Fab and IgE $F(ab')_2$ have been produced by proteolysis (Parham, 1986).

The production of other types of antibody fragment by proteolysis is more difficult. There has been much interest in the production of smaller antigen-binding fragments which may be more effective at penetrating tissues. The Fv fragment comprises only the paired heavy and light chain variable domains and is thus approximately half the size of a Fab fragment. Fv fragments were first prepared by pepsin digestion of the murine IgA myeloma protein, MOPC315 (Inbar *et al.*, 1972). However, attempts to produce Fv fragments of other antibodies have proved largely unsuccessful, being restricted to a small number of cases of digestion of IgM (Lin and Putnam, 1978), IgA myeloma protein (Sharon and Givol, 1976), IgG containing λ light chain (Ornatowska and Glasel, 1991) and IgG2a with a deleted CH1 domain (Takahashi *et al.*, 1991).

Although they are of great utility for experimental applications, the problems of expense, heterogeneity of product, difficulty of purification and limited transferability from one antibody to another have limited the applications of antibody fragments produced by proteolytic means. Also, the range of fragments which can be produced in this way is restricted, and thus much effort has recently been invested to investigate the potential of recombinant antibody fragments.

2.4.2 *Recombinant antibody fragments*

The direct expression of recombinant antibody fragments allows a range of antibody fragments to be produced using any monoclonal antibody specificity without the restrictions imposed by proteolysis of the native antibody structure. $F(ab')_2$, Fab′, Fab and Fv can all be readily produced by expression of the appropriate gene segments in a suitable host, as can a range of novel fragment types. Antibody fragments can be expressed in mammalian cells and also in microbial systems, particularly *E. coli* (see Chapter 5). Two modes of expression are available for use with *E. coli*, intracellular expression and secretion to the periplasm. In common with other types of protein expressed in *E. coli*, intracellular expression of antibody fragments usually results in the accumulation of the expressed protein in a denatured form as an insoluble inclusion body which requires solubilisation and refolding to obtain active material (e.g. Bird *et al.*, 1988; Field *et al.*, 1989). This is often an expensive and low yielding procedure. Alternatively, attachment of a signal sequence to each protein chain can result in secretion of the antibody fragment to the periplasm, from which soluble active protein can be recovered (Skerra and Pluckthun, 1988; Better *et al.*, 1988). The amount of antibody fragment secreted varies between antibodies and between different types of antibody fragment. In general smaller antibody fragments are produced more efficiently than larger fragments, and therefore *E. coli* has become the most widely used expression host for Fv-based fragments, whereas with Fab fragments the most efficient expression system appears to depend on the particular antibody studied.

Fab-based fragments

Fab and Fab′ fragments are produced by co-expression of the light chain and Fd or Fd′ fragment of the heavy chain. These vary only in the presence of the hinge region in Fd′. Fab fragments are relatively stable due to extensive non-covalent interactions between the protein chains and also the presence of a disulphide bond anchoring the chains together. Assembly of expressed light chain and truncated heavy chain (Fd or Fd′) has generally been efficient. However, expression of Fd′ with light chain results in Fab′ production, with little or no $F(ab')_2$ produced. Expression in both mammalian cells and *E. coli* has been shown to result in high yields of Fab and Fab′ fragments and may result in more homogeneous molecules than those obtained by proteolysis.

For many applications the increased avidity of the bivalent molecule may be important and therefore attempts to produce bivalent antibody fragments have been made. Direct expression of $F(ab')_2$ can be achieved, but requires design of a hinge region containing multiple disulphide bonds. Hinge regions with single cysteine residues have resulted in very little $F(ab')_2$ production (Carter *et al.*, 1992a; King *et al.*, 1994) and hinge regions containing two cysteine residues results in only 10–30% $F(ab')_2$ formation (King *et al.*, 1992a; Better *et al.*, 1993). Engineering a hinge region with three potential disulphide bonds has resulted in the production of approx. 70% $F(ab')_2$ from Fab′ expressed in *E. coli* (Rodrigues *et al.*, 1993) and inclusion of five disulphide bonds in the hinge region has resulted in quantitative $F(ab')_2$ formation (King *et al.*, unpublished data). Alternatively, recombinant $F(ab')_2$ can be produced by *in vitro* re-oxidation of the hinge thiols of the expressed Fab′ (King *et al.*, 1992a) or chemically cross-linked $F(ab')_2$ can be produced by specifically cross-linking the hinge thiols (Carter *et al.*, 1992a; King *et al.*, 1994) (Figure 2.4). If chemical cross-linking is used a single hinge thiol is desirable to reduce side-reactions and allow high cross-linking yields. This is easily designed into the Fab′ construct to be expressed. Cross-linking is achieved using a cross-linker with two thiol reactive groups, usually maleimide groups, and can be used to generate $F(ab')_2$-like molecules with two Fab′s of the same antibody or using two different Fab′s to produce bispecific antibody fragments (Glennie *et al.*, 1987; see Section 2.5). Chemical cross-linking also allows the possibility of incorporating extra moeities into the cross-linking reagent and therefore the site-specific attachment of reagents away from the antigen-binding site (Figure 2.4).

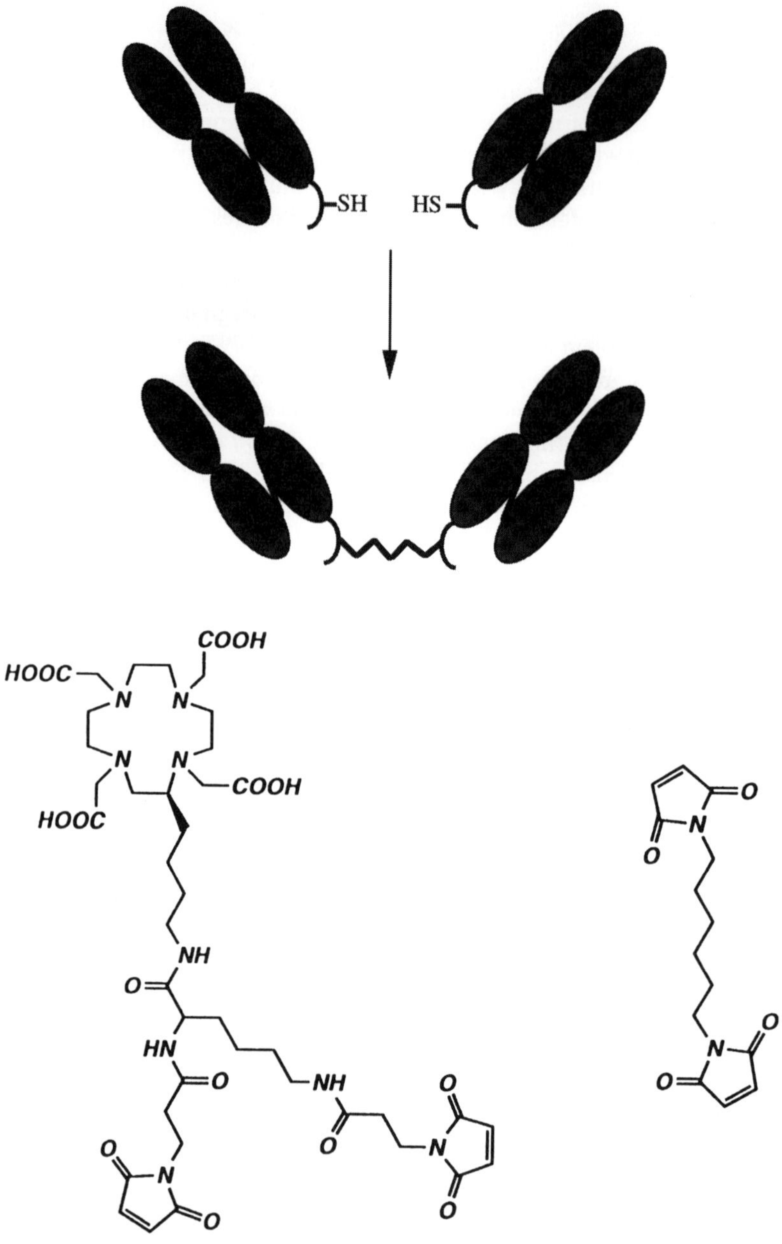

Figure 2.4 Chemical cross-linking of Fab′ fragments to form chemically cross-linked di-Fab, using either two Fab's of the same specificity or two of different specificity, in which case a bispecific fragment is produced. Also shown are the structures of two linkers which can be used for this purpose. Hexane bis-maleimide is a simple linker which allows di-Fab formation through cross-linking cysteine thiol groups. Alternatively linkers can be used which contain sites for site-specific attachment. In this case a 12N4 macrocycle is incorporated for site-specific labelling with radioactive metals (King *et al.*, 1994).

For example, cross-linkers containing macrocyclic ligands for the stable attachment of radioisotopes have been used to allow formation of radiolabelled antibody fragments without the need for further derivatisation of the antibody (King *et al.*, 1994).

Fv-based fragments

The two chains of the Fv fragment are less stably associated than the Fd and light chain of the Fab fragment with no covalent bond and less non-covalent interaction, but nevertheless functional Fv fragments have been expressed for a number of different antibodies. Characterisation of Fv fragments has revealed that although the antigen-binding ability of the Fab fragment is retained, dissociation of the two domains takes place at low protein concentration or under physiological conditions. Thus strategies for stabilising Fv fragments have been of great interest. Two strategies have been investigated with a number of different antibodies: firstly, mutating a selected residue on each of the V_H and V_L chains to a cysteine to allow formation of a disulphide bond between the two domains; and secondly, the introduction of a peptide linker between the C-terminus of one domain and the N-terminus of the other, such that the Fv is produced as a single polypeptide chain known as a single-chain Fv (Figure 2.5).

The production of single-chain Fv′s (scFv′s) was first described in 1988 (Bird *et al.*, 1988; Huston *et al.*, 1988). Since then a large number of scFv′s have been produced which retain antigen-binding characteristics, although in some cases antigen-binding affinity is reduced. Expression in *E. coli* has been the most frequently used production method, with both intracellular expression and secretion enabling high yields of scFv to be made. The production of scFv molecules requires the identification of a suitable peptide linker to span the 35–40 Å distance between the C-terminus of one domain and the N-terminus of the other and allow correct folding and assembly of the Fv structure. Several different types of linker have been used and shown to result in functional scFv. Linkers have been selected from searching existing protein structures for protein fragments of the appropriate length and conformation, or have been designed *de novo* based on simple, flexible structures with perhaps the most commonly used being the 15 amino acid sequence

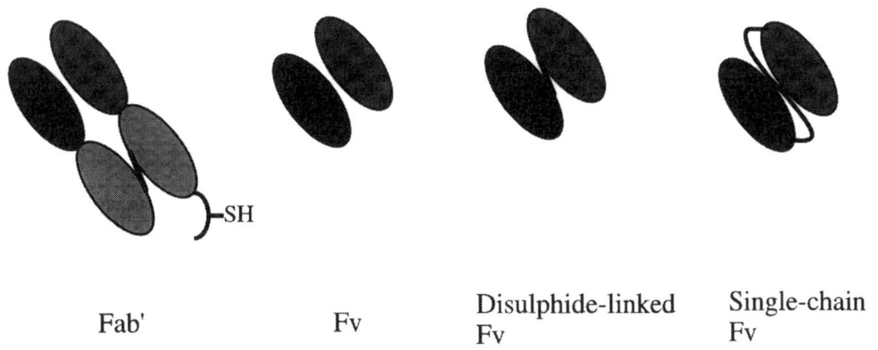

Figure 2.5 The organisation of domains in a Fab′ fragment compared to Fv, disulphide-linked Fv and single-chain Fv fragments: disulphide-linked Fvs are formed when a selected amino acid residue on both V_L and V_H is mutated to cysteine such that a disulphide bond can form between the variable domains; single-chain Fvs are formed as a result of the introduction of a peptide linker into the sequence, linking the C-terminus of one variable domain to the N-terminus of the second

$(Gly_4Ser)_3$. Active single-chain Fv molecules have been produced in both of the two possible orientations, VH–linker–VL or VL–linker–VH; however, for some antibodies one particular orientation may be preferable as a free N-terminus of one domain, or C-terminus of the other, may be required to retain the native conformation and thus full antigen binding. For example, VL–linker–VH scFv′s of the anti-tumour antibody B72.3 have been shown to be more active than VH–linker–VL with a range of different linker lengths (Desplancq *et al.*, 1994).

Many scFv′s have been shown to be susceptible to aggregation with dimers and larger species formed in a concentration-dependent manner (reviewed by Raag and Whitlow, 1995). This is because the interactions between VH and VL are relatively weak and there appears to be an equilibrium set up between assembled and unfolded scFv in solution. This leads to domain interchange with neighbouring molecules taking place at high concentrations (Figure 2.6). This is particularly prevalent if strain is introduced into the conformation of the scFv molecule by the use of short linker peptides. Conversely, the use of longer peptide linkers relaxes any strain in the conformation and encourages monomer formation (Whitlow *et al.*, 1993). For example, the B72.3 scFv is in the form of mostly dimers and high molecular weight aggregates with linker lengths of 0, 5 and 10 amino acids but mostly monomeric with linkers of 15, 20, 25 and 30 amino acids (Desplancq *et al.*, 1994). In addition, the tendency towards dimer formation at high concentration decreases as the linker length increases (Figure 2.6). The potential of forming dimers with very short linkers, or no linker at all, can be exploited to produce stable dimeric structures which have been termed ‘diabodies’ (Holliger *et al.*, 1993). Such an approach can also be used to create molecules with two different binding specificities, bispecific diabodies, by fusing the VH of an antibody of one specificity to the VL of another and vice versa (see Section 2.5). The crystal structure of a diabody recognising a phospholipase enzyme has been solved and it has been demonstrated that the VH and VL domains on the two polypeptide chains associate as expected, although in this case there is also non-covalent association between two diabody molecules to form a tetrameric structure (Perisic *et al.*, 1994).

Fv′s stabilised by disulphide linkages appear to be more stable structures than scFv′s (Glockshuber *et al.*, 1990). The introduction of a disulphide bond between the VH and VL domains to form a disulphide-linked Fv requires the identification of residues in close proximity on each chain which are unlikely to affect directly the conformation of the binding site when mutated to cysteine, and will be capable of forming a disulphide bond without introducing strain into the structure of the Fv. Sites have been identified in both CDR regions and framework regions which appear to result in the formation of such disulphide bonds and allow the production of stabilised Fv fragments which retain antigen-binding characteristics (Glockshuber *et al.*, 1990; Reiter *et al.*, 1994). The use of sites in the framework region is likely to be of more general utility, and several different antibody Fv fragments have now been stabilised in this way (reviewed by Reiter and Pastan, 1996).

A different approach would be to use a single immunoglobulin domain for antigen binding. In a small number of cases a single-antigen binding domain may have antigen-binding activity. For example, VH domains with affinity for lysozyme in the 20 nM range have been isolated which have been termed single domain antibodies or dAbs (Ward *et al.*, 1989). Light chain domains may also have antigen-binding activity in some cases (Masat *et al.*, 1994). However, single domains have been difficult to express and are poorly soluble due to an exposed hydrophobic surface which normally associates with the other V domain, and thus such small antigen binding proteins have not been widely used.

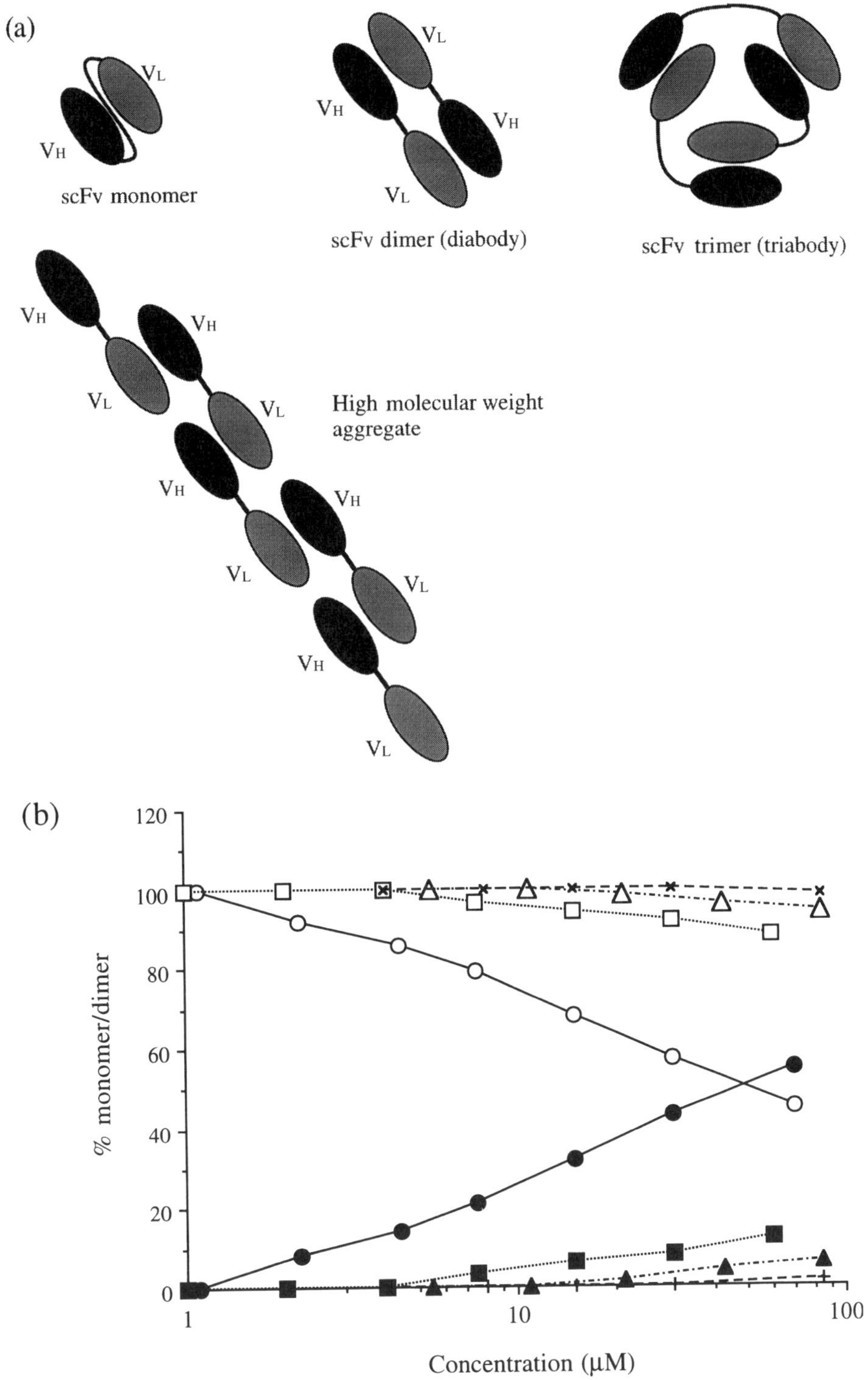

Figure 2.6 Aggregation of single-chain Fv: (a) representation of scFv monomer, dimer (diabody), trimer (triabody) and high-molecular weight aggregate; (b) extent of dimer formation with variation in linker length for scFv of the antibody B72.3, adapted from Desplancq *et al.* (1994). scFvs in the orientation V_L–V_H with linker lengths of 15, 20, 25 and 30 amino acids are shown. (○) Monomer, (●) dimer V_L–15–V_H; (□) monomer, (■) dimer V_L–20–V_H; (△) monomer, (▲) dimer V_L–25–V_H and (x) monomer, (+) dimer V_L–30–V_H.

Camels have been found to have naturally occurring antibodies without light chains, and thus attempts have been made to engineer the interface of the VH domain to mimic camel VH domains which may be more soluble (Davies and Riechmann, 1995). Although these attempts have been partially successful, in that antigen-binding specificities could be obtained from a library of such engineered VH domains, only low affinity binding has been achieved to date. This is probably a consequence of the reduction in available CDR regions for antigen binding from 6 to 3 compared to authentic Fv. In some cases it may be possible to increase the number of antigen contact residues, for example by engineering an extra antigen-binding loop which can be inserted into the framework 3 region of VH domains to increase binding affinity (Simon and Rajewsky, 1992).

Multivalent antibody fragments

As monovalent binding entities, Fv fragments suffer from relatively low avidity binding and therefore there has been much interest in producing multivalent Fv fragments which still retain a relatively small size. The production of 'diabodies' is one approach as described above, and some diabodies can be produced to high levels in the form of stable dimers (Zhu *et al.*, 1996). However, the stability of such dimeric species varies from one antibody to another (Whitlow *et al.*, 1994). In some cases, particularly with direct fusions of VH and VL, stable trimeric species are produced which have been termed 'triabodies' (Iliades *et al.*, 1997). Some triabodies show improved avidity for antigen as expected from the increased number of binding sites, although this is not always the case. A crystal structure of a non-functional triabody produced from the variable domains of two anti-hapten antibodies has been solved and demonstrated a dramatic alteration in the conformation of the VH CDR3 antigen-binding loop (Pei *et al.*, 1997).

Several alternative approaches have been examined in attempts to produce dimeric species (Figure 2.7). A peptide linker can be introduced between two scFv fragments to produce a single polypeptide chain di-scFv molecule. However, attempts to produce such dimers have often resulted in reduced antigen binding for the two binding sites, even when anti-hapten antibodies were used (Mallender and Voss, 1994). One of the most successful strategies to produce dimeric species has been to introduce a cysteine residue which can be used to introduce a disulphide bond or a chemical cross-link between two scFv fragments. Either this can be introduced directly at the C-terminus of an scFv, via a linker sequence, or a natural antibody hinge region can be attached to form an scFv′ fragment (Cumber *et al.*, 1992; Adams *et al.*, 1993; King *et al.*, 1994). Dimeric scFv can then be formed in a similar manner to that described for Fab′ fragments above, and offers the same potential for site-specific attachment of radiolabels or other moieties (King *et al.*, 1994).

An alternative approach has been the fusion of Fv fragments with peptides that naturally form dimeric structures. Amphipathic helices can be fused to the C-terminus of scFv fragments to allow spontaneous dimerisation (Pack and Pluckthun, 1992; Pack *et al.*, 1993). These 'dimerisation domains' can be taken from, for example, a four-helix bundle structure or a leucine zipper from a naturally occurring protein. Leucine zippers are peptides of approximately 30 amino acids which naturally form amphipathic helices with leucine residues lining up on the hydrophobic face of the helix. The hydrophobic faces of two zipper peptides associate with each other through hydrophobic interactions. Therefore the association of these molecules occurs through non-covalent interactions, which require further stabilisation to form stable dimeric species. Stabilisation is achieved through engineering a disulphide bond to be formed between the two dimerisation domains when

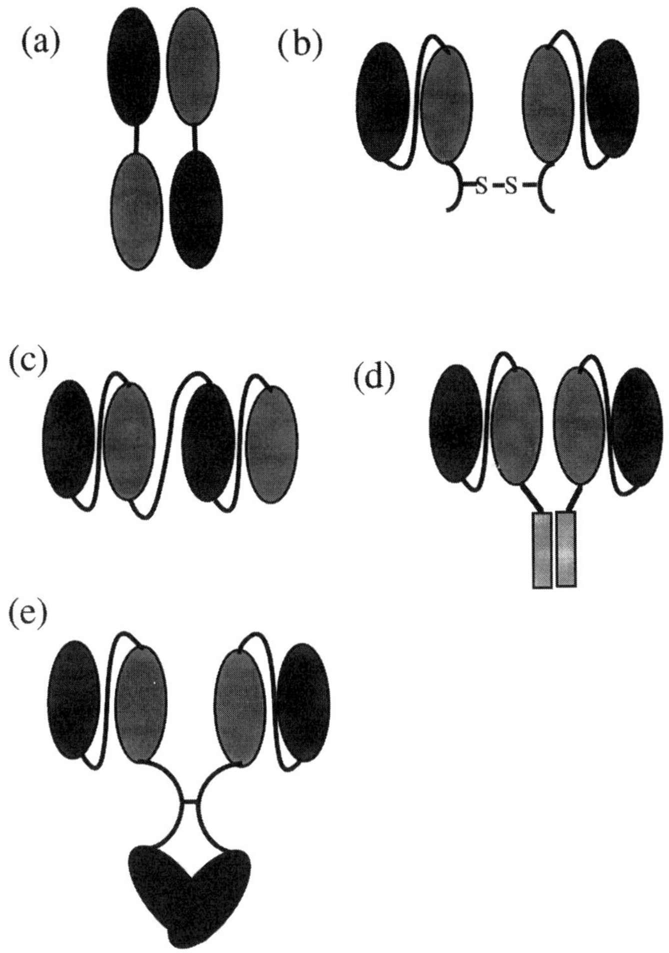

Figure 2.7 Strategies for formation of dimeric scFv species: (a) diabodies formed spontaneously from scFv with short peptide linkers; (b) formation of a disulphide bond or chemical cross-linking through attached hinge region peptides; (c) direct fusion of one scFv to another through an extra linking peptide; (d) the use of dimerisation domains such as leucine zippers; (e) minibody formed from the use of IgG CH3 domains and a hinge region peptide to drive, and stabilise, dimerisation

associated. This approach to dimer formation has the advantage that no *in vitro* manipulation is required after expression and purification of the molecule. However, for *in vivo* applications of Fv fragments, the added dimerisation domains may increase the immunogenicity of the molecule, thus limiting its usefulness. An alternative dimerisation domain is the IgG CH3 domain (Hu *et al.*, 1996). As described in Chapter 1, there are extensive non-covalent interactions between IgG CH3 domains which help drive assembly of the native paired heavy chains in IgG. This interaction can also be used to promote dimerisation of scFv–CH3 fusion proteins which can be stabilised by inclusion of a hinge region containing cysteine residues to allow disulphide bond formation (Figure 2.7).

The use of self-assembling structures can be extended to produce higher avidity antigen-binding proteins such as tetrameric scFv's. A modified amphipathic helix can result in tetrameric fragments (Pack *et al.*, 1995), as can fusions to the core region of streptavidin

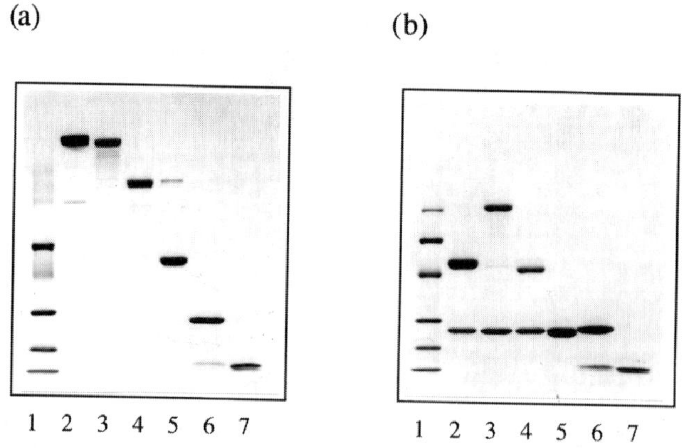

Figure 2.8 SDS-PAGE analysis of chemically cross-linked di-Fab and tri-Fab compared to other immunoglobulin forms: (a) (non-reducing conditions) lane 1, molecular weight markers; lane 2, IgG; lane 3, chemically cross-linked tri-Fab′; lane 4, chemically cross-linked di-Fab′; lane 5, Fab′ containing trace of $F(ab')_2$; lane 6, scFv; lane 7, Fv; (b) SDS-PAGE under reducing conditions, samples as for the non-reduced gel

which forms a natural tetrameric structure (Kipriyanov *et al.*, 1996). These fusion proteins can be difficult to express to high levels directly in soluble form, although procedures have been developed to enable recovery of active material from insoluble protein expressed in *E. coli* (Kipriyanov *et al.*, 1996). The tetramerisation domain of the human protein p53 has also been used to produce fusion proteins which assemble to tetrameric scFv's (Rheinnecker *et al.*, 1996). The use of human protein domains may have advantages for *in vivo* use as the human protein may be less immunogenic than entirely synthetic sequences or bacterial proteins such as streptavidin, although as a normally intracellular protein, p53 may not be the optimal choice for such reagents.

Trimeric and tetrameric antigen-binding proteins can also be produced by chemical cross-linking of Fab′ or scFv′ with tri- or tetra-maleimide cross-linking reagents (Schott *et al.*, 1993; King *et al.*, 1994), or by polyoxime chemistry (Werlen *et al.*, 1996). This approach can be easily used for a range of fragments from different antibodies, either recombinant or produced by proteolysis, and is perhaps the most attractive method for the production of multivalent fragments at present. Such cross-linked fragments can be produced in high yield and purified to homogeneity (Figure 2.8). Tri-scFv's and tri-Fab′s produced in this way show increased avidity of binding to antigen and favourable *in vivo* properties (see Chapter 4).

2.5 Antibodies with multiple specificities

Bispecific antibodies have two different specific antigen-binding sites, one on each arm of the antibody molecule, and are potentially useful for a range of diagnostic or therapeutic applications. For example, a bispecific antibody can be prepared with one binding specificity to a target tissue antigen and a second specificity to peroxidase for use in immunohistochemistry (Milstein and Cuello, 1983). Alternatively, for therapy, a bispecific antibody could be used with one arm binding to a tumour cell antigen and one to a cell killing

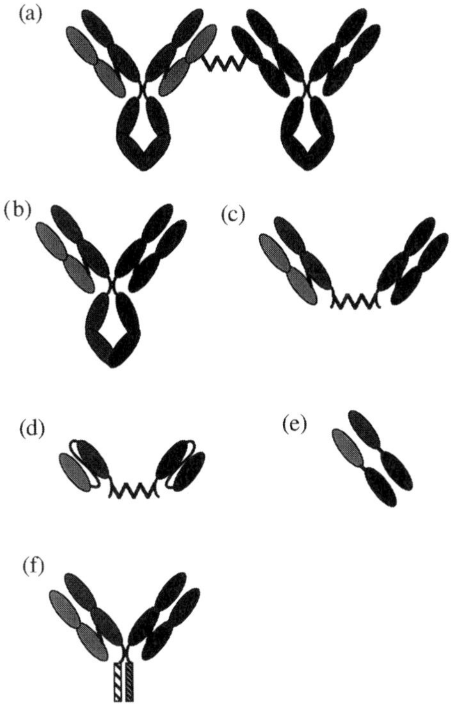

Figure 2.9 Approaches to bispecific antibody production: (a) chemical cross-linking of IgG, (b) production of IgG from a hybrid hybridoma, (c) chemical cross-linking of Fab's, (d) chemical cross-linking of scFvs, (e) bispecific diabody, (f) association through leucine zippers

agent such as a drug, toxin or effector cell (reviewed by Fanger and Guyre, 1991). There are now several approaches to the production of bispecific antibodies (Figure 2.9). All of these approaches have advantages and disadvantages with no one method being used predominantly at the present time.

Hybrid hybridoma cell lines, also known as quadromas, can be produced by the fusion of two antibody-secreting cells such that the resultant hybrid produces two individual heavy and light chains. This can be achieved using two hybridoma cells, or by direct fusion of a hybridoma cell and an immune spleen cell (triomas). Fusion of two existing hybridoma cell lines is usually preferable as the antibodies to be used can be characterised individually and thus the outcome is more predictable. The heavy and light chains associate with each other to produce IgG using the normal cellular mechanisms, but without any selectivity in the pairing of the chains ten combinations are possible, with the bispecific IgG making up only a small proportion of the total (Figure 2.10). Attempts to improve the proportion of bispecific produced have been made, for example, by using rat/mouse hybrid hybridomas wherein the light chains will only pair with the heavy chain of the same species (Lindhofer *et al.*, 1995). However, purification of the correct bispecific antibody from a mixture of very similar IgG molecules remains a considerable problem. If both antigens are available these can be used for sequential affinity purification although this is an expensive operation which is usually difficult to perform on a large scale. Alternatively, methods such as ion-exchange chromatography based on differences in charge of the individual IgG chains can be used (Allard *et al.*, 1992), or if the two

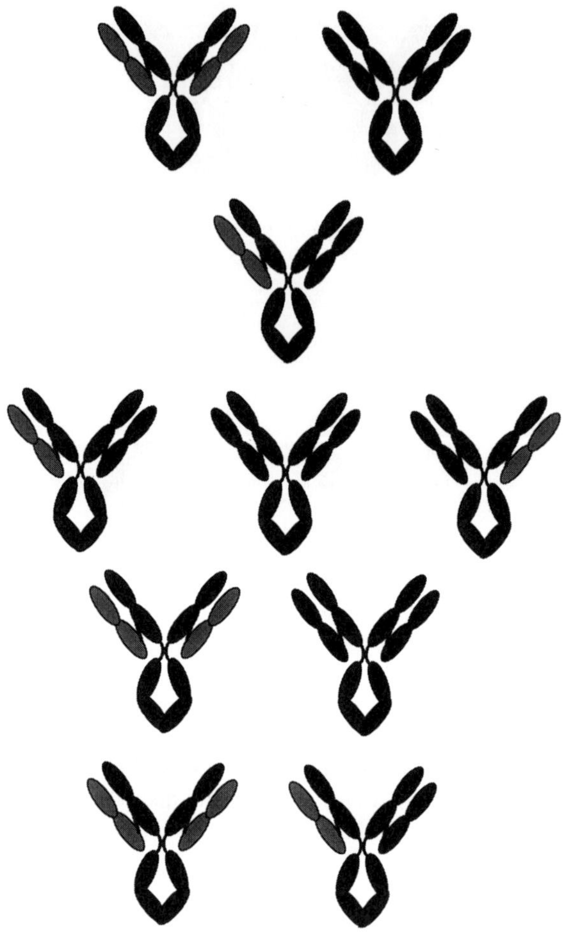

Figure 2.10 Combinations of heavy and light chains possible from a hybrid hybridoma

heavy chains are of different isotype, fractionation on protein A chromatography may be possible (Lindhofer *et al.*, 1995). A further problem with hybrid hybridomas is that they are usually unstable as a result of being polyploid due to carrying the chromosomes of each of the parent cells. Chromosome loss and loss of the ability to secrete bispecific antibody is a constant danger and maintenance of a productive cell line therefore requires frequent cloning to ensure that both sets of heavy and light chain genes are still present. An alternative method to overcome the instability problem is the introduction and expression of the genes for two heavy and light chains in the same cell by recombinant DNA techniques (Songsivilai *et al.*, 1989). This method also allows engineering of the antibody genes to introduce other desired features; for example, humanised antibody genes can be used.

A bispecific reagent can also be prepared simply by chemically cross-linking two IgG molecules (e.g. Karpovski *et al.*, 1984). This results in an undefined heteroaggregate of approx. 300 kDa, and one or both of the antigen-binding sites may very well be damaged during the non-specific chemistry required. This approach does have the advantage, however, that bivalent binding for both antigens may be retained in some cases resulting in high avidity binding to both antigens.

The production of bispecific IgGs by protein engineering techniques has also been investigated by designing specific protusions, 'knobs' on one CH3 domain and a corresponding 'hole' in the CH3 domain of the second antibody heavy chain (Ridgeway *et al.*, 1996). This is achieved by swapping a small amino acid at the interface with a large one to make the 'knob' and vice versa to produce a complementary 'hole'. Assembly of the two engineered heavy chains is thus favoured while assembly of the homodimers containing two 'knobs' or two 'holes' is unfavourable. Coexpression of the two heavy chains resulted in mostly heterodimer produced. However, pairing of the correct light chains may require further engineering of the interface between the heavy and light chains, possibly using further 'knobs' and 'holes'.

A more attractive alternative is to prepare Fab′ fragments from the two antibodies required and then to cross-link the two Fab′ fragments via the hinge cysteine residues to produce bispecific $F(ab')_2$. This can be achieved by use of a thiol reactive cross-linker (Figure 2.4). Methods have been devised for the preparation of bispecific molecules without any homodimers produced, and such bispecific reagents have been shown to maintain the full functionality of each binding site (Brennan *et al.*, 1985; Glennie *et al.*, 1987). Chemically cross-linked $F(ab')_2$ produced with bis-maleimide reagents, which therefore contain thioether bonds between the two Fab′s, are produced in higher yields and are more stable than disulphide-linked molecules (Glennie *et al.*, 1987). This approach can be used with Fab′ molecules produced by proteolysis or with recombinant Fab′ or other antibody fragments engineered to contain a free thiol group, such as scFv′. If Fab′ with multiple hinge thiols is used, or alternatively tri-functional linkers, it is also possible to produce bispecific tri-Fab with two antigen-binding arms to one of the antigens, allowing higher avidity binding (Tutt *et al.*, 1991a). This approach can be modified to allow the production of trispecific antibody fragments, which allows cross-linking of two independent antigens on an effector cell as well as target antigen (Tutt *et al.*, 1991b).

The formation of bispecific $F(ab')_2$ molecules can also be accomplished using 'dimerisation domains' (Kostelny *et al.*, 1992). Leucine zipper peptides have been taken from the transcription factor proteins Jun and Fos which preferentially associate with each other to form a heterodimer. These zipper peptides have been fused to the C-terminus of the Fd′ chains of two Fab′ fragments and expressed as fusion proteins. When co-expressed *in vivo* little heterodimer was produced, but when the individual Fab′-Jun and Fab′-Fos proteins were expressed and purified separately, they could be reduced and recombined *in vitro* to produce the bispecific molecule in high yield. The design of these molecules also incorporates hinge disulphides to stabilise the bispecific $F(ab')_2$ produced. The Jun and Fos peptides attached may increase the immunogenicity of the $F(ab')_2$, but further design could incorporate proteolytic cleavage sites between the Fab′ and the dimerisation domains to allow their subsequent removal once the disulphide bonds were formed.

Bispecific scFv molecules can also be produced by several means using techniques similar to those described for the formation of homodimers above (Section 2.4.2). Direct fusion of two scFv molecules to form single polypeptide molecules containing two different specificities has been attempted by several groups. Initial attempts to express such bispecific molecules in *E. coli* resulted in insoluble material, even if secretion to the periplasm was attempted (Gruber *et al.*, 1994; Mallender and Voss, 1994). Refolding of functional molecules was achieved but resulted in reduced antigen-binding activity of one or both antigen-binding sites. Subsequent expression of single-chain bispecific constructs in mammalian cells resulted in molecules in which both binding sites appeared to be functional (Mack *et al.*, 1995; Jost *et al.*, 1996).

Fusion of the VL of one antibody to the VH of a second antibody with a short linker and coexpression with the VL of antibody two fused to the VH of antibody one allows bispecific diabodies to be produced (Figure 2.9). Expression of high yields of bispecific diabodies in *E. coli* is possible, with yields of 935 mg/l reported after purification with approximately 75% as functional heterodimers (Zhu *et al.*, 1996). Engineering of the domain interfaces has also been investigated to improve assembly of dimeric molecules (Zhu *et al.*, 1997). Improvements in the proportion of functional heterodimer to 92% were reported by engineering the interface to contain 'knobs' and 'holes' as described for the pairing of two heavy chains above. Bispecific diabodies can also be combined with phage display technology to allow selection of diabodies with improved binding and association properties (McGuinness *et al.*, 1996). Although it is easier to produce in *E. coli,* non-specific association of diabodies has been demonstrated (Perisic *et al.*, 1994), and further work examining both the physical properties of these molecules and their *in vivo* characteristics is required.

Genetic fusions of anti-dansyl scFv's to the C-terminus of anti-dextran antibody, either as IgG or as $F(ab')_2$, has been described resulting in tetravalent bispecific molecules (Coloma and Morrison, 1997). Although produced at relatively low levels (0.5–5 mg/l), proteins capable of binding both antigens resulted. It was not clear however, whether the potential for bivalent binding by both specificities was achieved.

2.6 Engineering effector functions

For many applications a simple blocking or neutralising effect of antibody is not enough, and it is desirable to use the antibody to target a diagnostic or therapeutic effector mechanism. Therapeutics can utilise natural effector functions such as complement activation, phagocytosis or ADCC, or entirely novel functions can be introduced. For example, the antibody can be used to target radioisotopes, drugs or toxins to kill tumour cells. Novel functions can be introduced by chemical coupling to antibody molecules or in many cases by recombinant DNA techniques. Similarly, reagents useful as reporter molecules for diagnostic agents, such as enzymes or peptides, can be attached to antibodies either chemically or by means of recombinant DNA. Antibody engineering can thus be used to tailor the effector functions required for individual applications.

2.6.1 Engineering natural effector functions

Use of different human isotype constant regions for chimeric and humanised antibodies can confer different biological properties as expected. For example, Shaw *et al.* (1988) produced a set of mouse/human chimeric antibodies from the mouse MAb 17-1A which recognises a tumour-associated antigen. It was demonstrated that the chimeric molecules could activate the effector functions expected for each particular human IgG isotype. This has also been shown for several other sets of matched chimeric or humanised antibodies. Thus the choice of isotype for a recombinant antibody allows the selection of the desired effector function profile. For example, human IgG1 and IgG3 are the most active IgG isotypes in binding to Fc receptors and eliciting ADCC responses, whereas IgG2 or IgG4 can be chosen to minimise Fc binding and ADCC effects. The ability of human antibodies to mediate complement-dependent lysis is highly dependent on a number of poorly understood factors, including antigen access and density, antigen size and proximity to the cell membrane. However, comparisons of sets of recombinant antibodies with identical variable regions have revealed that human IgG1 and IgG3 are also the most active isotypes for complement activation (see Table 1.1) and thus when an active isotype is required one of these two should be chosen. IgG3 molecules are difficult to purify and handle; their long hinge region results in susceptibility to proteolysis and aggregation, and therefore IgG1 is a common choice when an active antibody is required.

In some cases isotypes are chosen for chimeric and humanised antibodies to mimic the functional properties of the original rodent antibody. The effector functions of murine and human isotypes are compared in Table 1.1. Most mouse and rat antibodies are poorly effective at activation of human effector functions, although mouse IgG2a and rat IgG2b are more efficient than other isotypes. Active human isotypes can confer human effector functions to humanised mouse antibodies previously unable to mediate such effects (e.g. Junghans *et al.*, 1990; Caron *et al.*, 1992a), although this is not always the case due to restrictions of the antibody/antigen interaction mentioned above, and also due to the existence of specific cellular protective mechanisms. As more is understood of the molecular basis of Fc receptor binding and complement activation, specific mutations can be made to alter effector functions of a particular isotype. For example, in many *in vivo* situations antibodies are used as blocking or neutralising molecules and effector functions are undesirable as these could lead to unnecessary toxicity. In these cases human isotypes such as IgG2 or IgG4 have been used, although even these relatively neutral isotypes have low level functionality for both ADCC and complement lysis, which may be undesired. The recognition site for Fc receptor binding on the lower hinge/CH2 domain has been mapped to residues Leu234–Ser329, and mutations in these residues have been shown to reduce Fc receptor binding (Lund *et al.*, 1991; Canfield and Morrison, 1991). The residue Leu235 in this motif has been changed to Glu in a humanised IgG4 and shown to remove the residual Fc receptor binding of this isotype (Alegre *et al.*, 1992). In engineering human antibodies for specific effects of this type, however, it must be remembered that mutations from the natural sequence may have unexpected effects on other functional properties or may increase the immunogenicity of the molecule, and therefore careful testing of the engineered molecule is required.

Other considerations must also be borne in mind when choosing an antibody isotype. The flexibility of the hinge region varies between the human IgG isotypes, and this may affect binding properties, particularly to cell surface antigens. For example, the isotype choice for a chimeric antibody against intercellular adhesion molecule 1 (ICAM-1) affected the antigen-binding properties of the antibody (Morelock *et al.*, 1994). In this case a chimeric IgG1 variant had equivalent antigen-binding ability to the murine antibody, yet less active molecules were produced with human IgG4 or IgG2 constant regions. The affinity of the individual binding sites was shown to be the same for all the chimeric antibodies, the difference being in the avidity due to differences in the ease of bivalent binding to this cell surface antigen. This correlated with the different flexibility of the hinge region in these antibodies, IgG1 being the most flexible with highest avidity and IgG2 the least flexible with the lowest avidity. Similarly, differences in antigen-binding properties between antibodies with identical variable regions have been identified in other antibody–antigen systems with both murine and human antibodies (Cooper *et al.*, 1993; McCloskey *et al.*, 1996).

Human IgG4 is a common isotype choice when an antibody with few effector functions is required. Natural human IgG4 contains a proportion of molecules in which the inter-heavy chain disulphide bonds do not form although the two halves of the molecule are still assembled through non-covalent interactions (King *et al.*, 1992a). During analysis by

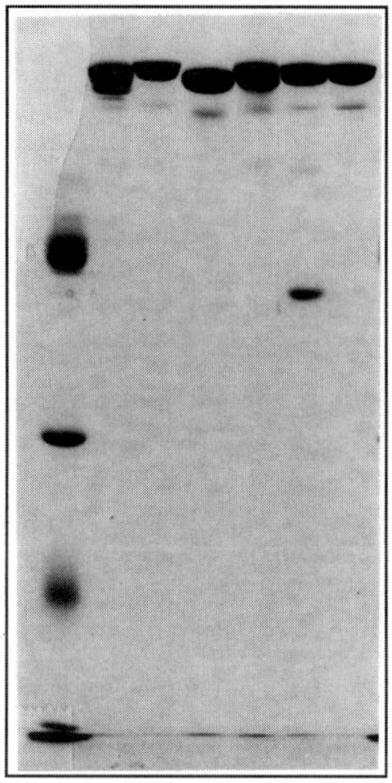

Figure 2.11 SDS-PAGE of an isotype series of the mouse:human chimeric antibody B72.3 under non-reducing conditions; the presence of a band of non-disulphide-linked molecules is observed at approx. 80 kDa for IgG4: lane 1, molecular weight markers; lane 2, murine B72.3; lane 3, chimeric IgG1; lane 4, chimeric IgG2; lane 5, chimeric IgG3; lane 6, chimeric IgG4; lane 7, chimeric IgG4P, Ser-Pro mutant (see text for details)

SDS-PAGE non-covalent interactions are broken such that two bands are seen under non-reducing conditions, one of disulphide bonded IgG at approx. 155 kDa and one representing IgG with no inter-heavy chain disulphide bond at approx. 80 kDa (Figure 2.11). This can lead to confusion during analysis of purified antibody as the 80 kDa band (which is present at 5–25% depending on the particular IgG4 antibody) may be viewed as an impurity, although it is present in all natural IgG4 antibodies studied as well as recombinant antibodies (King *et al.*, 1992a). The core hinge region of human IgG4 contains the sequence CPSC, compared to IgG1 which contains the sequence CPPC. The serine residue present in the IgG4 sequence leads to increased flexibility in this region and therefore a proportion of molecules form disulphide bonds within the same protein chain (an intrachain disulphide) rather than bridging to the other heavy chain in the IgG tetramer to form the interchain disulphide (Bloom *et al.*, 1997). Changing the serine residue to a proline to give the same core sequence as IgG1 allows complete formation of inter-heavy chain disulphides in the IgG4 molecule, hence removing the 80 kDa band seen on SDS-PAGE (Angal *et al.*, 1993). This altered isotype, termed IgG4P, has been used for several chimeric and engineered human antibodies to overcome the double-banded gel pattern (e.g. Owens and Robinson, 1995).

It is also possible to produce recombinant antibodies of other classes which are functional in Fc receptor binding. Mouse:human chimeric IgM, IgA1, IgA2 and IgD molecules have all been expressed and have shown functional binding to their respective receptors (Bruggemann *et al.*, 1987; Morton *et al.*, 1993; Shin *et al.*, 1992). The properties of these constant regions are therefore also available for incorporation into recombinant

Figure 2.12 Dimeric IgG formed through mutation of a CH3 domain residue to cysteine (see text for details)

antibodies if required. In some cases these isotypes may offer advantages, for example a chimeric IgA2 has been used to translocate across an epithelial cell layer via the polymeric Ig receptor for IgA and therefore target tumour cells by reaching an antigen more abundant on the apical cell surface (Terskikh *et al.*, 1994).

Although the amino acids involved in the interaction sites for Fc receptor binding and complement activation are beginning to be identified, at present the structural basis by which effector functions are elicited is not fully understood. Mutations in recombinant antibody molecules are playing a large role in elucidating recognition sites for individual functional effects. When the molecular basis of these interactions is understood it may be possible to tailor antibodies to bind to a chosen set of receptors and so design their *in vivo* properties.

The role of carbohydrate in maintaining the conformation of the IgG CH2 domains must also be considered. Removal of carbohydrate by mutation of the normal attachment site for N-linked carbohydrate at position 297 in the CH2 domain results in severe reductions in the ability of antibody to mediate effector functions (Tao and Morrison, 1989). This is probably a consequence of altered conformation of the antibody molecule (Lund *et al.*, 1990). Recent studies have shown that mutation of amino acids involved in structural contacts with antibody carbohydrate can influence effector functions as well as the profile of sugar residues attached to the antibody itself (Lund *et al.*, 1996). Manipulation of antibody carbohydrate structure may, therefore, represent another route towards engineering antibody effector functions.

Another approach to alter effector functions is the production of polymeric IgG antibodies. As described in Chapter 1, the ability of IgG to elicit complement activation is dependent on multiple molecules attaching to antigen to allow Fc clustering. Pentameric and hexameric IgM molecules contain multiple attachment sites for C1q binding which are revealed by a conformational change upon antigen binding and thus IgM molecules in general are more active at complement activation than IgG. Simple chemical cross-linking can be used to produce dimeric or polymeric species with increased antigen-binding avidity (Tsai *et al.*, 1995). However, antibody engineering allows more controlled and reproducible site-specific linking strategies to be considered. Two approaches to engineering IgG molecules for increased effector function via multimerisation have been investigated. Substitution of a serine residue near the C-terminus of the CH3 domain (Ser444) to cysteine allowed the production of IgG dimer of a chimeric human IgG1 (Figure 2.12). This mutation resulted in 50% of the molecules forming dimeric IgG, reminiscent of IgA dimers, which were approx. 200-fold more potent in complement-mediated lysis (Shopes, 1992). The same mutation introduced into a humanised IgG1, huG1-M195, also resulted in the formation of IgG dimers which were more potent at both complement activation and ADCC (Caron *et al.*, 1992a). In addition, dimers of an antibody to an internalising antigen such as huG1-M195 internalise into cells more rapidly than the IgG, which may offer advantages for delivery of cytotoxic agents to cells *in vivo* (see Chapter 4). The second approach investigated has been the expression of IgG molecules with a tailpiece from the IgM μ chain which results in the production of IgM-like IgG polymers with up to six IgG units (Smith and Morrison, 1994; Smith *et al.*, 1995). All four human IgG isotypes have been shown to result in polymers using this approach, and as expected the ability of such multimeric molecules to mediate complement activation is increased, as is their ability to interact with the FcγRII receptor. Disadvantages include the heterogeneity of polymers produced, with species from monomers to hexamers present, activation of complement in the absence of antigen for some isotypes, and altered *in vivo* tissue distribution and clearance properties, with rapid loss of polymeric species from the circulation.

Another approach is the use of bispecific antibody fragments to redirect effector functions. Bispecific diabodies have been produced with one specificity for target antigen and the second for serum immunoglobulin (Holliger *et al.*, 1997). Such diabodies can retarget serum Ig to the target antigen and elicit a range of effector functions including complement activation, phagocytosis and ADCC. This potentially exciting strategy offers the potential to develop reagents capable of retargeting specific immunoglobulin types and thus their particular effector functions, or all immunoglobulins via a common epitope such as light chain constant domain (Holliger *et al.*, 1997). Similarly, more specific reagents may be produced through retargeting individual effector molecules. For example, a bispecific diabody with one specificity towards complement component C1q and the other against lysozyme as a model target antigen was capable of activating complement resulting in specific target cell lysis (Kontermann *et al.*, 1997).

2.6.2 *Attachment of diagnostic or therapeutic agents*

Chemical conjugates

Chemical conjugates of antibodies with drugs, radioisotopes, proteins or other molecules have been widely investigated, and a range of chemical approaches are available and widely used. Chemical conjugation is often the quickest route to the preparation of conjugates for many purposes, and in some instances conjugates with superior properties to those produced as fusion proteins may result (see below). Indeed, chemical conjugates of antibodies with enzymes or other reporter groups are a mainstay of the diagnostics industry (see Chapter 3). Chemical conjugation is, of course, the only option for the attachment of non-proteinaceous materials such as drugs to antibodies, and the ability to use a range of different linker chemistries allows variation in the properties of the reagents produced to be tested until the optimum is identified. The major issues in the production of such conjugates are the degree of substitution which can be achieved, the stability of the linkage and the biological activity of the resulting conjugate. Loss of some or all of the antigen-binding ability of the antibody is a common consequence of non-specific conjugation methods which may result from the modification of amino acids at or close to the CDR amino acid residues involved in antigen binding. This is more pronounced when antibody fragments are used, as there is more chance of randomly modifying a residue important for function simply due to the smaller number of amino acids present. Similarly, care must also be taken that the effector or reporter molecule attached is not damaged by the conjugation methodology. Therefore, choice of the conjugation method and control of the conjugation process to minimise such effects is a major consideration. Other drawbacks of this approach include the heterogeneous nature of the conjugates produced which can lead to difficulties with batch-to-batch variation.

A variety of amino acid residues on the surface of antibody molecules can be used for chemical modification and coupling. Residues with useful functional groups for attachment include tyrosine (phenolic hydroxyl group), aspartic and glutamic acid (carboxyl group), lysine (amino group) and cysteine (thiol group). The tyrosine phenolic hydroxl group is reactive towards electrophilic agents, and its major role as a modification site is as a site for labelling of proteins with radioactive iodine. The carboxyl group of the acidic aspartate and glutamate residues can be used for coupling with carbodiimides such as EDC which react with both carboxyl and amine groups. However, such reagents are relatively non-specific and often lead to antibody aggregation and precipitation. The most useful and widely used sites for attachment to antibodies are the terminal amino group of surface lysine residues and the free thiol group of cysteine residues which can be generated by reduction of disulphide bonds in the hinge region. A large variety of reagents are available as linkers for attachment to these two groups under mild conditions which minimise damage to the antibody (Figure 2.13). Immunoconjugates which incorporate low molecular weight ligands are usually prepared using a derivative of the ligand with a suitable amino acid reactive group, whereas for protein conjugates a linker is required. Both homobifunctional linkers with two identical reactive groups and heterobifunctional linkers with reactive groups for different moieties are commercially available.

The most useful amine reactive groups include imidoesters and *N*-hydroxysuccinimide esters. Both of these groups react with the terminal amino group of lysine residues under relatively mild conditions (Figure 2.13), but many imidoesters require high pH (>9) for selective reactivity and can be reversible. Therefore *N*-hydroxysuccinimide esters have become the most popular reagents for linking to lysine residues due to their efficient coupling at physiological pH and the stability of the linkage. The amine groups at the N-terminus of the protein chains can also react with these linkers. Other lysine-reactive groups are not useful for protein–protein linkage but are commonly used for attachment of small chemical moieties such as drugs, or fluorescent reporter groups onto antibodies. For example, fluorescein and rhodamine isothiocyanates are commonly used for the attachment of these fluorescent groups to antibodies, and several amine-reactive groups have been used for polyethylene glycol attachment to antibodies (see Figure 2.2).

The cysteine thiol group is the most reactive amino acid side chain and is thus a useful site for specific conjugation. Thiol-reactive groups for antibody modification include maleimide, vinylsulphone, haloacetyl or pyridyl disulphide groups (Figure 2.13). Maleimides react selectively with cysteine residues at neutral pH, although there is reactivity with amine groups at higher pH values. A stable thioether bond is generated. Haloacetyl groups, usually containing iodoacetyl, are also useful reagents for introducing stable thioether bonds into immunoconjugates. These reagents are less selective than maleimides and at high concentrations can also react with histidine groups. Also, free iodine is generated under some conditions which may react with tyrosine, tryptophan or histidine residues causing unnecessary antibody damage. Pyridyl disulphides react with thiols to form a disulphide bond and are useful in situations when a less stable chemical linkage is required. A major advantage of coupling via pyridyl disulphides is the release of pyridine-2-thione which can be used to monitor the reaction spectrophotometrically at 343 nm. Antibody molecules do not normally contain a free thiol group as all of the available cysteine residues form disulphide bonds. Free thiols can be liberated by reduction of the hinge region of IgG, or of $F(ab')_2$ fragments to form Fab′ which is a method of site-specific labelling (see below). Alternatively, surface lysine residues can be modified to

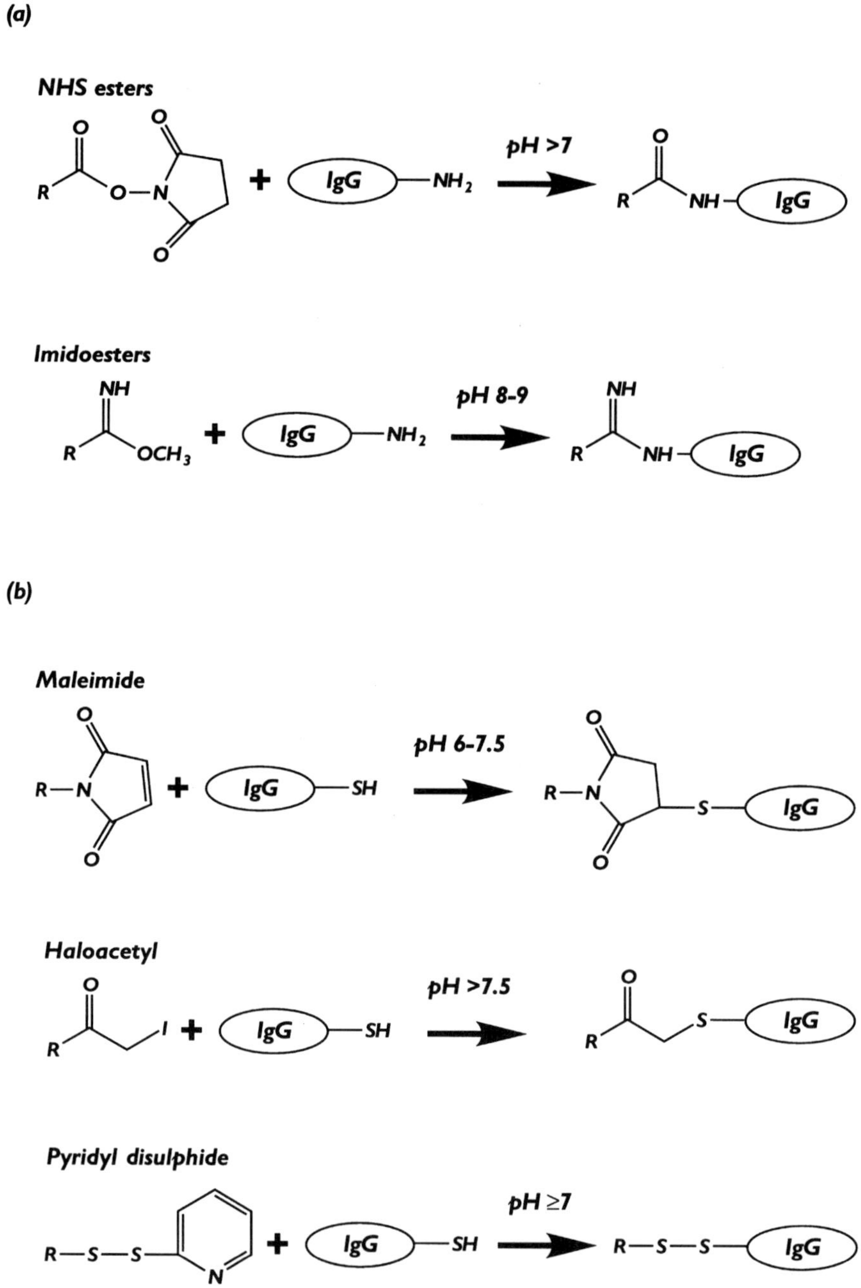

Figure 2.13 Examples of amine and thiol reactive linkers commonly used for chemical modification of antibodies: (a) lysine reactive linkers such as NHS esters and imidoesters; (b) thiol reactive linkers such as maleimide, haloacetyl and pyridyl disulphide

introduce a thiol group using reagents such as 2-iminothiolane, SATA or SPDP (Figure 2.14), which in turn can be used as an attachment point. One advantage of modifying lysine residues with thiol groups in this way is to achieve good control over the conjugation reaction, and thus minimise loss of biological properties of the conjugate. While

(a) Homobifunctional

Glutaraldehyde

DSS

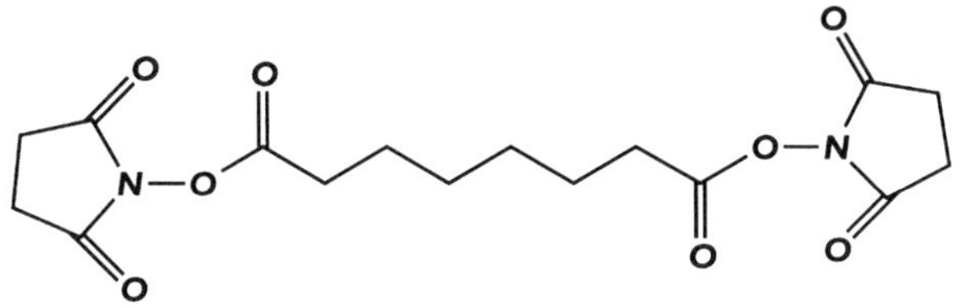

BMH

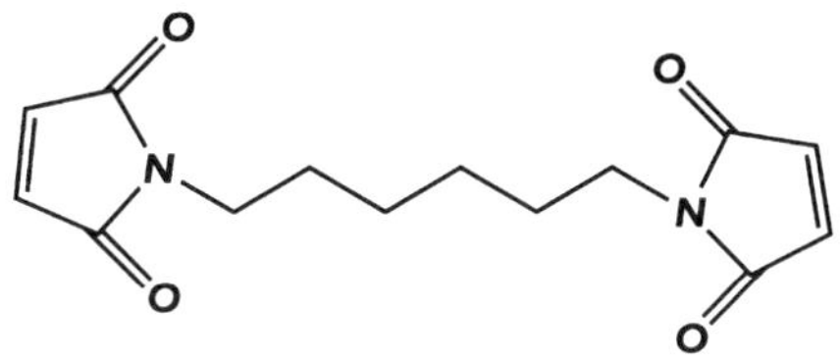

(b) Heterobifunctional

SMCC

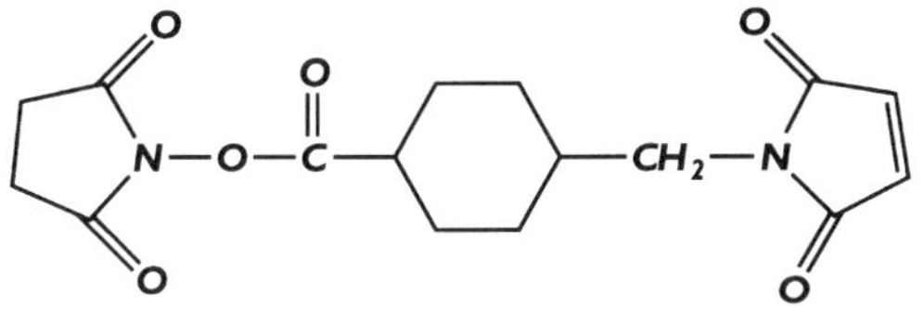

MBS

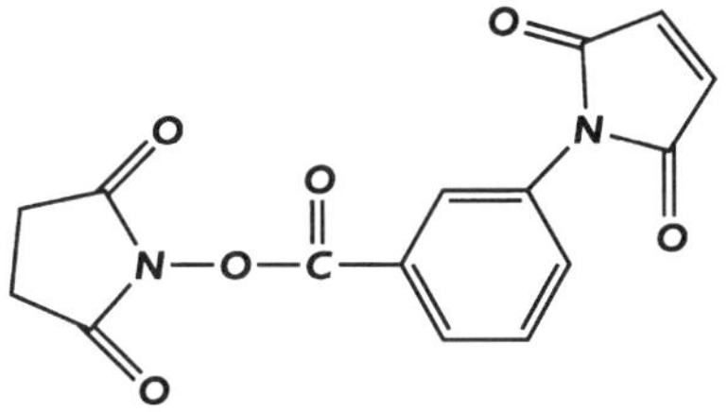

SPDP

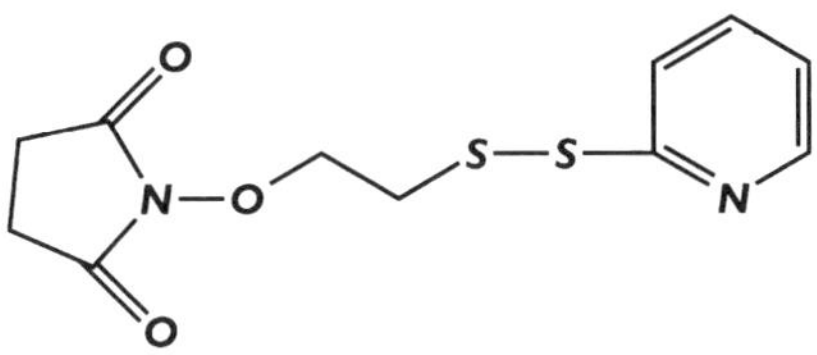

Figure 2.14 Some useful commercially available protein cross-linking reagents: (a) homobifunctional reagents glutaraldehyde, disuccinimidyl suberate (DSS) and bis-maleimidohexane (BMH); (b) heterobifunctional reagents, succinimydl 4-*N*-maleimidomethylcyclohexane-1-carboxylate (SMCC), maleimidobenzoyl-*N*-hydroxysuccinimide ester (MBS) and *N*-succinimidyl-3-[2-pyridyldithio]propionate (SPDP)

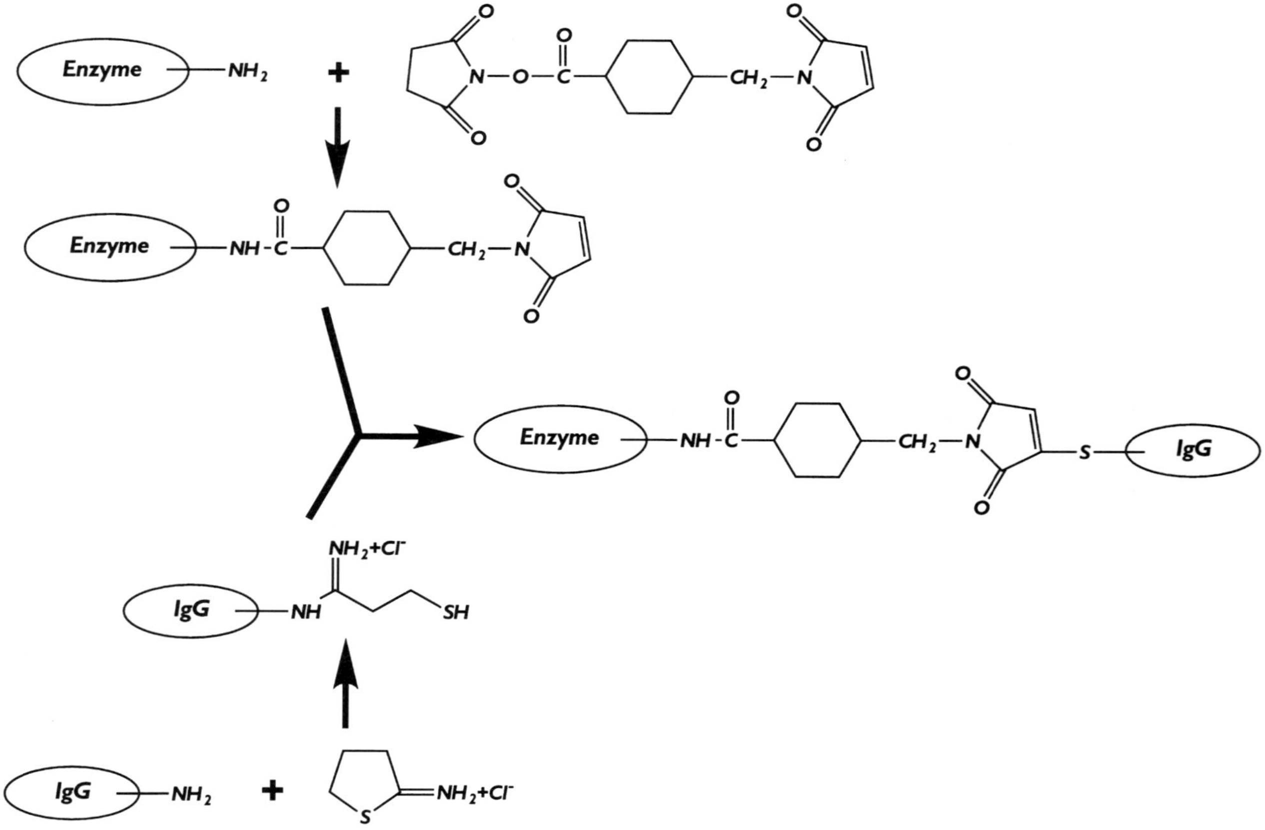

Figure 2.15 Example reaction scheme for attachment of an enzyme to antibody using SMCC attachment to enzyme and 2-iminothiolane activation of IgG

titration of lysine groups to determine the level of substitution is an insensitive procedure, the presence of one thiol group per antibody can be easily measured using reagents such as dithionitrobenzoic acid (DTNB) or dithiodipyridine (DTDP).

Both homobifunctional and heterobifunctional reagents are readily available (Figure 2.14). Homobifunctional cross-linkers which are either lysine-reactive or thiol-reactive can be used for IgG conjugation, although in some cases their use can result in unacceptable aggregation. The use of heterobifunctional linkers often allows more control of the conjugation reaction. Heterobifunctional reagents with one thiol-reactive and one amine-reactive group are particularly useful for the production of stable conjugates with minimum aggregation, such as antibody–enzyme conjugates. A typical reaction strategy is shown in Figure 2.15. The introduction of such a stable linkage is desirable in many cases though for some applications, for example in the construction of immunotoxins (see Chapter 4), a reducible linkage has been shown to be useful. In these cases linkers such as SPDP can be used to form a disulphide linkage after reaction of SPDP with each reagent individually, or alternatively heterobifunctional reagents with a disulphide group between the two functional groups can be used. In a similar way, sites for attachment of other functionalities can be designed into the linkers used. For example, sites for radiolabelling either with iodine or by chelation of radioactive metals can be incorporated.

Site-specific attachment

As a consequence of the loss of antigen-binding avidity often seen on conjugation, several methods for attachment of reagents at specific sites away from the antigen-binding site have been investigated (Table 2.5). As mentioned above, the generation of free cysteines by selective reduction of the hinge region has been used for both IgG and antibody fragments, for example to attach fluorescent groups (Packard *et al.*, 1986) or for site-specific radiolabelling (King *et al.*, 1994). The Fc region carbohydrate also provides a natural specific attachment site for IgG molecules. Carbohydrate is usually modified by periodate oxidation to generate reactive aldehydes which can then be used to attach reactive amine containing compounds by Schiff base formation. As the aldehydes can react with amine groups, reactions are carried out at low pH so that the lysine residues are protonated and unreactive. Hydrazide groups are most suitable for attachment to the aldehydes generated since they are reactive at low pH to form a hydrazone linkage. The linkage can then be further stabilised by reduction with sodium cyanoborohydride to form a hydrazine linkage (Figure 2.16). Comparisons of conjugates made by attachment to

Table 2.5 Approaches to site-specific labelling of antibodies and antibody fragments

Strategy	Reference
Modification of hinge cysteine residues	Packard *et al.*, 1986
Introduction of surface cysteines in constant region domains by protein engineering	Lyons *et al.*, 1990
Modification of Fc carbohydrate	Rodwell *et al.*, 1986
Introduction of glycosylation site to light chain by protein engineering	Leung *et al.*, 1995
Introduction of extra lysine residues by protein engineering	Hemminki *et al.*, 1995
Addition to C-terminus of antibody fragments by reverse proteolysis	Fisch *et al.*, 1992

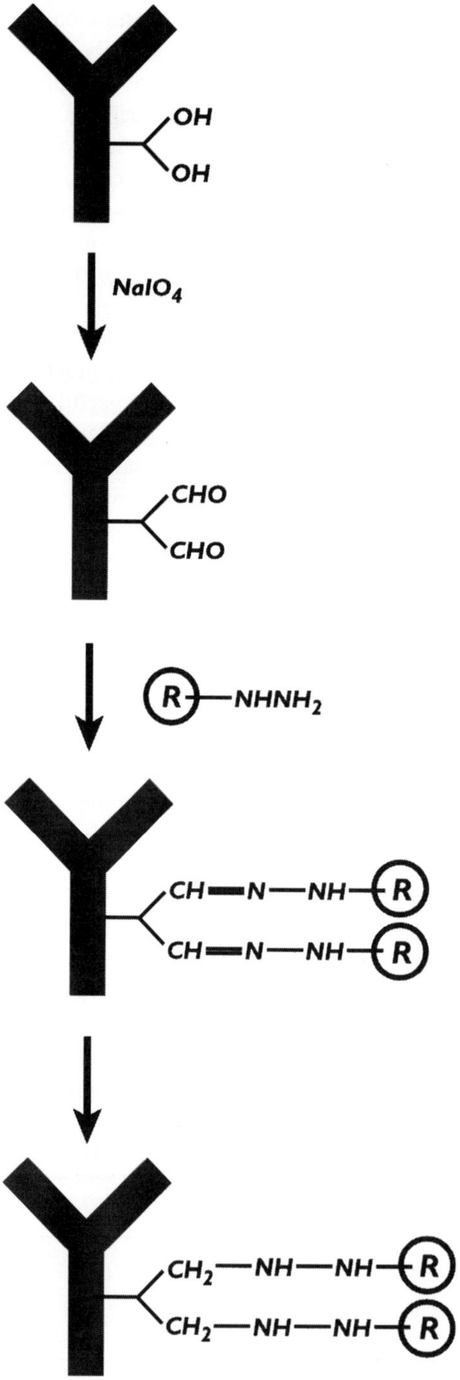

Figure 2.16 Reaction scheme for site-specific attachment to antibody carbohydrate via periodate oxidation

Fc carbohydrate with conventional conjugates with ligands randomly attached to lysine residues have demonstrated significantly improved antigen-binding properties (Rodwell *et al.*, 1986). The disadvantages of this approach are the relatively harsh conditions required, which can damage and aggregate some antibody molecules. Methionine residues present in some antibody variable regions may be particularly susceptible to oxidation by periodate, which can lead to loss of antigen-binding avidity. In some cases histidine or tryptophan residues might also be affected.

Antibody engineering can be used to introduce specific attachment sites into antibody molecules, and this can be incorporated as part of the design of an engineered molecule. Extra cysteine residues can be introduced onto the surface of antibody constant domains to provide a specific attachment site without the need to disrupt native disulphide bonds. Introduction of specific cysteine sites in the CH1 domain of the IgG heavy chain has been shown to result in sites to which ligands can be attached without any loss of antigen binding (Lyons *et al.*, 1990). Alternatively, extra carbohydrate sites can be engineered into the molecule to allow attachment via periodate oxidation. Some antibody light chains have an unusual natural glycosylation site, and thus the light chain has been used as a site to introduce a glycosylation site into antibodies which do not normally have carbohydrate attached to the light chain (Leung *et al.*, 1995). A third engineering strategy is to introduce extra lysine residues into the surface of the constant region domains (Hemminki *et al.*, 1995). Although this approach does not introduce a unique labelling site, lysine-reactive reagents are more likely to modify the antibody at the increased concentration of lysine residues in the constant region, resulting in the retention of more antigen-binding reactivity.

A more specialised approach is the use of reverse proteolysis to attach reagents specifically at the C-terminus of Fab′ heavy chains (Fisch *et al.*, 1992). After production of a $F(ab')_2$ fragment by the protease lysyl endopeptidase, experimental conditions can be altered such that the same protease working in reverse is capable of the specific attachment of carbohydrazide groups to the C-terminus of the $F(ab')_2$ heavy chains. These carbohydrazide groups could then be used as an attachment point for a radiolabelled chelator reacting via an aldehyde group to form a hydrazone linkage. A similar approach has also been used to attach an aldehyde-modified enzyme to the C-terminus of $F(ab')_2$ (Werlen *et al.*, 1994).

Fusion proteins

An attractive alternative to chemical conjugation for antibody–protein conjugates is the direct expression of fusion proteins. This approach can be viewed as another method of site-specific attachment, as the protein to be fused is attached away from the antigen-binding site and no further modification is necessary. The antibody or antibody fragment is fused to the protein at the gene level, usually via attachment of the sequence to either full length or a truncated heavy chain. Expression of functional fusion proteins was first shown by fusion of a Fab fragment to Staphylococcal nuclease (Neuberger *et al.*, 1984). Since then the production of active antibody and antibody fragment fusion proteins with a variety of different enzymes, cytokines, and other proteins has been achieved (Table 2.6). Fusion proteins overcome problems with loss of antigen-binding ability on chemical conjugation and are also well-defined, homogeneous reagents. Disadvantages are that the flexibility of using linkers with varying chemical stability is lost, and in some cases fusion proteins have proved difficult to express in a functional form.

Table 2.6 Some examples of antibody fusion proteins

Antibody form	Fusion	Reference
Enzyme fusions		
F(ab′)$_2$	Staphylococcal nuclease	Neuberger *et al.*, 1984
scFv, dsFv	β-lactamase	Goshorn *et al.*, 1993 Rodrigues *et al.*, 1995
Fab′	β-glucuronidase	Bosslet *et al.*, 1992
scFv, F(ab′)$_2$	Alkaline phosphatase	Wels *et al.*, 1992 Ducancel *et al.*, 1993
scFv	Urokinase	Holvoet *et al.*, 1991
Toxin fusions		
scFv, dsFv	Pseudomonas exotoxin	Brinkmann *et al.*, 1991 Reiter *et al.*, 1996
scFv, Fab, F(ab′)$_2$	Gelonin	Better *et al.*, 1995
Fab′, scFv	Angiogenin	Rybak *et al.*, 1992 Newton *et al.*, 1996
scFv	Eosinophil-derived neurotoxin	Newton *et al.*, 1994
Cytokines and growth factors		
scFv, F(ab′)$_2$	IL-2	Fell *et al.*, 1991 Dorai *et al.*, 1994
IgG, Fab	IL-2, GM-CSF TNFα, TNFβ	Gillies *et al.*, 1992 Gillies *et al.*, 1993
Fab, F(ab′)$_2$, IgG	IGF-1, IGF-2	Shin *et al.*, 1994
Other proteins		
F(ab′)$_2$	Aequorin	Casadei *et al.*, 1990
Fab, F(ab′)$_2$, IgG	Avidin	Shin *et al.*, 1997
Fab	Biotin carboxyl carrier protein	Weiss *et al.*, 1994
Fab, F(ab′)$_2$, IgG	Transferrin	Shin *et al.*, 1995
Fab	Staphylococcal enterotoxin A	Dohlstein *et al.*, 1994
scFv	Protein A fragment	Tai *et al.*, 1990
F(ab′)$_2$	Metallothionein	Das *et al.*, 1992

The choice of antibody form for fusion purposes depends on the other properties of the molecule required, e.g. overall size, pharmacokinetics and valency. As shown in Table 2.6, fusion proteins with single-chain Fv, disulphide-linked Fv, Fab, F(ab′)$_2$ and IgG have all been constructed. However, the choice of construct can have unexpected effects on the assembly and activity of the fusion protein. A series of fusions of interleukin-2 (IL-2) to the heavy chain of the chimeric anti-ganglioside antibody ch14.18 have been made with addition directly after the Fab such that the entire Fc is replaced by IL-2, after CH2 such that the CH3 domain is replaced with IL-2, and fusion to the end of the entire heavy chain (Gillies *et al.*, 1992). The Fab fusion was inactive in antigen binding and the fusion to CH2, although capable of binding antigen, did not assemble correctly to the divalent species even though a full-length hinge region was present. The fusion to the end of the entire heavy chain resulted in enhanced antigen binding of the antibody and the molecule was shown to be fully assembled. The IL-2 fused to the antibody remained active although the antibody effector functions were slightly reduced.

In some case spacer peptides have been included between the antibody and fused protein which may assist folding or activity of the fused protein. For example, a linker between an scFv toward the transferrin receptor and the RNAse angiogenin was shown to be required for optimal activity of both the scFv and the enzyme (Newton *et al.*, 1996). Also, spacers can be designed to allow proteolytic release of the fused protein when internalised into cells. Other choices in the nature of the construct also need to be made, as the fusion can be made at the C-terminus or N-terminus of the antibody chain and to heavy or light chain. In a study of antibody fragment fusions to the plant toxin gelonin, fusions to Fab, $F(ab')_2$ and scFv were made at either the N- or C-terminus (Better *et al.*, 1995). Of these immunotoxins, the divalent $F(ab')_2$ variants were the most potent, as might be expected due to divalent antigen binding and possibly more effective internalisation into target cells. scFv and Fab conjugates with gelonin fused at the N-terminus or C-terminus of the light or heavy chain were all approximately equipotent. Linkers were examined with sites for trypsin-like proteases or with sites for the lysosomal enzymes cathepsin B and D. These are expected to be cleaved intracellularly and may assist gelonin release and hence cytotoxicity. However, in this study there was little difference between the cytotoxicity of the constructs with and without the cleavable linkers, suggesting that the fusion proteins themselves may be active without the need for release of free gelonin.

The types of protein used in fusion proteins includes those used to induce dimerisation (Section 2.4.2), peptide or protein tags used to simplify antibody fragment purification (see Chapter 5), enzymes for both therapeutic and diagnostic purposes, cytokines and growth factors, and many others as shown in Table 2.6. Fusions can also be used to couple some non-proteinaceous materials to antibodies via specific binding proteins. Fusion of the metal-binding protein metallothionein to antibody fragments can be used to allow subsequent binding of the radionuclide ^{99m}Tc (Das *et al.*, 1992). A small peptide which can chelate ^{99m}Tc has also been used effectively (George *et al.*, 1995). Another approach to radiolabelling is to attach a peptide sequence which can be enzymically labelled with ^{32}P (Neri *et al.*, 1996). Non-proteinaceous materials which are naturally bound by other proteins can be attached by fusion of a suitable ‘acceptor’ sequence to the antibody fragment followed by expression in *E. coli*. In this way lipid-tagged scFv fragments have been produced by fusion of the amino terminal part of the *E. coli* major lipoprotein, which is sufficient for attachment of lipid during expression (Laukkanen *et al.*, 1993). Biotin can be attached via the biotin-carboxyl carrier protein (BCCP) subunit of *E. coli* acetyl-CoA carboxylase (Weiss *et al.*, 1994). In this case a Fab-BCCP fusion was expressed in *E. coli* and biotin was attached during secretion of the fusion protein.

2.7 Engineering pharmacokinetics and biodistribution

For therapeutic applications of MAbs, the pharmacokinetics and biodistribution of the agent to be used in humans is of great importance. The degree of binding to the target antigen *in vivo* is dependent not only on binding characteristics of the antibody such as affinity and avidity which can be measured *in vitro*, but also on a variety of other factors such as the accessibility of the antigen, the concentration, and the time of exposure of the antibody to the target antigen. Access to antigens in the vascular compartment is relatively straightforward when antibodies are injected intravenously, but access to other body tissues is usually more difficult. Such extravascular targets require the antibody to pass through the endothelial cell layer lining the blood capillaries to reach interstitial fluid compartments. The permeability of blood vessels to proteins varies according to the size of the protein, with smaller proteins able to reach interstitial spaces more quickly. Therefore, large proteins such as intact antibody molecules penetrate into tissues relatively slowly, while antibody fragments penetrate tissues more quickly. However, the amount of antibody to reach the target tissue is also dependent on the concentration in circulation and the rate of clearance from the blood, and therefore the study of antibody pharmacokinetics has an important role to play.

In both human and animal systems, antibody clearance from the blood follows a classical two-compartment kinetic model. Initially, a short distribution phase, or alpha phase, takes place during which the antibody is distributed throughout body tissues, i.e. equilibrates between intravascular and extravascular compartments. This is followed by an elimination or clearance phase (beta phase) in which the antibody is metabolised and excreted. The rates of these two phases are commonly calculated and quoted as half-life values ($t_{\frac{1}{2}}\alpha$ and $t_{\frac{1}{2}}\beta$). Alternatively, a single-compartment kinetic model may be used, in which case a single half-life can be calculated.

2.7.1 Pharmacokinetics of IgG

Human IgG molecules have a long circulating half-life in humans, with $t_{\frac{1}{2}}\alpha$ of 18–22 hours and $t_{\frac{1}{2}}\beta$ of 21–23 days for human IgG1, 2 or 4 (reviewed by Mariani and Strober, 1990). Human IgG3 has a shorter half-life with a $t_{\frac{1}{2}}\beta$ of approximately 9 days. This may be due to the extended hinge region of IgG3 which makes the molecule more susceptible to proteolytic cleavage. Human IgM is cleared more quickly, with a half-life of approximately 5 days. This very large protein remains largely in the vascular compartment and thus pharmacokinetic data are normally described using a one-compartment kinetic model. The pharmacokinetics of human IgG are unusual in that the half-life varies with concentration. At very high IgG levels, such as those seen in myeloma patients which can approach 100 mg per ml of serum, the half-life is decreased to similar values to IgM, while in patients with hypogammaglobulinemia with reduced plasma IgG levels a prolonged half-life is observed. It has been postulated that a receptor-mediated event is thus responsible for maintaining the long circulating half-life for IgG, through IgG binding, preventing degradation and resulting in recirculation to the plasma (Brambell *et al.*, 1964). This hypothesis has been supported by subsequent experimental investigations. It is possible to calculate the amount of IgG which is protected from degradation by this receptor-mediated process in humans at full saturation. This is approximately 150 mg per kg body weight per day (Mariani and Strober, 1990).

It has been known for some time that this receptor-mediated recycling is mediated by the Fc region of the antibody, as isolated Fc has a long half-life, whereas Fab and $F(ab')_2$ are cleared relatively rapidly. Also, co-administering Fc with IgG can result in reduced IgG half-life in a similar manner to increasing the IgG dose. Recently the amino acids in the Fc fragment important for maintaining IgG in circulation in mice have been elucidated by site-directed mutagenesis studies (Kim *et al.*, 1994; Ghetie *et al.*, 1996). The receptor binding site is located at the interface region between the CH2 and CH3 domains and overlaps with the binding site for Staphylococcal protein A, commonly used for purification of IgG (see Chapter 5). Thus protein A complexes are rapidly cleared due to

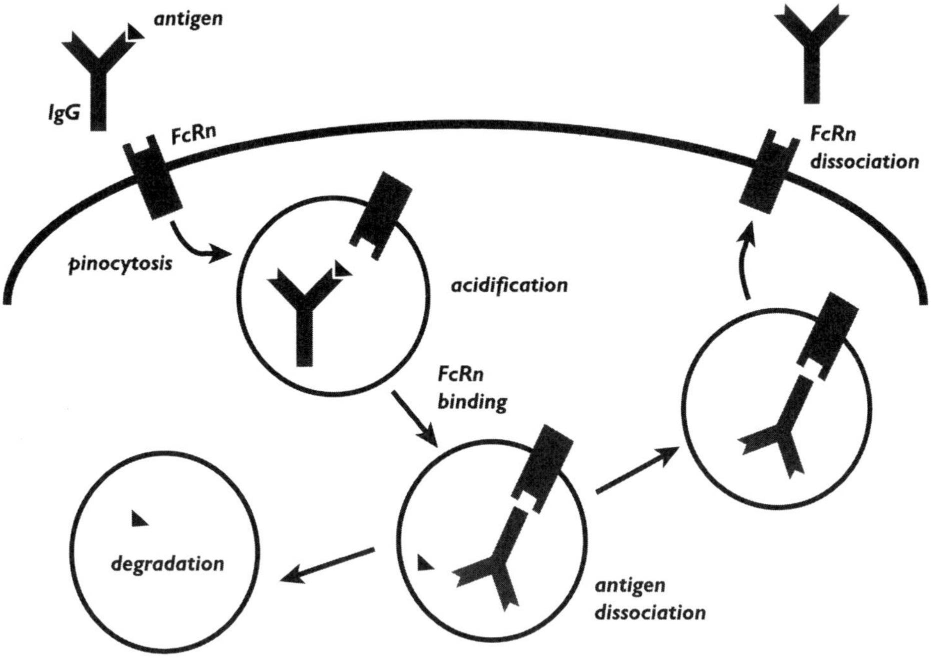

Figure 2.17 Mechanism of FcRn-mediated salvage of IgG from catabolic degradation: the low pH of the endosome results in tight binding of IgG to the FcRn receptor, leading to protection from degradation and return to the cell surface where the higher pH allows re-release of the IgG into circulation

competition with the receptor-mediated process, and isolated Fc fragments have the same $t_{\frac{1}{2}}\beta$ as the whole IgG molecule. The cellular receptor responsible for maintaining half-life has been identified as the neonatal Fc receptor, FcRn. This is the same receptor which is responsible for transfer of IgG from the mother to the neonate which provides immunity for the first few weeks after birth. Mice deficient in this receptor are unable to acquire IgG from the mother, have reduced plasma IgG levels and clear administered IgG or Fc with an abnormally short half-life (Ghetie *et al.*, 1996). The FcRn-mediated pathway for maintaining high IgG levels is thought to operate through its pH dependence (Figure 2.17). IgG is strongly bound to FcRn at low pH and dissociates at higher pH. IgG is internalised into cells and salvaged from the endosome during acidification due to binding to FcRn at low pH. Low pH may also strip any bound antigen from the antibody which would then be degraded in the lysosome. Antibody is then recycled to the cell surface where the higher pH of the extracellular fluid promotes dissociation from FcRn and return of the IgG to the circulation.

Many murine MAbs have been examined in human studies in a variety of diagnostic and therapeutic situations. The half-lives of such murine antibodies vary both from antibody to antibody and from patient to patient, and depend on a number of factors including isotype and the presence of circulating antigen. In general, $t_{\frac{1}{2}}\beta$ values for murine antibodies in humans have been in the range of 15–48 hours on the first administration. Such studies are complicated by human immune responses to murine antibodies (HAMA, see Section 2.3) which result in rapid clearance of murine antibodies on repeat injection. There is now a growing collection of data on the pharmacokinetics of chimeric and humanised antibodies in man, some of which is summarised in Table 2.7. These data are

Table 2.7 Pharmacokinetics of some chimeric and humanised antibodies in humans

Antibody	Indication	$t_{\frac{1}{2}}\alpha$	$t_{\frac{1}{2}}\beta$	Reference
c17-1A (γ1)	Colorectal carcinoma	18 ± 2 hours	101 ± 16 hours	LoBuglio *et al.*, 1989 Meredith *et al.*, 1991
cB72.3 (γ4)	Colorectal carcinoma	18 ± 7 hours	224 ± 66 hours	Khazaeli *et al.*, 1991 Meredith *et al.*, 1992a
c14.18 (γ1)	Melanoma	24 ± 1 hours	181 ± 73 hours	Saleh *et al.*, 1992
cL6 (γ1)*	Carcinoma		54–109 hours	Goodman *et al.*, 1993
C2B8 (γ1)*	B-cell lymphoma		106 (38–252 hours)	Maloney *et al.*, 1994
cAnti-CEA (γ4)*	Colorectal carcinoma	7 (1.4–18 hours)	91 (30–292 hours)	Buchegger *et al.*, 1995
hu2PLAP (γ1)*	Carcinoma		73.1 ± 30.2 hours	Hird *et al.*, 1991
Humanised anti-Tac(γ1)*	Graft v. host disease		88† hours (44–360)	Anasetti *et al.*, 1994
Humanised M195 (γ1)*	Myeloid leukemia	0.3 ± 0.4 hours	38 ± 9 hours	Caron *et al.*, 1994
CDP571 (γ4)	Healthy volunteers	19.6 hours (1–49)	225† hours (87–537)	Stephens *et al.*, 1995
hMN-14 (γ1)*	CEA producing cancer	5.3 ± 6.2 hours	56 ± 32 hours	Sharkey *et al.*, 1995
rhuMAb HER2 (γ1)	Breast cancer		199 ± 120 hours	Baselga *et al.*, 1996
hCTM01 (γ4)*	Ovarian cancer	4.2 ± 6.1 hours	64.7 ± 20.2 hours	van Hof *et al.*, 1996

*In these cases circulating antigen, or large pools of readily accessible antigen, are known to be present which may reduce antibody half-life in some cases.
†Increase in $t_{\frac{1}{2}}$ with increasing dose.

Table 2.8 Approaches used to manipulate antibody pharmacokinetics

Use of particular isotypes	
Clearing regimes:	anti-antibody *in vivo*
	avidin–biotin system
	extracorporeal immunoadsorption
Chemical modification: galactosylation	
Engineering antibody domain deletions	
Antibody fragments	

difficult to compare due to different antibodies being used in a variety of different clinical conditions with, in some cases, the presence of pools of circulating antigen, or because some of the antibodies used target readily available antigen on blood cells. However, it is apparent that the $t_{\frac{1}{2}}\beta$ values for chimeric and humanised antibodies are consistently higher than those for murine antibodies. In some instances the same antibody has been administered to man in both murine and chimeric forms. In all of these cases the antibody half-life is significantly longer for the chimeric antibody, for example the $t_{\frac{1}{2}}\beta$ of 17-1A was increased from 16 hours for the murine antibody to 101 hours for the chimeric IgG1 version, and that of B72.3 from 24–48 hours for the murine antibody to 224 hours for the chimeric IgG4 (LoBuglio *et al.*, 1989; Khazaeli *et al.*, 1991). In the one report to date of the pharmacokinetics of a humanised antibody in healthy volunteers, the anti-TNF antibody CDP571 had a half-life which approached the 21 day half-life of normal human IgG4 (Stephens *et al.*, 1995), suggesting that in healthy subjects at least, the normal mechanisms for regulation of IgG half-life described above also apply to humanised antibodies.

The long half-life of humanised antibodies is ideal for some applications, for example in providing protection against infection, or providing long-term neutralisation of a cytokine. However, antibodies have also been widely investigated for their ability to deliver diagnostic or therapeutic agents such as drugs, radionuclides or protein toxins to sites such as tumour cells in attempts to detect and/or destroy them. In these cases a long circulating half-life can result in unacceptable toxicity. For example, a circulating antibody–radionuclide conjugate will irradiate other tissues and limit the amount of radioactivity that can be delivered to a tumour. The ideal molecule would target the toxic agent to the tumour cells and clear rapidly from the rest of the body. Therefore there has been much interest both in strategies to remove antibody from circulation more rapidly and in designing molecules with shorter *in vivo* half-lives. Some approaches to this problem are summarised in Table 2.8.

The choice of a particular isotype for an engineered antibody can influence the pharmacokinetic properties of the antibody. The γ1 and γ4 isotypes of chimeric B72.3 have been compared directly for their pharmacokinetics in cynomolgous monkeys (Hutzell *et al.*, 1991). The $t_{\frac{1}{2}}\beta$ was 90 hours for the γ1 and 262 hours for the γ4 isotype. Although direct comparisons have not been done in humans, the same trend has been observed (see Table 2.7), with γ1 antibodies usually clearing faster than γ4. However, the clearance of γ1 antibodies is still slow, and other strategies are required to reduce the half-life further.

2.7.2 *Pharmacokinetics of antibody fragments*

As the catabolic site for IgG lies within the Fc domain, there is no specific mechanism for maintaining the serum level of $F(ab')_2$ or smaller antigen-binding fragments and these are cleared from circulation more rapidly than IgG. One of the most useful approaches to developing molecules with short *in vivo* half-life has thus been the use of antibody fragments. Small antibody fragments are filtered through the kidney, from the glomerular capillaries to the renal tubules, and thus rapidly excreted. The nominal molecular weight cut-off for kidney filtration is believed to be approx. 60–70 kDa, and thus Fab and Fv fragments are rapidly cleared through this route. During this process antibody fragments are often re-absorbed by the kidney tubules and metabolised. As well as shorter $t_{\frac{1}{2}}\beta$ values, antibody fragments also distribute to, and penetrate, tissues more rapidly, which is reflected in a shorter α phase. For example, the ability to penetrate into tumour tissue has been shown to be size-related, with smaller fragments demonstrating faster penetration (Yokota *et al.*, 1992).

There have been many studies comparing the pharmacokinetics of IgG with $F(ab')_2$ and Fab fragments in animal models. In the majority of studies Fab clears faster than $F(ab')_2$, which is considerably faster than IgG. For example, analysis of the murine antibody MOPC21 revealed the residence time in the body for IgG, $F(ab')_2$ and Fab to be 8.5, 0.5 and 0.2 days respectively (Covell *et al.*, 1986). A similar pattern was observed with B72.3 Fab, $F(ab')_2$ and IgG labelled with either of two different radioisotopes (Brown *et al.*, 1987). In several human studies $F(ab')_2$ has been shown to be cleared more rapidly than IgG. For example, Buist *et al.* (1993) found a plasma elimination half-life of 70 hours for chimeric MOv18 and 20 hours for $F(ab')_2$. However, this is not the case for all studies. In a trial of ^{131}I-radiolabelled anti-CEA antibody a similar half-life was observed for IgG and $F(ab')_2$ (Lane *et al.*, 1994).

$F(ab')_2$ molecules from different IgG isotypes have different pharmacokinetic properties. A comparison of chimeric $F(ab')_2$ of the human IgG isotypes γ1, 2 and 4 revealed that the γ2 version was retained in circulation longer than the other isotypes (Buchegger *et al.*, 1992). This pattern may reflect the number of disulphide bonds in the hinge region, 4 for γ2 , 2 for γ1 and γ4, as $F(ab')_2$ with more disulphide bonds between the two heavy chains may take longer to be metabolised, either reductively or proteolytically, *in vivo*. As $F(ab')_2$ is above the molecular weight cut-off for rapid kidney clearance, metabolism to smaller fragments probably plays an important role in $F(ab')_2$ pharmacokinetics. Recombinant $F(ab')_2$ of the same isotype with different numbers of disulphide bonds inserted in the hinge region also demonstrated that more disulphide bonds results in a longer half-life (Rodrigues *et al.*, 1993). Similarly, $F(ab')_2$ with a chemically cross-linked hinge region has been shown to have a longer residence time in the blood than $F(ab')_2$ linked by a single disulphide bond (Rodrigues *et al.*, 1993; King *et al.*, 1994).

Fv and scFv fragments have also been examined in experimental systems and shown to clear very rapidly (Milenic *et al.*, 1991; King *et al.*, 1992b). A comparison of scFv with other forms of the antibody CC49 showed very rapid plasma and whole body clearance for the scFv in both mouse and monkey studies, with 80% of scFv lost from the plasma in 15 minutes in mice and 95% removed from the plasma of monkeys by 30 minutes (Milenic *et al.*, 1991). Calculation of $t_{\frac{1}{2}}\alpha$ and $t_{\frac{1}{2}}\beta$ values can be misleading due to not taking account of the duration of each phase which may differ between antibody fragment forms. However, results for comparisons of some fragment forms of two antibodies, one murine and one chimeric, are demonstrated in Table 2.9. Differences in overall values probably reflect differences in the method of calculation used in each case, but it is easily seen that the overall trends are maintained between fragments. The pharmacokinetics of divalent scFv fragments prepared by a number of techniques including disulphide linkage, chemical cross-linking and the use of dimerisation domains have also been examined (Adams *et al.*, 1993; King *et al.*, 1994; Haunschild *et al.*, 1995). In

Table 2.9 Pharmacokinetics of antibody fragments in mice: data for CC49 from Milenic *et al.* (1991) and for cB72.3 from King *et al.* (1994); all values in hours

	CC49		cB72.3	
	$t_{\frac{1}{2}}\alpha$	$t_{\frac{1}{2}}\beta$	$t_{\frac{1}{2}}\alpha$	$t_{\frac{1}{2}}\beta$
IgG	0.65	113	11.1	56.4
$F(ab')_2$	0.43	12	0.87	26.1
Cross-linked $F(ab')_2$			3.14	21.2
Fab′	0.15	1.5		
scFv	0.06	1.5	0.23	4.0

all cases the half-life of the dimeric species is increased only slightly, the di-scFv still being rapidly removed from circulation. The preparation of tri-scFv results in a further small increase in half-life, though these multivalent molecules are still cleared rapidly (King *et al.*, 1994).

Further control of half-life of antibody fragments can be exerted by PEGylation as described in Section 2.3.3. PEGylation of many different proteins, including antibody fragments, has resulted in conjugates with increased half-life. Comparison of the *in vivo* properties of IgG, $F(ab')_2$ and Fab′ fragments with and without PEG attached revealed that PEGylated IgG was relatively unaffected, whereas the half-life of $F(ab')_2$ and Fab′ was increased by PEGylation (Pedley *et al.*, 1994). Application of this technology to smaller Fv-based fragments allows the preparation of a series of molecules with a range of different half-lives which can then be selected for a particular application.

Selective engineering of the IgG molecule can also be carried out to affect half-life. Removal of antibody carbohydrate in the CH2 domain can be achieved by substitution of the Asn residue at position 297 which is normally glycosylated. Such aglycosyl antibodies may have a reduced half-life, although this appears to be dependent on isotype. An aglycosylated mouse:human chimeric IgG4 and chimeric IgG3 antibody cleared faster than the native IgG, whereas aglycosyl chimeric IgG1 was relatively unaffected (Tao and Morrison, 1989; Rhind *et al.*, 1990). Aglycosyl mouse IgG2b also clears more rapidly than the glycosylated antibody (Wawrzynczak *et al.*, 1992). This may reflect differences in the role of the antibody carbohydrate in maintaining the conformation of the two CH2 domains, which in turn would affect FcRn interaction. Alternatively, differing conformational effects between isotypes may result in increased susceptibility to proteolysis for those isotypes which clear rapidly in aglycosyl form. More subtle variation in the carbohydrate structure can also affect half-life. A chimeric IgG1 antibody produced in a mutant CHO cell line, which is incapable of processing a high-mannose intermediate through the terminal steps of the glycosylation pathway, has a shorter half-life than the same antibody produced in a wild-type CHO line (Wright and Morrison, 1994). This is probably a result of enhanced clearance through mannose receptors in the liver, as proteolytic stability of the antibody was unaffected.

The serum persistence of Fc has been increased by introducing mutations into the FcRn binding region (Ghetie *et al.*, 1997b). Random mutagenesis of the sequence around the FcRn binding site, followed by selection using phage display, allowed variants with higher affinity for FcRn to be isolated with serum half-life increased from 120 to 150 hours. Specific engineering to disrupt FcRn binding in an IgG has not yet been reported, although mutations in the CH2 and CH3 domains which have this effect in Fc fragments have been described (Kim *et al.*, 1994). Mutation of these key residues may allow the production of antibodies with a range of serum half-lives. Removal of the entire CH2 domain of the IgG heavy chain has been reported (Mueller *et al.*, 1990). The resulting truncated heavy chain still allows assembly of a tetrameric structure with full antigen-binding ability. The pharmacokinetics of such CH2 deleted variants are similar to those of stabilised F(ab′)$_2$ fragments with a $t_{\frac{1}{2}}\alpha$ of 1.5 hours and a $t_{\frac{1}{2}}\beta$ of 12 hours reported in mice (Mueller *et al.*, 1990). This again demonstrates disruption of the FcRn receptor binding site.

As described in Section 2.5, bispecific diabodies can be produced with one specificity for target antigen and one for serum immunoglobulin. This results in not only the ability to retarget effector functions, but also a significant increase in half-life compared to a control diabody (Holliger *et al.*, 1997). A β-phase half-life of 10 hours was reported for serum Ig targeted diabody, compared to less than 1 hour for the control diabody. This was still significantly less than the IgG alone, possibly as a result of dissociation of the diabody from the IgG during the low pH values experienced during FcRn-mediated recycling (Figure 2.17).

2.7.3 Clearance

A further approach to remove antibody from circulation has been to allow antibody to bind to its target for a period of time and then attempt to remove the residual antibody. Such 'clearance' strategies have been investigated particularly with radiolabelled anti-tumour antibodies. Clearance can be achieved using a second antibody with specificity for the anti-tumour antibody. After the desired time interval, injection of the second antibody results in binding to the radiolabelled antitumour antibody to form immune complexes. These are then rapidly removed by clearance through the liver and to a lesser extent other clearance organs. Initial work used liposomally entrapped second antibody to achieve this, although equivalent results were later generated with second antibody alone (Begent *et al.*, 1982; Pedley *et al.*, 1989). Effective clearance of the radiolabelled antibody has been observed using such systems with enhanced tumour : blood ratios of radioactivity allowing improved imaging of tumours, and resulting in less toxicity associated with blood-borne activity (Blumenthal *et al.*, 1989). A variation of this approach uses the high-affinity interaction between avidin and biotin to remove antibody from circulation. The initial radiolabelled antibody is conjugated to biotin before use and then can be cleared through administration of avidin or streptavidin (Paganelli *et al.*, 1991). A comparison of these two approaches found that superior clearance could be achieved with second antibody, with a larger and faster reduction in blood levels achieved (Marshall *et al.*, 1994).

Problems in using this approach include the relative complexity and expense of preparing and using a clearing agent. Also, the clearance system may result in the sudden deposition of large amounts of immune complex and consequently radioactivity into the liver, or other clearance organ, which may result in toxic effects. This can be overcome using a related method known as extracorporeal immunoadsorption. In this procedure blood is pumped outside the body through an extracorporeal shunt in which the radiolabelled antibody is removed by binding to an affinity column before being returned to the patient (Johnson *et al.*, 1991b). The affinity column can be prepared from either antigen or second antibody attached to a solid support. Clinical trials with this approach have proved that most of the circulating activity can be removed from the patient efficiently, although the complexity and expense of such a procedure has meant that it has not been widely used.

2.7.4 Chemical modification

Chemical modification of antibody has also been used to alter half-life of antibody. Attachment of galactose, galactosylation, allows recognition of the conjugate by the asialoglycoprotein receptor in the liver resulting in very rapid clearance. This procedure was developed for use with antibodies administered locally, for example intra-peritoneally for ovarian cancer, such that when antibody leaked out from the peritoneal cavity to the blood, the circulating activity was rapidly removed (Mattes, 1987). However, penetration of antibody from the peritoneal cavity to ovarian tumours is not very effective and hence galactosylated antibody has also been evaluated for intravenous cancer therapy (Ong *et al.*, 1991). In this case it was necessary to inhibit the action of the asialoglycoprotein receptor for 2–3 days by injection of large amounts of asialo-bovine submaxillary mucin which initially saturates the receptor. Inhibition for 2–3 days allowed the galactosylated antibody to bind to experimental tumours in nude mice, after which time the inhibitor was cleared and antibody removed from the circulation. In common with the *in vivo* clearance mechanisms described above, this procedure can result in the rapid deposition of antibody in the liver, and thus is restricted to use with therapeutic agents which will not be toxic to liver cells. Radioiodinated antibody appears to be rapidly metabolised by the liver with the excretion of radio-iodine, and thus may be useful for therapy with such galactosylated conjugates (Ong *et al.*, 1991). But perhaps the most useful application of galactosylated antibody is as a second antibody clearing agent. Use of galactosylated second antibody increases the rate of removal of the first antibody from the circulation and has been applied to the removal of antibody–enzyme conjugates from circulation prior to prodrug therapy (Sharma *et al.*, 1990; see Chapter 4).

2.7.5 Fc region to extend half-life

The finding that antibody half-life is dependent on the Fc region as described above, and that isolated Fc regions have the same extended $t_{\frac{1}{2}}\beta$ as IgG molecules, has led to the use of the Fc region to extend the half-life of other proteins. CD4 is a receptor for the HIV virus, and has been of interest for AIDS therapy. However, soluble CD4 has a very short plasma half-life and was therefore difficult to examine *in vivo*. Fusion proteins between CD4 and IgG Fc, termed immunoadhesins, have been prepared which use the IgG Fc region to increase dramatically the half-life of the soluble CD4 protein (Capon *et al.*, 1989). Since then a similar approach has been used to produce Fc fusion proteins with a range of different proteins in attempts to improve half-life (Charnow and Ashkenazi, 1996). An additional property of many immunoadhesins is the retention of some Fc effector functions, and a range of applications are open to such molecules both as research reagents and as potential human therapeutics. The CH2 domain alone can also be used to extend half-life, though the increase in $t_{\frac{1}{2}}\beta$ is less than can be achieved with whole Fc fusions. However, the use of this single domain allows the design of molecules with an internal CH2 domain. For example, a CD4–CH2–PE40 toxin molecule has been produced with extended plasma half-life (Batra *et al.*, 1993).

3

Monoclonal Antibodies in Research and Diagnostic Applications

3.1 Introduction

Antibodies have proved to be invaluable reagents for the detection and quantitation of many types of substances both *in vitro* and *in vivo* and have, therefore, found wide application in both the research laboratory and the diagnostics industry. Many of the techniques used in these two situations are similar, and thus it is helpful to review them together. In a chapter of this length it is impossible to cover all the ways that MAbs have been used in these areas; thus the emphasis in this chapter is on applications which are particularly widely used and those which are likely to benefit most from the ability to design engineered or modified antibodies.

3.2 Immunoassays in diagnostics and research

Immunoassays are carried out to detect and quantitate the presence of a particular antigen or antibody in a test fluid. The development of immunoassays began with polyclonal antisera and thus precedes the introduction of MAbs. However, today MAbs are widely used, and many assays routinely used in, for example, hospital clinical biochemistry laboratories which rely on them. In fact MAbs have found widespread use in laboratories of all types, in applications from drug discovery to detecting drugs of abuse in athletes and horses. In contrast to polyclonal antisera, production of MAbs allows a potentially unlimited supply of identical reagent which can be selected from a number of different clones to have the optimal characteristics for the intended assay. For some assay formats such as radioimmunoassay, however, polyclonal antisera remain the best reagents, as they are frequently of higher apparent affinity than MAbs. Therefore, it is useful to examine the types of immunoassays to appreciate the role of monoclonal antibody reagents. Some of the properties of MAbs of interest for immunoassay development are given in Table 3.1.

Table 3.1 Properties of monoclonal antibodies important for immunoassay design

Unlimited quantity of identical antibody
Antibodies to distinct epitopes can be selected from panels of clones
Easily purified to reduce background binding
Readily digested to produce fragments with equivalent specificity
Can be produced to impure antigen
Often of lower apparent affinity than polyclonal antisera

3.2.1 *Radioimmunoassay*

Radioimmunoassay (RIA) methodology was first developed in the 1960s and represented the first sensitive method for immunoassay (Yallow and Berson, 1960). RIAs are competitive assays based on the competition of radiolabelled antigen with antigen present in the sample to be assayed for binding to antibody (Figure 3.1). The antibody complex is then separated from the free antigen and the amount of bound material determined by counting the radioactivity. Separation can be achieved by a number of methods, although it is most convenient to use antibody immobilised onto a solid phase directly, or to

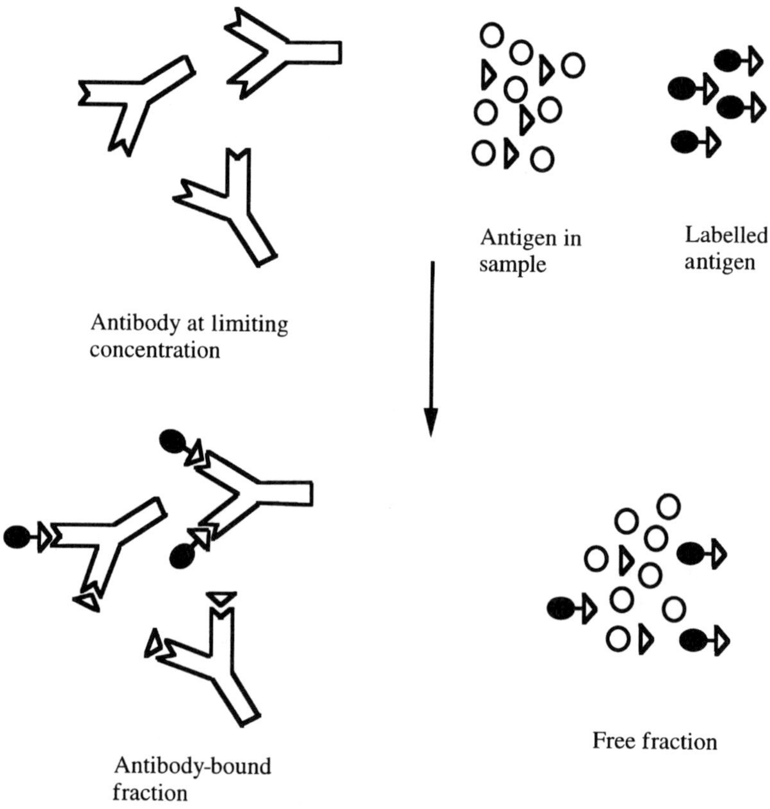

Figure 3.1 Format of radioimmunoassay: antibody at limiting concentration is added to sample containing antigen and radiolabelled antigen; the antibody-bound fraction is then separated and either free or antibody-bound counts (or both) are measured

remove the antibody–antigen complex by a second immobilised antibody which recognises the first antibody used. Antibody can be immobilised to many types of solid supports, e.g. microparticles such as cellulose, sepharose and paramagnetic particles, or to large solid phase plastic surfaces such as beads, tubes or microwells. The RIA assay format requires excess antigen over antibody and for high sensitivity small amounts of high affinity antibody are required. Polyclonal antisera often have very high apparent affinity or avidity, probably due to a small proportion of the total antibody present having high affinity and to cooperativity between antibody molecules. The mixture of antibodies present in a polyclonal antiserum may allow attachment of several antibody molecules to the antigen at different sites and, as antibody molecules are bivalent, this can lead to the formation of multimeric complexes, which lead to a low off-rate and higher apparent affinity. Therefore, polyclonal sera have remained the ideal reagents for RIA, and although MAbs can be used their performance is often not optimal. However, RIA techniques are relatively long and tedious to carry out and in many cases have been superseded by MAb techniques based on other assay formats.

3.2.2 ***Immunoradiometric assay***

The immunoradiometric assay (IRMA) uses an excess of labelled antibody rather than antigen to quantitate the antigen present. This means that sensitivity is not dependent on antibody affinity. The IRMA was initially developed as a single-site assay in which excess labelled antibody was added to the sample containing antigen to form immune complexes (Miles and Hales, 1968). The remaining antibody was then removed by incubation with solid-phase antigen and after separation of the solid and liquid phases, counting of either phase could be carried out to determine the amount of antigen present (Figure 3.2). This form of IRMA is not widely used due to the requirement for large amounts of both labelled antibody and immobilised antigen. A variation of the IRMA, the two-site IRMA, has become a much more useful assay format. In this assay antigens are quantitated using two antibodies which bind to different epitopes (Woodhead *et al.*, 1974). Antibody is immobilised onto a solid phase and incubated with sample containing antigen. After binding of the antigen to the solid-phase antibody, excess labelled second antibody is added which binds to a second epitope on the antigen. The excess labelled antibody is then washed away and the amount of bound antibody quantitated by counting (Figure 3.3). Requirements for this type of assay are an antigen which is large enough to contain two epitopes, and supplies of two different specific antibodies. This type of assay format is very difficult to achieve with polyclonals, and requires extensive purification of the antisera to isolate preparations binding to individual epitopes. MAbs are ideally suited, however, as different clones can be tested until two non-competing antibodies are identified. These assays are much more rapid, sensitive and robust than competitive binding assays such as RIA and are now widely used.

Iodine 125 (^{125}I), a high-energy gamma emitter, is usually used as the isotope for antibody labelling in IRMA applications. It allows the development of sensitive assays due to the ability to achieve high specific activity labelled antibody preparations. ^{125}I is attached to antibodies via two main methodologies, either by attachment to tyrosine residues or by attachment to lysine residues. Attachment to tyrosine is achieved by oxidation of sodium iodide ($Na^{125}I$) to form molecular iodine which reacts rapidly with tyrosine side chains (Figure 3.4). The reaction is then stopped by the addition of a reducing agent (e.g. sodium metabisulphite) to stop generation of reactive iodine, or by

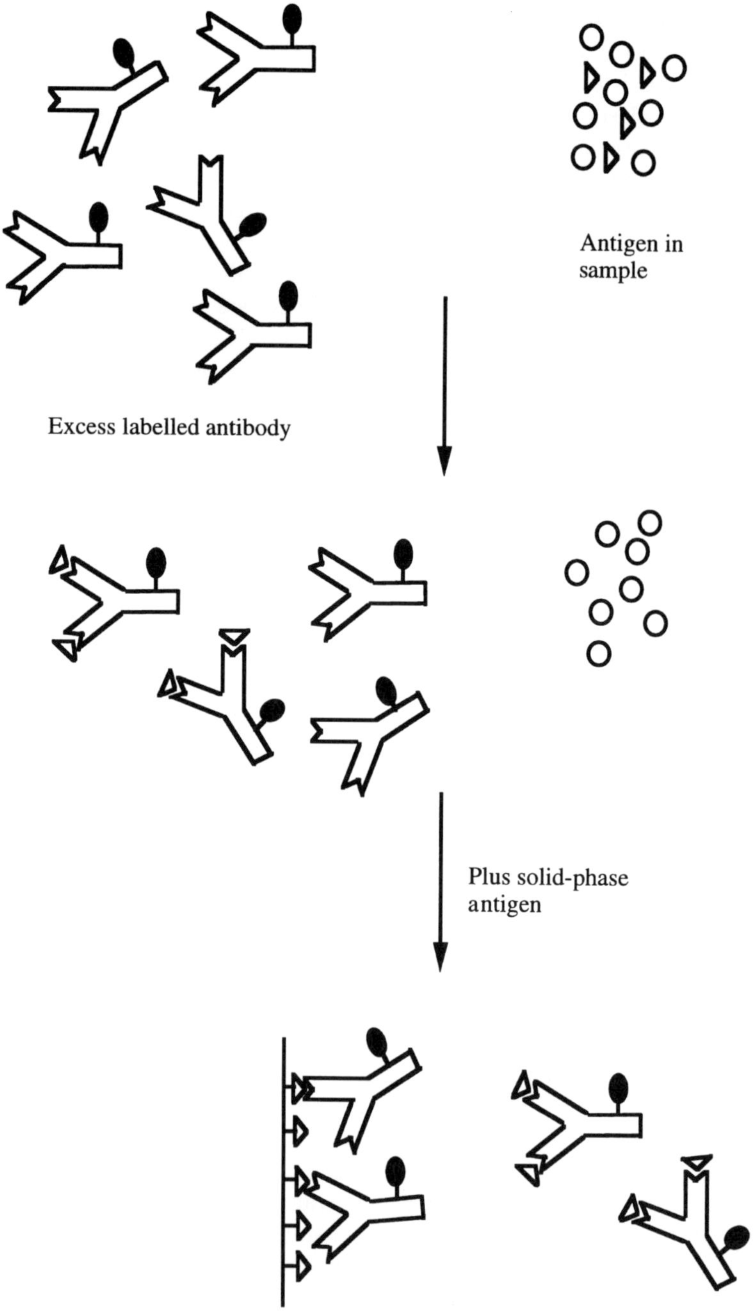

Figure 3.2 Format of immunoradiometric assay: excess labelled antibody binds antigen in the sample and is then added to solid-phase antigen; determination of bound or free counts allows quantitation of the amount of antigen present in the sample

adding excess free tyrosine to mop up all of the reactive iodine. High specific activities can be generated using this methodology, although care must be taken that oxidative damage to the antibody is minimised. Commonly used oxidants are chloramine T or solid-phase reagents such as iodogen coated onto a plastic bead or tube. The use of solid-phase

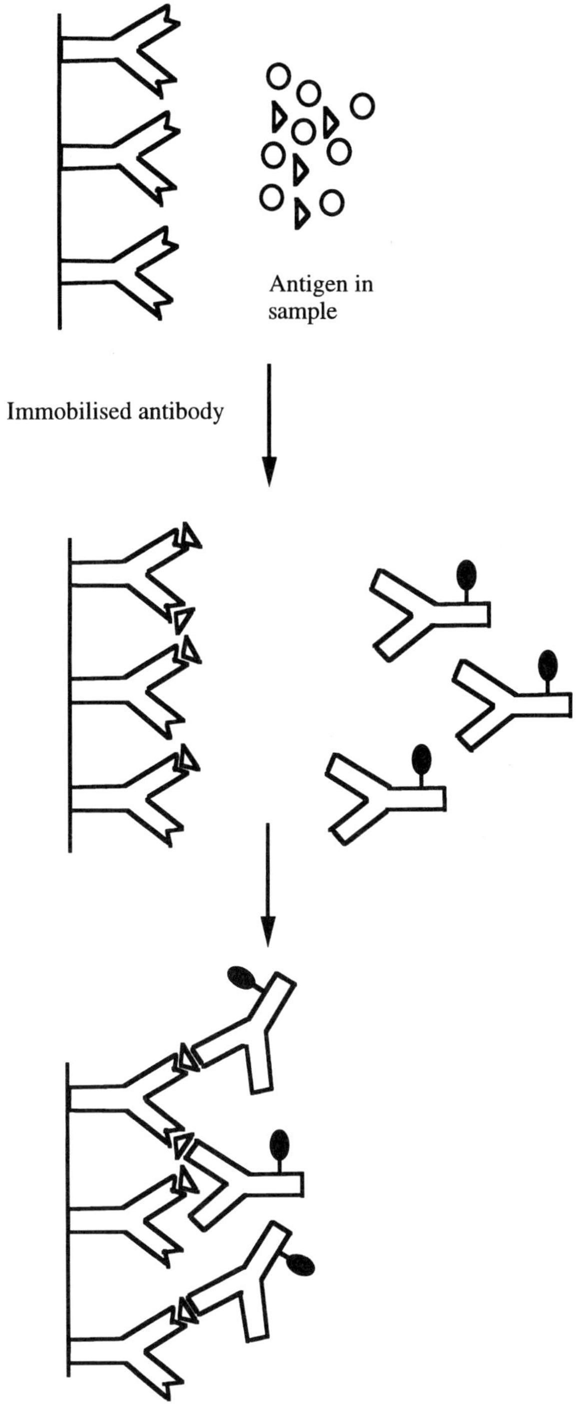

Figure 3.3 Format of two-site immunoradiometric assay: immobilised antibody captures antigen in the sample; a second antibody reactive with a second epitope on the antigen is then used in labelled form to bind to antigen and reveal the presence of antigen

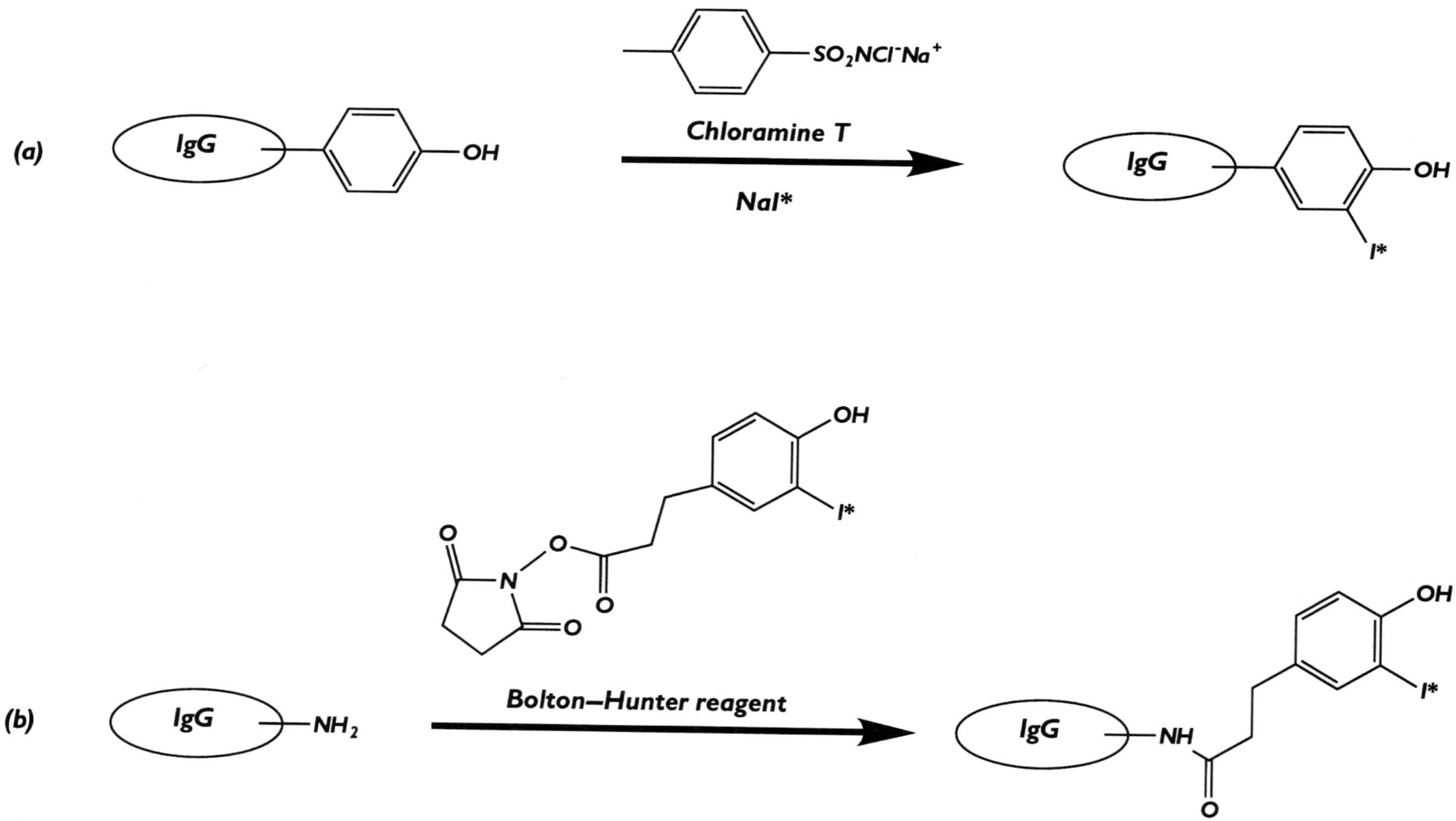

Figure 3.4 Radioiodination of MAb by (a) the use of chloramine T and sodium iodide and (b) Bolton–Hunter reagent (Bolton and Hunter, 1973)

reagents allows rapid removal of oxidant from the system and thus can result in less damage to the antibody. Attachment to lysine residues is accomplished via indirect labelling using amino reactive compounds pre-labelled with ^{125}I. This approach was first described using the ^{125}I-labelled *N*-hydroxysuccinimide ester of 3-(4-hydroxyphenyl) propanoic acid (Bolton and Hunter, 1973), subsequently known as Bolton–Hunter reagent. This and other similar reagents are commercially available in radiolabelled form and allow simple labelling of antibodies under mild conditions (Figure 3.4). Specific activities achieved with Bolton–Hunter reagent are usually lower than with oxidative methods as the incorporation efficiency is lower, but for antibodies sensitive to oxidative damage and in cases where relatively low specific activity is required, this method is particularly useful.

The limitation in sensitivity of the two-site IRMA is the specific activity of the labelled antibody, the affinity of the first antibody and the level of non-specific binding. MAbs can be readily purified, and antigen-binding fragments such as $F(ab')_2$ easily produced, which can reduce levels of non-specific binding to surfaces. Low specific activity is often due to antibody damage during radiolabelling. The availability of large amounts of MAb allows the screening of several radiolabelling methods and the development of a suitable method to minimise such damage.

3.2.3 Non-isotopic immunoassays

A major disadvantage with the use of the two-site IRMA is the need to use radiolabelled antibody. ^{125}I has a half-life of 60 days and thus new labelled preparations need to be continually produced. Also, the use of radiolabelled materials requires stringent precautions in the laboratory to prevent radioactive contamination. Several alternative labels have been investigated, which are replacing radioisotopes. These include enzymes, fluorescent and chemiluminescent labels, and indirect methods (Table 3.2).

Suitable enzymes can be attached to antibodies without loss of activity, and can convert substrate to a coloured product which is then quantitated spectrophotometrically. Enzyme detection has become the standard method for most laboratory immunoassays and two-site (or sandwich) ELISAs (enzyme-linked immunosorbent assays) are very

Table 3.2 Commonly used non-isotopic labels for immunoassays

Enzymes	Horseradish peroxidase Alkaline phosphatase Urease β-galactosidase Xanthine oxidase
Fluorescent labels	Fluorescein Phycoerythrin Europium chelates and other lanthanide chelates
Chemiluminescent	Luminol/isoluminol Acridinium esters
Indirect methods	Biotin Streptavidin/avidin Protein A, protein G

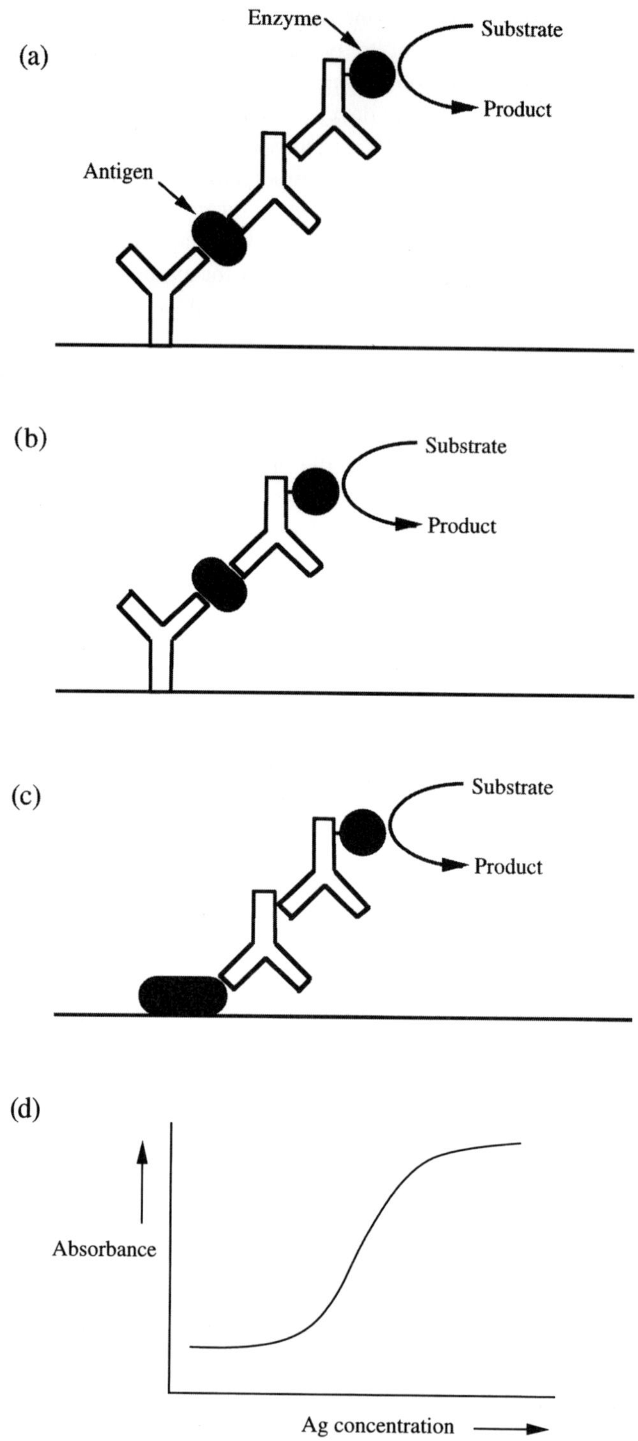

Figure 3.5 Common ELISA formats: (a) two-site ELISA for antigen quantitation using enzyme conjugated anti-antibody to detect; (b) two-site ELISA for antigen quantitation using directly conjugated second antibody; (c) determination of antigen-specific antibody; (d) expected form of standard curve for such direct binding ELISA formats

widely used. These are similar in format to immunoradiometric assays, although often an extra antibody layer is included to help amplify the signal generated by the enzyme (Figure 3.5). The use of such formats on microtitre plates allows assays to be carried out simply and quickly. Commonly used enzymes such as alkaline phosphatase and horseradish peroxidase can convert substrate to product with a high turnover number to allow the generation of a large amount of coloured product per enzyme molecule bound, resulting in amplification of the signal.

ELISAs are commonly used for measuring antibody as well as antigen. In this case ELISAs with immobilised antigen can be used to measure the amount of antigen-specific antibody by detecting with an anti-antibody reagent after binding. Alternatively, the amount of total antibody in a sample can be quantitated in a two-site ELISA using, for example, immobilised antibody to the light chain and detecting with enzyme conjugated to antibody directed to the heavy chain or vice versa. ELISAs can also be used in a number of competitive formats, for example in the detection of soluble antigens (Figure 3.6), and can also be carried out with whole cells as the antigen (Feit *et al.*, 1983). ELISAs are usually carried out on plastic microtitre plates on which the 'capture' reagent is immobilised. These plates are usually made of polystyrene which is optically clear and are commonly used in the standard 96-well format. Such plates are particularly advantageous as they allow dilutions of standards and samples to be run under identical conditions on the same plate, which is ideal for an accurate result. Antibody or antigen is normally immobilised to the plate by passive adsorption through hydrophobic interactions, although it has been suggested that covalent binding may improve assay performance (Douglas and Monteith, 1994).

Antibody–enzyme conjugates are often produced by chemical coupling (see Section 2.6.2). This is usually a satisfactory procedure although there are often variations from batch to batch and in some cases it can lead to loss of some of the antigen-binding activity of the antibody or to loss of enzyme activity. Site-specific conjugation to Fc region carbohydrate has been investigated, and alkaline phosphatase conjugates prepared in this way (Section 2.6.2) have been shown to retain full binding activity (Husain and Bienarz, 1994). Recently, attempts have been made to generate suitable reagents by expression of antibody–enzyme fusion proteins. Alkaline phosphatase fusions to both scFv and $F(ab')_2$ fragments have been prepared, expressed in *E. coli* and shown to be fully functional for both antigen binding and enzyme activity (Wels *et al.*, 1992; Ducancel *et al.*, 1993). For application in immunoassays a general reagent consisting of anti-human IgG scFv or $F(ab')_2$ fused to alkaline phosphatase has been produced and shown to be capable of detecting human IgG to hepatitis B antigen in an ELISA (Carrier *et al.*, 1995). An additional advantage of this approach is that the use of antibody fragments has been shown to reduce background binding in ELISAs compared to intact IgG molecules. The ease of recombinant techniques for the direct expression of antibody fragment–enzyme fusion proteins suggests that this route of production may become increasingly important in immunoassay development.

3.2.4 Improving sensitivity

Several approaches have been investigated to improve the sensitivity of immunoassays, to develop ultrasensitive assays, by manipulation of the detection system (Kricka, 1993). Increases in enzyme amplification of the signal generated have been described in which the product of the antibody–enzyme conjugate reaction is used to set up a cycle in which

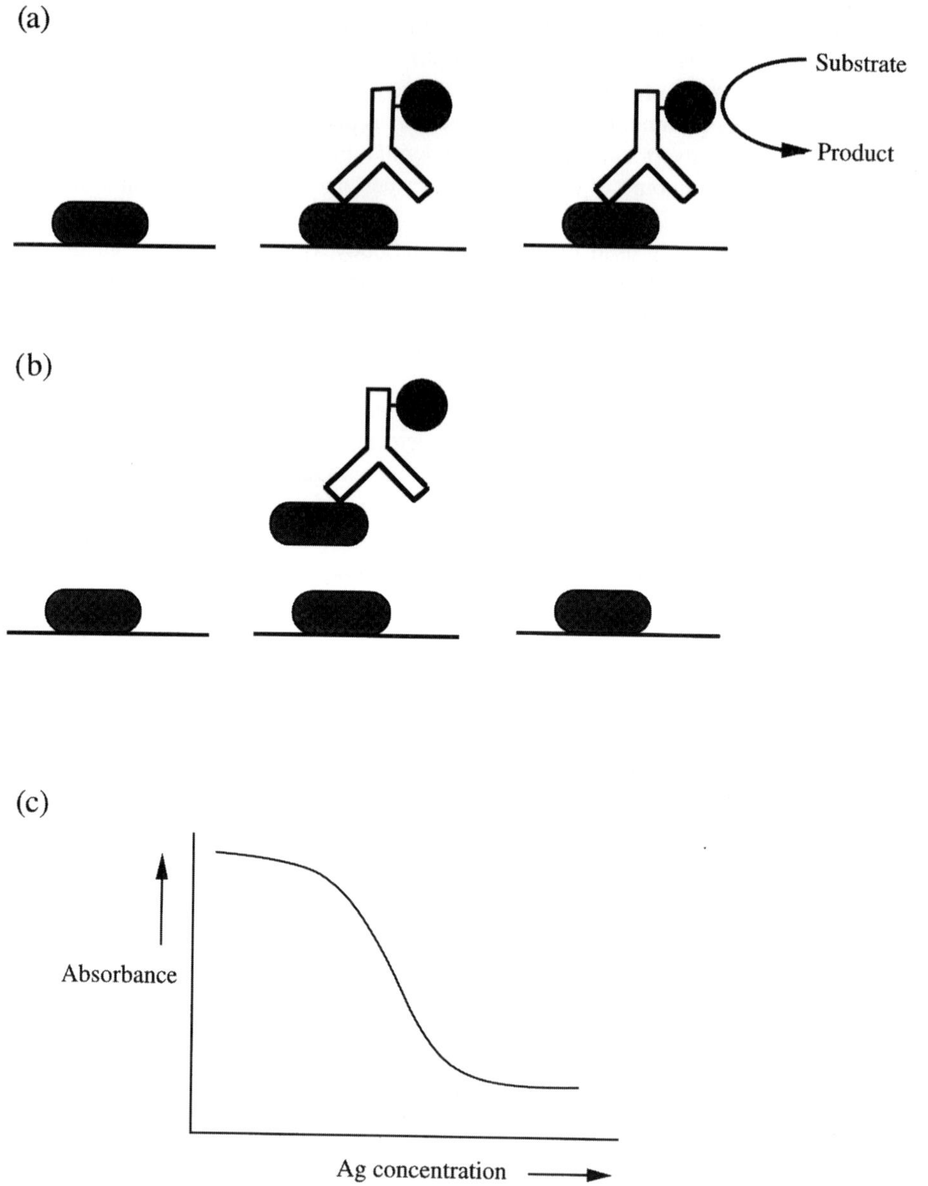

Figure 3.6 Competitive ELISA format assay: (a) in the absence of antigen in the sample, conjugated antibody binds to immobilised antigen and gives rise to the production of coloured product; (b) in the presence of antigen in the sample, conjugated antibody is prevented from binding to immobilised antigen leading to the absence of signal; (c) expected form of standard curve from such a competitive ELISA

a second enzyme system generates large amounts of coloured product as shown in Figure 3.7 (Stanley *et al.*, 1985). A high degree of signal amplification results as each NAD molecule generated can be responsible for the production of many hundred formazan molecules, and this has led to the development of several immunoassays with reported sensitivities over 100-fold higher than conventional alkaline phosphatase substrates such

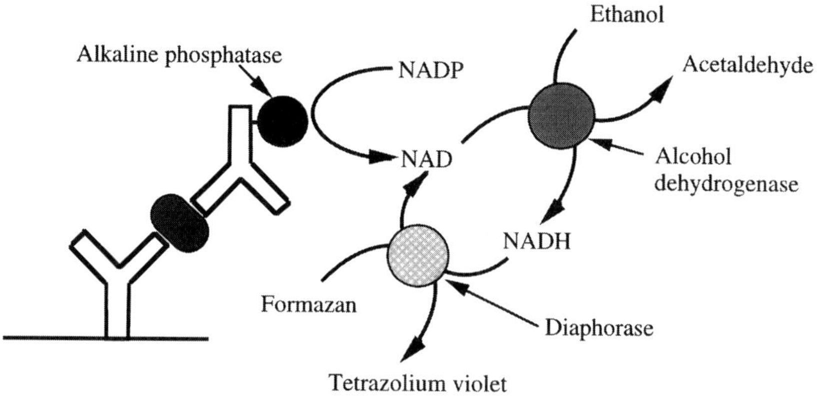

Figure 3.7 Example of an amplification mechanism for enhancing signal in ELISA

as *p*-nitrophenyl phosphate (Clark and Price, 1986). In many cases such highly sensitive assays may have problems with the level of non-specific background and thus careful optimisation of the assay is usually required.

Enzyme labels can also be made more sensitive by the use of substrates, resulting in the generation of fluorescent or chemiluminescent substances, which can be detected with greater sensitivity than colorimetric reactions. Substrates resulting in both fluorescent and chemiluminescent assays are available for the most commonly used enzymes, horseradish peroxidase and alkaline phosphatase (e.g. Albrecht *et al.*, 1994; Akhavan-Tafti *et al.*, 1995). Enzyme amplification cycles have also been combined with fluorescent readouts to allow determination of very small amounts of alkaline phosphatase, down to one thousandth of an attomole (one zeptomole or 350 molecules) when applied to an immunoassay of proinsulin (Cook and Self, 1993).

Fluorescent labels can also be directly attached to antibody and used as a detection system for immunoassays. The major problem with the use of fluorescent labels, however, is the background level of fluorescence generated by many biological substances and by plastics used for immobilisation. Hence conventional fluorescent compounds used in immunocytochemistry (see Section 3.4) such as fluorescein are of relatively little use. This can be overcome by the use of time-resolved fluorescent techniques using lanthanide chelates as the fluorescent reagents (Diamandis, 1988). In this technique pulses of light are used to excite the fluorescent material present and then the emitted light is measured after a short time interval of a few hundred microseconds. As background fluorescence is due to very rapid events, with fluorescence lifetimes of 100 nanoseconds or less, the interference can be removed by using fluorescent compounds which emit light for longer periods (1000 microseconds or more) and measuring emitted light after the background fluorescence has died away. Measurement of light emitted by such systems requires instruments which can pulse and measure fluorescence many times per second to build up a strong signal, and several such instruments have been commercialised. Chelates of lanthanides such as europium or terbium are suitable reagents as they have ideal properties of long fluorescence lifetimes, large Stokes shifts (the difference between the wavelength of light used for excitation and that emitted) and narrow emission wavelength bands. Many sensitive assays have been developed using this technology and they are widely used in clinical diagnostics. Attempts have also been made to produce recombinant molecules capable of binding lanthanides directly. A fusion protein between a single-chain Fv and an engineered lanthanide binding protein has been produced and shown to bind terbium (MacKenzie *et al.*, 1995).

Chemiluminescent reactions have also been explored as a sensitive means of detection. Chemiluminescence results from a reaction in which the product is in an electronically excited state. Electrons in the excited state then relax to their ground state with the emission of light. Most useful chemiluminescent labels for direct attachment to antibodies have been the acridinium esters, which are capable of coupling to antibodies or antigens without a large loss in their quantum yield (Weeks *et al.*, 1983). They are usually coupled to antibody or antigen as *N*-hydroxysuccinimide esters which allows high coupling efficiencies, and chemiluminescence is initiated simply by the addition of hydrogen peroxide. An alternative luminescent reagent for attachment to antibody is the calcium-activated photoprotein aequorin, initially isolated from the jellyfish *Aequorea victoria*. Aequorin conjugated antibody has been used for the development of several sensitive immunoassays for hormones such as thyrotrophin, chorionic gonadotrophin, lutrophin and follitrophin (Rigl *et al.*, 1995). As a protein, aequorin can also be attached to antibody by recombinant means, and the construction and expression of a functional Fab′–aequorin fusion protein has been described (Casadei *et al.*, 1990).

The use of an antibody-linked peroxidase system to generate a chemiluminescent acridinium ester has been reported to result in longer duration of light than that generated from directly conjugated acridinium esters (Akhavan-Tafti *et al.*, 1995). Peroxidase also catalyses the oxidation of luminol by hydrogen peroxide which can be used to generate a strong signal in the presence of a suitable enhancer such as 4-iodophenol, 1,6-dibromo-2-naphthol or 6-hydroxybenzothiazole (Thorpe and Kricka, 1986). Such enhanced chemiluminescent assays have been shown to be useful in many situations, for example in the measurement of thyroxine (Christofides and Sheehan, 1995). Other enzymes can also be used, and xanthine oxidase has been shown to be useful in a luminol-based immunoassay system for interleukin-5 (Rongen *et al.*, 1997). Several substrates for alkaline phosphatase conjugates have been developed which give rise to chemiluminescent compounds after enzyme action (Bronstein *et al.*, 1989). The use of the substrate adamantyl 1,2-dioxetane aryl phosphate (AMPPD) has been shown to result in particularly low detection levels, down to one zeptomole, when applied to immunoassay of thyrotrophin (Bronstein *et al.*, 1989).

A simple method to increase sensitivity of immunoassays might be simply to put more of the label on to each detecting antibody molecule. This is limited by loss of immunoreactivity of the antibody when too many labelling molecules are attached. One way to increase this is to attach a carrier molecule, which can be loaded up with labels. Poly-lysine has been used in this way for the attachment of multiple fluorescent groups (Exley and Ekeke, 1981). However, the most useful strategy to date has been the use of the biotin:avidin or biotin:streptavidin interaction. Each streptavidin or avidin molecule can bind four biotin groups with high affinity. The detecting antibody can be labelled with several biotin molecules until the limit for retention of antibody-binding activity is reached. When bound in an assay, biotinylated antibody can then be revealed with streptavidin that is labelled with multiple numbers of signal generating molecules such as enzymes or fluorescent or chemiluminescent reagents, hence leading to a considerable amplification in signal. For example, thyroglobulin molecules labelled with up to 480 europium chelate groups can be attached to streptavidin and used to detect biotinylated antibody resulting in an amplification of 4500- to 6750-fold (Diamandis, 1991). Biotinylated antibody can also be produced by recombinant means by use of Fab fusion proteins to the biotin–carboxyl carrier protein (BCCP) subunit of *E. coli* acetyl-CoA carboxylase (Weiss *et al.*, 1994). However, production of fully biotinylated material is difficult due to the requirement for sufficient biotin ligase enzyme to allow *in vivo* incorporation of biotin during expression of the Fab–BCCP fusion protein. Only approximately one sixth of the fusion protein produced was able to react with streptavidin (Weiss *et al.*, 1994).

Another approach to amplify signal is to attach the detecting antibody to a liposome, into which have been entrapped many detectable molecules, such as enzymes, fluorescent or chemiluminescent reagents. Liposomes are closed spherical structures in which a phospholipid bilayer encloses an aqueous compartment in which it is possible to entrap 10^3–10^6 water-soluble molecules in a stable manner. When used in an immunoassay, the liposome can be lysed at the end of the assay and the contents quantitated. Immunoliposomes can be produced by coupling antibody to lipid via a heterobifunctional crosslinker, either randomly or site-specifically via antibody carbohydrate (Torchilin, 1994; Ansell *et al.*, 1996). However, immunoliposomes often show a lower association rate than free antibody and may also increase non-specific binding levels. Nevertheless, direct comparisons of immunoassays for thyrotrophin with antibody conventionally labelled with an acridinium ester (10 labels per antibody) with an immunoliposome containing approx. 250 000 labels per liposome resulted in an improvement in signal to background ratio of approx. 3-fold for the immunoliposome assay (Law *et al.*, 1989). Lipid-tagged antibodies have also been produced by expression of scFv fragments fused to part of the *E. coli* major lipoprotein which is tagged with lipid when expressed in *E. coli*. The resulting lipid-tagged antibody fragments can be incorporated into liposomes with high efficiency and have been used to generate europium chelate loaded liposomes for a use in fluoroimmunoassay (Laukkanen *et al.*, 1995). A higher signal was generated using liposomally entrapped europium chelate compared to conventionally labelled scFv, resulting in a more sensitive assay.

The polymerase chain reaction (PCR – see Chapter 1) has also been adapted for use as a sensitive detection system for immunoassay, termed immuno-PCR (Sano *et al.*, 1992). Antibody is coupled to DNA through the avidin–biotin system. In the first system described, streptavidin is first bound to immobilised antibody through a protein A–streptavidin conjugate. This is then used to capture biotinylated DNA which is then detected and amplified through PCR with the resulting products analysed by electrophoresis. The resulting assay has been shown to be extremely sensitive, capable of detecting as few as 600 molecules in an immunoassay of bovine serum albumin (Sano *et al.*, 1992). PCR has also been applied to the simultaneous determination of multiple analytes (Hendrickson *et al.*, 1995). In this case human thyroid stimulating hormone, human chorionic gonadotropin and *E. coli* β-galactosidase were measured by a combination of three specific MAbs, each conjugated to a unique oligonucleotide which were designed to result in different size PCR products from the same primer sequence. The different size products could be resolved by electrophoresis, resulting in an assay which was 2–3 orders of magnitude more sensitive than a comparable ELISA.

3.2.5 Assay formats

The majority of immunoassay formats rely on the separation of bound and free antibody before detection. Homogeneous assays in which there is no need to separate bound and free antibody have been investigated in attempts to produce simplified assays. Several formats have been successful as competitive assays which are easy to use and automate, although sensitivity is often relatively low. The enzyme multiplied immunoassay technique, known as EMIT, uses glucose-6-phosphate dehydrogenase–antigen conjugates which catalyse the conversion of substrate to coloured product. When bound by antibody, enzyme activity is inhibited and therefore the amount of coloured product is proportional to the amount of competing antigen present (Armbruster *et al.*, 1993). Later versions of this system use a recombinant form of the enzyme which is more suitable for automation (Vogl *et al.*, 1996). A similar system, cloned enzyme donor immunoassay, known as CEDIA, also operates as a competitive assay. In this system two inactive fragments of β-galactosidase, enzyme donor and enzyme acceptor, are produced by recombinant means which when mixed together associate to produce active enzyme. When the enzyme donor is conjugated to antigen, antibody binding prevents association and formation of the active enzyme. Any free antigen competes for binding to the antibody, hence allowing enzyme activity to be generated by association of the antigen–enzyme donor with the acceptor (Henderson *et al.*, 1986). Such homogeneous systems have been developed for immunoassay of many small molecular weight substances such as cortisol, ferritin, digoxin, folate, vitamin B12 and drugs of abuse (Van der Weide *et al.*, 1992; Armbruster *et al.*, 1995). Other homogeneous assay formats have also been developed with fluorescent endpoints, including fluorescence polarisation (Colbert *et al.*, 1985), fluorescence energy transfer (Calvin *et al.*, 1986) and fluorescence quenching (Barnard *et al.*, 1989). Many other formats have been investigated, including some based on light scattering which may be simple to automate (Armbruster *et al.*, 1993). Overall the advantages of homogeneous assays are in speed and convenience. However, the sensitivity of homogeneous immunoassays remains limited, largely because the use of a separation step in other assay formats allows washing, which reduces or removes interfering substances and non-specific binding events.

For most large diagnostic laboratories there is now a requirement for a high degree of automation in performing immunoassays, and many automated assays are in use and continue to be developed. Both heterogeneous and homogeneous assay formats have been automated using a wide variety of detection systems, including enzyme reactions, fluorescence, chemiluminescence and light-scattering techniques (Chan, 1996). Attempts are being made to develop all-in-one formats suitable for automation (Lovgren *et al.*, 1996). Whether any of these systems will eventually dominate the future market for automated assays or whether the current range of immunoassay techniques will continue to be extended remains to be seen.

While large diagnostics centres require specialised instrumentation, there has also been much interest in developing ‘point of need’ assays which can be performed simply in the clinic, doctor’s office or home to give a simple + or – readout. Applications include home pregnancy testing (measurement of chorionic gonadotrophin, hCG) and fertility testing (lutrophin measurement). Several useful formats have been developed and commercialised based on immunochromatographic test strips and membranes. For example, the hCG test strip, known as Clearblue One Step (Unipath) has been developed for home use. In this test coloured submicron latex particles are used to indicate a result by formation of a blue line in the large window and the small window (positive) or the small window alone (negative). A sample of urine is captured on a porous applicator which is then held in contact with the test membrane which has three zones of antibody (Davidson, 1992). Urine is used as the mobile phase and as it passes along the membrane it reaches the first zone consisting of blue microparticles with a Mab to the alpha subunit of hCG attached. The urine picks up the particles and carries them to the second zone, under the large window, which has a MAb to the beta subunit of hCG immobilised on the membrane.

If hCG is present the coloured particles will bind resulting in a blue line under the large window. The urine continues to migrate along the membrane and reaches the third zone, under the small window, when excess or unreacted particles will be trapped resulting in a blue line whether hCG is present or not, allowing a control on the assay format. This assay can be carried out in three minutes with an overall performance (sensitivity and specificity) of 99% (Davidson, 1992). Many variants of such tests are now available which have successfully extended the applications of MAbs in the home for such applications as pregnancy testing (hCG assay), ovulation prediction (luteinising hormone, follicle stimulating hormone) and contraception (luteinising hormone and oestrone-3-glucuronide). Similar tests are under development or in use for applications such as identification of infectious diseases, drugs of abuse screening and monitoring disease markers.

3.2.6 Advantages of monoclonal antibodies in immunoassay

Although many immunoassays are still carried out with polyclonal antibody reagents, MAbs have clearly demonstrated an important role as described above, and have allowed a whole new generation of assays to be developed of importance in the research laboratory, the diagnostics industry and even the home. In many cases the combination of monoclonal and polyclonal reagents may allow powerful assays to be developed, and the design of a particular assay can take account of the properties of the individual reagents available.

By selection of an appropriate MAb, the interference of closely related substances in an assay can often be removed so that specificity is increased. For example, a MAb has been raised to the C-terminal part of human brain acetylcholinesterase which can distinguish between brain acetylcholinesterase and that present in red blood cells (Boschetti *et al.*, 1996). Brain acetylcholinesterase leaks into the amniotic fluid in prenatal neural tube defects as well as neurological disorders. Samples of amniotic fluid from amniocentesis are often contaminated with blood and thus an assay which can distinguish between the two forms could be important in allowing accurate diagnosis of such serious conditions, reducing any false positive results. Closely related compounds may also be problematic in therapeutic drug measurements. Monoclonal reagents have shown superior specificity in the measurement of digoxin (Datta *et al.*, 1996), and can be used to discriminate between a drug itself and its metabolites, as shown in the case of cyclosporin monitoring to allow adjustment of individual dosing regimens (Quesniaux, 1991). Another common form of interference in immunoassays is the presence of autoantibodies to the sample antigen. The use of MAbs in such cases also allows reduced interference, as demonstrated in the case of an IRMA for the measurement of serum thyroglobulin (Marquet *et al.*, 1996).

Assay sensitivity may also be increased with MAbs. Simple substitution of a polyclonal capture antibody with a monoclonal has resulted in increased sensitivity in a fluorescent assay for salmon calcitonin (Rong *et al.*, 1997). Similarly, sensitivity for the human cancer marker, prostate specific antigen, has been increased in another immunofluorometric assay which may be of utility in the diagnosis and staging of prostate and breast cancers (Ferguson *et al.*, 1996).

One problem with the use of MAbs has been in the detection of small molecular weight materials, such as many drugs, which are often not large enough to contain two distinct epitopes for use in two-site assays. Thus the measurement of such substances has relied largely on competitive techniques, using detection either with radioisotopes

(RIA) or with non-isotopic techniques such as those described above. Non-competitive approaches are now being developed. Two-site ELISAs have been developed based on the interaction of a small molecular weight analyte, such as digoxin, with an antibody which can then be detected by an antibody which recognises the new epitope formed by the bound analyte in the immune complex (Self *et al.*, 1994). Alternatively, assays have been developed using one second antibody which sees the antibody–analyte immune complex and one which recognises the primary antibody only when the analyte is not bound (Mares *et al.*, 1995).

MAbs have found applications in the measurement of a wide range of substances including hormones, metabolites, disease markers, therapeutic drugs, drugs of abuse and food and environmental contaminants such as mycotoxins, microorganisms, herbicides and pesticides. Antibody engineering technology is also beginning to have an impact on the design of immunoassays. Direct expression of fusion proteins may result in improved, homogeneous reagents in both enzyme-based and fluorescent assays. Antibody specificities can be directly expressed as fragments which may exhibit fewer problems of non-specific binding, and the power of phage selection of antibodies (see Chapter 1) allows rapid selection of useful binding specificities. Phage libraries may also allow the generation of useful antibodies which are difficult or impossible to obtain by other means, such as for antigens which are highly conserved between species, and it has been demonstrated that phage-bound antibody fragments may even be used directly in immunoassays (Navarra-Teulon *et al.*, 1995). The applications of bispecific antibodies are also likely to increase as their production becomes simpler (see Section 2.5). For example, bispecific antibodies can be produced with specificities for two epitopes on the same target which may offer advantages in specificity and affinity of binding and extend applications in immunoassay development (Cook and Wood, 1994).

Bispecific antibodies are also finding application in simple immunoassay formats. Rapid diagnosis of pulmonary embolism (PE) in the clinic may be crucial to further treatment and it has been suggested that high levels of the fibrin degradation product, D-dimer, is associated with PE (Bridey *et al.*, 1989). A rapid assay has been developed as an exclusionary test for PE which can allow exclusion of patients from potentially hazardous anti-coagulant therapy (Ginsberg *et al.*, 1995). This assay uses a bispecific antibody in which one arm recognises D-dimer and one red blood cells. The antibody causes agglutination of the patient's own red blood cells in the test tube, hence allowing a simple positive or negative test to be performed within a few minutes of taking a blood sample.

Entirely novel assay formats are also possible based on recombinant antibody fragment production. It has been shown that the interaction between the two variable domains of an antibody (VH and VL) is considerably strengthened in the presence of antigen (King *et al.*, 1993). This interaction has been used to develop an open sandwich ELISA method for the detection of antigen in which VL immobilised on an ELISA plate is used to bind antigen in the presence of phage displayed VH. The resulting complex is then detected by use of a peroxidase labelled anti-phage antibody (Ueda *et al.*, 1996). If such antigen-promoted association of VH and VL is generally applicable then an exciting new class of immunoassays could result for both large and small antigens.

3.3 Immunosensors

An alternative to conventional immunoassays is the use of a biosensor based on antibody:antigen interaction, termed an immunosensor. Immunosensors are useful analytical

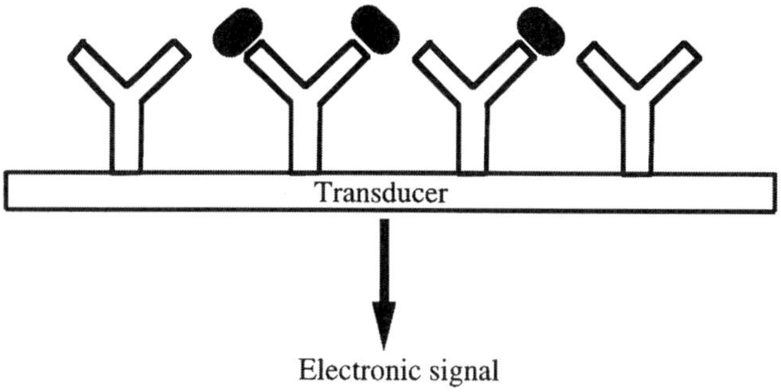

Figure 3.8 Diagram of a general immunosensor: antigen binding is detected by generation of signal which is converted to electronic format by a transducer

tools for monitoring antibody:antigen reactions in real time, often without the need for labelling either of the components. As such there has been great interest in developing immunosensors for analytical applications in the clinic, as well as for environmental monitoring and for food analysis. Immunosensors have been produced which are capable of detecting medical diagnostic markers such as hormones, drugs (therapeutic and abused), microorganisms and environmental pollutants such as pesticides. Under appropriate conditions immunosensors can also be used to analyse the kinetics of antibody:antigen binding.

An immunosensor is a solid-state device in which the antibody:antigen reaction is detected via a transducer which provides a signal that the binding reaction has taken place (Figure 3.8). Three main types of transducer have been used in biosensor technology; these exploit changes in mass (piezoelectric or acoustic wave), electrochemical properties (potentiometric, conductimetric or amperometric) or optical properties (fluorescent, reflective, luminescent, surface plasmon resonance or waveguide properties).

3.3.1 Mass-detecting immunosensors

Piezoelectric materials, such as quartz crystals, can be made to resonate by application of an alternating electric field with the frequency of the resulting oscillation being determined by the mass of the crystal. In piezoelectric biosensors, quartz crystals are coated with antibody. Binding of antigen results in a change in mass which is monitored through the change in frequency of the crystal oscillation. One problem with the use of such immunosensors is non-specific binding of sample components to the crystal, but sensitive devices have been produced for such applications as insulin measurement (Raman-Suri *et al.*, 1995), metamphetamine detection in human urine (Miura *et al.*, 1993), herbicide detection in drinking water (Yokoyama *et al.*, 1995) and detection of viruses and bacteria (Konig and Gratzel, 1993).

A further variant of the piezoelectric sensor uses surface acoustic waves to monitor binding to the crystal surface. In this application the oscillation of the crystal is at higher frequency and an acoustic wave is generated by application of an alternating voltage across interlaced electrodes, known as an interdigital transducer. A second interdigital transducer detects the acoustic signal a few millimetres away. Binding of sample to the crystal slows the acoustic wave, with the change in velocity being proportional to the analyte concentration. Such devices have the potential for higher sensitivity but may suffer more interference from factors such as temperature, pressure and conductivity, all of which may affect the properties of the acoustic wave. An acoustic wave sensor for measurement of the pesticide atrazine enabled sensitive detection in a competitive assay, with the ability to re-use the device 48 times with only a 30% loss in response (Tom-Moy *et al.*, 1995).

3.3.2 Electrochemical immunosensors

Potentiometric immunosensors are based on the change in potential that results when antibody is immobilised on an electrode and its antigen binds to it. Antibodies, like all proteins, are polyelectrolytes and in many cases binding of antigen will alter the charge. Therefore the potential difference between an electrode with immobilised antibody and a reference electrode will depend on the concentration of antigen present. However, direct detection results in only small changes in potential (1–5 mV) and low signal : noise ratios because the the charge density on the antibody is low compared with background ions. Also, signal is dependent on variables such as pH and ionic strength of the sample and sensitive to interference from other ions. Attempts to improve such sensors have included the development of antibody field-effect transistor devices (immunoFET). ImmunoFET operates through measurement of conductivity through a channel region between source and drain electrodes. The conductivity is controlled by the strength of electric field generated by the gate, providing an amplification effect. Antibody is immobilised on a membrane which must be thin enough to allow redistribution of the small charge changes which occur on antigen binding. However, reliable immunoFET sensors have not been developed due, at least in part, to problems in manufacturing suitable membranes.

Conductimetric sensors monitor variations in the conductivity between two electrodes by measuring variations in the current across the electrodes. Many reactions lead to variations in ion concentration through either the consumption or production of ionic species, and thus such sensors have been developed with many enzyme systems. Application of conductimetric sensor technology to immunosensors has been limited to date.

Amperometric sensors measure the current generated when electroactive species are reduced or oxidised at the electrode. Antibodies are not directly electroactive and thus enzyme labels are used to generate electroactive species which can then be measured. Most often such sensors use oxygen or hydrogen peroxide electrodes, with the current produced being directly proportional to the amount of oxygen or hydrogen peroxide reduced or oxidised. Catalase was the first label used to construct an immunosensor in this way. MAb to human chorionic gonadotropin (hCG) was immobilised on an oxygen electrode to which hCG and catalase-labelled hCG could compete for binding (Aizawa *et al.*, 1979). Subsequently alternative labels, such as alkaline phosphatase, have been used in the detection of several antigens (Treloar *et al.*, 1994). Substrates for alkaline phosphatase which can be used to generate electroactive compounds include *p*-aminophenol phosphate or *N*-ferrocenyl-4-aminophenol phosphate. Alternative labels have also been developed for such applications as measurement of apolipoprotein E in serum (Meusel *et al.*, 1995) and detection of herbicides such as 2,4-dichlorophenoxyacetic acid using a disposable immunosensor (Kalab and Skladal, 1995). The development of amperometric immunosensors has also been extended into *in vivo* use for the measurement of corticosteroids (Cook, 1997). This system used an immunosenor with antibodies immobilised on the electrode surface surrounded by a dialysis membrane which allows corticosteroids to equilibrate across. When the corticosteroid to be measured has equilibrated, a solution of corticosteroid conjugated to horseradish peroxidase is introduced into the internal cavity of the electrode, which then competes for antibody binding. The electrode surface can then be washed and peroxidase substrate added which is detected amperometrically. The probe can be regenerated with 1 mM HCl, allowing re-use every three minutes. The small size of this probe allowed implantation into the circulatory system of animals for monitoring hormone levels in real time, with the probe viable for several hundred measurements over 48 hours. The development of such systems may allow real-time, *in vivo* monitoring of a range of substances in the future.

3.3.3 Optical immunosensors

Optical immunosensors can be used with or without labels. Some sensors use labels such as enzymes or fluorescent or luminescent reagents to generate the optical signal, whereas in other cases binding is detected through the use of techniques such as total internal reflection, surface plasmon resonance or dielectric waveguides.

Labelled antibody immunosensors based on fluorescence or luminescence are attractive due to the relatively simple instrument design possible. Also, the use of optical fibre technology has enabled many devices to be miniaturised. These essentially operate via the generation or quenching of light at the surface of the optical fibre. Chemiluminescent assays have been developed based on the generation of light from the peroxidase catalysed oxidation of luminol, and fluorescent devices have been developed using different fluorescent labels, for applications such as the measurement of the cancer marker, prostate specific antigen in whole blood (Daniels *et al.*, 1995).

The most widely used optical immunosensors do not require labelled reagents. Surface plasmon resonance (SPR) uses light directed towards a layer of low refractive index from one of higher refractive index at an angle such that total internal reflection occurs. A metal film coated with a dextran layer is used to immobilise the antibody (or antigen). When light hits the metal film, surface plasmons in the film are excited and an evanescent wave is generated, a process known as SPR, which results in a decrease in the intensity of reflected light (Figure 3.9). When antigen binds to the immobilised antibody, the resulting change in refractive index causes a change in the angle at which the drop in reflected light intensity occurs. Continuous monitoring of the angle and intensity of reflected light therefore allows real-time analysis of binding events. A commercial system based on this principle, known as the BIAcore™, is now widely used for the analysis of biological interactions (Malmqvist, 1993). In this instrument SPR technology has been combined with a microfluidics system such that continuous monitoring is possible (Figure 3.9). This not only enables quantitative measurements to be made but also enables detailed kinetic analysis of binding interactions to be performed in real time. This instrument has therefore found many applications in the analysis of antibody:antigen interactions. BIAcore™ analysis is now used to determine the binding affinities of MAbs, and is widely used to monitor the outcome of antibody engineering experiments, for example in the selection of phage displayed antibodies, in the characterisation of antibody fragments and in epitope mapping (Malmborg and Borrebaeck, 1995).

Another commercially available type of optical biosensor, the IAsys™, is based on a combination of SPR technology with waveguide technology. This device has found applications in the area of bioprocess monitoring. The production of an Fv fragment could be monitored through fermentation and purification processes (Holwill *et al.*, 1996). Measurements were performed in five seconds, with an overall assay time of two minutes. This constant monitoring of the process allowed in-process decisions to be made to improve the process efficiency. A comparative study which used both BIAcore™ and the IAsys™

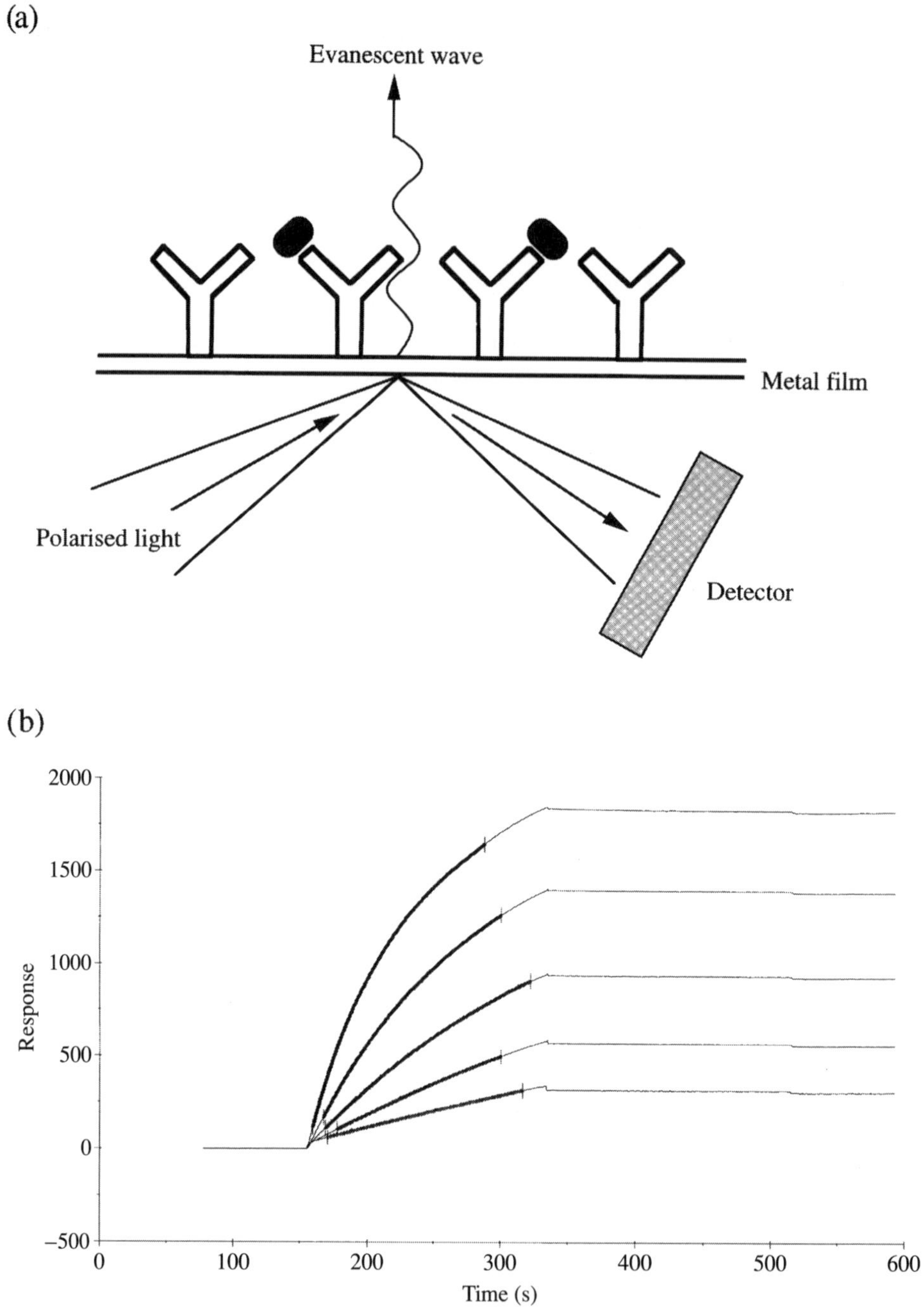

Figure 3.9 (a) Schematic diagram of BIAcore optical biosensor. Polarised light is reflected in the gold film of the sensor chip and detected on a diode array. Surface plasmon resonance is observed as a decrease in light intensity for a specific angle of incidence. The angle changes as the refractive index in the vicinity of the surface changes due to binding of large molecules on the immobilised ligand on the sensor chip. (b) Example of BIAcore sensorgrams for the binding, and slow dissociation, of IgG to an antigen-coated chip at five different concentrations. Measurement of inital on-rate and off-rate can be made to allow estimation of binding affinity.

system to monitor the kinetics of binding of lysozyme to anti-lysozyme antibody found good agreement between results obtained from the two systems (Yeung *et al*., 1995).

3.4 Immunocytochemistry

Immunocytochemistry allows the detection and location of antigens within cells and tissues by the application of labelled antibody (either directly labelled or via a labelled second antibody reagent) followed by microscopy. Suitable labels to allow both light microscopy and electron microscopy applications are available. In the research laboratory immunocytochemistry is used to visualise antigens and their subcellular locations in tissues. Double-labelling techniques, in which antibodies to two different antigens are applied with different labels, allow comparison of the relative distributions of two antigens in the same tissue samples. This is compared with conventional histological stains and markers to build up a detailed picture of antigenic distribution. Immunocytochemistry is also widely used in diagnostic pathology, particularly in the identification of types of tumour cells, infectious organisms or inflammatory cells.

Detailed methodologies for immunocytochemistry procedures for light and electron microscopy have been described (Beesley, 1993; Polak and Priestley, 1992), and therefore the technique is summarised only briefly here. Immunocytochemistry requires firstly tissue preparation followed by antibody binding and then detection. Two major types of tissue sections are used in immunocytochemical studies: fixed and frozen. Frozen sections are more difficult to prepare but the technique is relatively gentle and leaves more of the immunological features of the tissue intact. Fixed tissue sections are prepared, often in formalin-based fixative, and embedded in paraffin prior to cutting slices. Such fixed sections are simple to prepare, retain cellular morphology and can be stored for long periods but the process of preparation is more likely to destroy antigenic epitopes within the tissue. Nevertheless such fixed sections are widely used in diagnostic pathology due to their ease of preparation, and the existence of extensive slide libraries which can be used for comparative purposes. Cell suspensions from cell culture experiments, and cell smears from, for example, needle aspirates or tissue scrapings, can also be used for immunocytochemistry after suitable fixing procedures. Such slides obviously do not preserve the morphology of the intact tissue but are useful in many subcellular studies, or in the identification of infectious organisms in tissue samples.

MAbs offer improved specificity compared to many polyclonal reagents also used in immunocytochemistry. However, the loss of a MAb epitope on tissue fixing is more likely than the loss of all the epitopes recognised by a polyclonal antiserum. Therefore, polyclonal reagents may be preferred for the detection of some antigens, particularly when using fixed tissue sections. Non-specific binding is generally reduced by use of monoclonal reagents. In some tissues non-specific binding via Fc receptors may also be a significant problem which can be overcome by use of antibody fragments such as Fab and $F(ab')_2$. In addition, antibody fragments may ease the penetration of the reagent into tissues. Smaller fragments such as engineered Fv fragments may also be beneficial and have also been developed for use in immunocytochemistry applications (Kleymann *et al*., 1995).

Antibodies can be directly labelled for immunocytochemistry, but if several antigens are to be detected, this requires preparation of several individual labelled reagents. The use of labelled second antibody allows the detection of several antibodies with a single reagent. For example, sections can be probed with mouse MAbs and then detected with

Table 3.3 Examples of antibody labels for immunocytochemistry

Enzymes	Substrates	Colour
Horseradish peroxidase	Diaminobenzidine	Brown
	Tetramethylbenzidine	Blue
	Aminoethylcarbazole	Red
	4-chloro-1-naphthol	Blue-black
Alkaline phosphatase	Naphthol AS phosphate + fast red, blue or violet	Red, blue or violet
	Bromochloroindolyl phosphate + nitroblue tetrazolium	Blue
Fluorescent labels	Fluorescein	Green
	Rhodamine	Red
	Phycoerythrin	Orange
	Texas Red	Red
	Cyanin 3 or 5	Red or orange
Metal labels	Colloidal gold 1–40 nm	Electron-dense
	Gold + silver enhancement	Dark brown or black

a labelled anti-mouse Ig reagent. Second antibody methods also allow some amplification on detection, and avoid the problem of developing labelling methods to minimise loss of antigen-binding properties with many different antibodies. Alternatively, biotinylated antibodies which can be detected with a labelled streptavidin conjugate are widely used.

The major types of label used in immunocytochemistry are enzymes used to generate insoluble coloured products, fluorescent labels and colloidal gold particles (Table 3.3). Both fluorescent and enzyme substrate labels are available in different colours such that two or more antigens may be visualised simultaneously. For example, fluorescein and phycoerythrin can be used to detect two different antigens under a fluorescent microscope with differential colour fluorescence, green and orange respectively. Horseradish peroxidase is the most widely used enzyme label for animal tissues, though not for plant tissues due to the presence of endogenous peroxidase activity which can lead to problems with non-specific staining. Conversely plant tissues can be stained with alkaline phosphatase conjugates, although endogenous activity makes this technique unsuitable for many animal tissues. In practice, however, both enzymes are suitable for most types of tissue, as endogenous activity is easily inhibited. Colloidal gold is suitable as a label for use in both light and electron microscopy. In light microscopy, the signal from gold-labelled antibodies is seen as a pinkish colour. This is usually intensified by the use of silver enhancement in which the gold catalyses the reduction of silver ions to metallic silver which results in a stable signal. Under the electron microscope, colloidal gold is seen as dense black round particles that can easily be distinguished from cellular structures. Different size particles are available, for example, 5 nm and 30 nm, which can be used to allow the detection and relative quantification of two antigens simultaneously.

Many of the problems observed with immunocytochemistry are similar to those seen with immunoassay development. Non-specific binding is also a problem which can be overcome in many cases by purification of suitable reagents, or the use of antibody fragments or alternative labelling techniques.

One of the major uses of immunocytochemistry is the identification and classification of human tumour types (Giovagnoli and Vecchione, 1996). Information gained in this way can be of direct benefit in patient care. MAbs can allow the identification of the primary site of metastatic carcinoma, give information on the malignancy of the tumour and in some instances be an aid to determining prognosis. For example, the breast cancer marker $p185^{HER2}$ has been associated with increased probability of relapse and poor outcome (Slamon *et al.*, 1987). Immunocytochemistry to detect expression of this marker in breast cancer tissue is thus useful in determining likely patient prognosis. Similarly, applications of MAbs in the detection and identification of microorganisms are important clinically such as in the demonstration of human cytomegalovirus infection (Jahn and Plachter, 1993). There is also a role for immunocytochemistry in other disease states in the detection of cellular infiltrates and the determination of their activation state, the localisation of adhesion molecules and identification of locally produced cytokines. In pulmonary diseases such as allergic asthma and sarcoidosis, the analysis of cytokines from cells present in bronchoalveolar lavage fluid using immunocytochemistry can give an insight into the disease process (Krouwels *et al.*, 1997).

3.5 Flow cytometry and cell sorting (FACS)

Flow cytometry and fluorescence activated cell sorting techniques (FACS) are widely used in cell biology research for both analytical and preparative purposes. Fluorescently labelled antibodies capable of binding to the cells of interest are used to tag these cells, such that the cells become fluorescent. FACS equipment resolves mixtures of cells based on this fluorescence. Within the instrument a stream of single cells passes through a laser light beam and fluorescent cells can be individually detected. The distribution of antigens within the population of cells can therefore be readily studied. Preparative isolation of populations of cells becomes possible as the instrument can selectively impart an electrostatic charge to labelled cells and not others. The cells then pass through an electric field when the cells carrying a charge are deflected into a different container from the uncharged cells. As such this technique can be considered as an immunopurification technique for whole cells, which has been used for both research and clinical applications. Applications include the separation of cell populations for functional studies and the isolation of rare transfected cells to simplify cloning.

FACS techniques can also be performed with two different fluorescent labels, allowing the simultaneous detection of two cellular antigens. Such analyses are widely used in characterising populations of cells, for example in the characterisation of hematopoietic stem cells used for autologous transplantation, using antibodies to CD34 and CD19 (Fritsch *et al.*, 1995). The use of FACS for preparing a purified population of CD34 positive bone marrow cells for use in transplantation studies has also been described (Rondelli *et al.*, 1996). The application of flow cytometry to detect cellular antigens is not restricted to cell surface markers: intracellular substances can also be detected. The use of flow cytometry in the detection of intracellular interleukin-4 and -5 and interferon-γ has been compared with immunocytochemistry (Krouwels *et al.*, 1997). In this comparison flow cytometry was shown to be more suitable for use with double staining techniques for detection of more than one antigen and allowed isolation of defined cell populations. However, the detection of these cytokines was apparently less sensitive than with immunocytochemistry.

3.6 Western blotting (immunoblotting)

The technique of western blotting, also known as immunoblotting, is used to detect protein antigens after polyacrylamide gel electrophoresis (PAGE). PAGE is used to separate proteins in an electric current, and in its most widely used form is carried out in the presence of the detergent sodium dododecyl sulphate (SDS). SDS denatures proteins and binds to them such that a uniform negative charge is imparted to each protein. Therefore, when proteins are run through a gel using an electric current in the presence of SDS (SDS-PAGE) they can be separated on the basis of molecular weight. The range of molecular weights which can be separated depends on the pore size in the gel used and this is normally controlled by the concentration of acrylamide and cross-linker used in formation of the polyacrylamide gel. Proteins separated on SDS-PAGE are then visualised using a general protein stain such as coomassie blue or, more sensitively, silver stain. However, general protein stains stain every protein, and it is often desirable to be able to identify specific proteins. This can be achieved by western blotting. After SDS-PAGE the proteins present in the gel are transferred laterally (blotted) onto a membrane of nitrocellulose or an alternative polymer such as polyvinylidene difluoride (PVDF). The protein (or proteins) of interest can then be detected using the desired antibody which is either conjugated to a suitable label or detected by a secondary labelled antibody reagent. Labels commonly used include the radioisotope iodine-125 followed by exposure to X-ray film, colloidal gold with silver enhancement (see Section 3.4), and the enzymes horseradish peroxidase and alkaline phosphatase either with insoluble coloured substrates or, in the case of peroxidase, to generate a chemiluminescent readout by oxidation of luminol in the presence of hydrogen peroxide. Chemiluminescent detection by HRP-conjugated antibodies is usually enhanced with chemical enhancers and is a particularly sensitive detection method. An additional advantage is that after detection the signal can be removed by stripping off the bound antibody with protein denaturants as first developed for use with radiolabelled antibodies (Kaufmann *et al.*, 1987). This allows the blot to be reprobed with another antibody and re-detected, thus allowing several different antigens to be examined sequentially on the same blot. Western blotting is not confined to SDS-PAGE but can also be used to identify proteins separated by electrophoresis under non-denaturing conditions or by isoelectric focusing.

An example of a western blot is shown in Figure 3.10. Both polyclonal and monoclonal antibodies are used in western blotting. MAbs offer excellent specificity, but as detection is carried out on denatured proteins, reactivity of the MAb with the native conformation of the protein does not necessarily allow the antibody to be used in western blotting. This is because the antibody may recognise an epitope not made up of a stretch of linear polypeptide sequence but of parts of the protein which are spatially close in the protein's native conformation due to folding. Polyclonal antisera which recognise many different epitopes are more likely to contain antibodies which recognise some linear polypeptide sequences, and thus are often more successful in western blotting applications. However, screening can be carried out to identify those MAbs which are suitable for blotting.

The applications of western blotting in the research laboratory are wide and include the detection and analysis of natural proteins of low abundance, the analysis of recombinant protein expression and the detection of contaminant proteins. The combination of immunological reactivity with size analysis gained from western blotting allows analysis of protein processing and post-translational modifications. The use of MAbs allows precise information to be gained which would not be possible otherwise. This is demonstrated by a recent study to determine the role of matrix metalloprotease enzymes in human arthritis

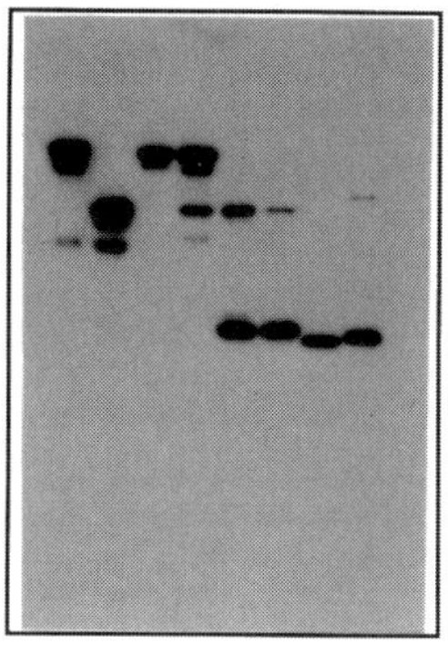

Figure 3.10 Example of a western blot. SDS-PAGE was carried out using a 4–20% polyacrylamide gel under non-reducing conditions. The blotted gel was probed with a murine monoclonal antibody to IgG Fd (Fab heavy chain) followed by a polyclonal rabbit anti-mouse IgG–horseradish peroxidase conjugate. The blot was then developed using enhanced chemiluminescence. Lane 1, recombinant human IgG4 (hIgG4); lane 2, $F(ab')_2$ derived from hIgG4; lane 3, hIgG1; lane 4, hIgG4 and $F(ab')_2$ mixture; lane 5, γ1 hFab′ and $F(ab')_2$ expressed in mammalian cells; lane 6, γ1 hFab′ and $F(ab')_2$ expressed in *E. coli*; lane 7, γ4 hFab′(a) expressed in mammalian cells; lane 8, γ4 hFab′(b) expressed in *E. coli*.

(Fosang *et al.*, 1996). Aggrecan is the major proteoglycan in cartilage and the molecule that allows the tissue to bear weight by deforming under compression. A MAb recognising the cleaved products of metalloprotease action on aggrecan allowed the detection of cleaved products in synovial fluid by western blot which could be quantitated with an ELISA using the same antibody. This analysis allows the protease activity involved to be characterised, leading to further insight into the degradative mechanisms involved in the arthritic process.

Western blotting can also be used in routine diagnostic procedures, for example in the confirmation of HIV infection (Gurtler, 1996). In this case anti-HIV antibodies are detected initially by ELISA tests, and reactive results are confirmed by western blotting which can allow differentiation of HIV types and subtypes. Western blotting has also been used to determine alcohol abuse by measurement of carbohydrate-deficient transferrins (Anton and Bean, 1994). Carbohydrate-deficient transferrins are produced in greater quantities than usual during periods of heavy alcohol intake due to decreased glycosylation of transferrin before release into the blood. Western blotting of serum samples allows separation and detection of the carbohydrate-deficient transferrin from the bulk of normal transferrin. In this case isoelectric focusing is used as the separation technique, where the difference in charge of the carbohydrate-deficient transferrin provides the basis for good separation from a large amount of normal transferrin.

Recombinant fusion proteins are also beginning to play a role in western blotting applications. Fusions of an scFv recognising the plant protein phytochrome have been made to both alkaline phosphatase and thc IgG binding protein staphylococcal protein A (Gandecha *et al.*, 1994). Western blotting of extracts from oat seedlings containing phytochrome demonstrated that both of these fusion proteins could be successfully used for western blotting. The alkaline phosphatase conjugate was visualised directly using the chromogenic substrates BCIP/NBT (bromochloroindolyl phosphate + nitroblue tetrazolium), whereas the protein A conjugate was used in a two-step method, detected by an IgG–alkaline phosphatase conjugate. The fusion proteins were expressed in *E. coli*, and detection on the western blots was possible using the fusion proteins directly from *E. coli* lysates without the need to purify the reagents. The functionality of such conjugates suggests that the role of recombinant fusion proteins in the immunoassay field in general will increase and allow the use of simple single-step procedures, which may be advantageous in reducing non-specific binding problems.

3.7 Immunopurification

Immunopurification uses the specificity of the antibody to bind to a substance of interest and isolate it in a purified form. Antibody is usually coupled to a solid phase, such as beaded agarose, and packed into a chromatography column for use in the procedure known as immunoaffinity chromatography. This is a type of affinity chromatography in which antibodies are used as the binding moiety to retain the protein, or other substance of interest, while contaminating material flows through the column. The bound protein is then eluted by a change in buffer conditions to those which promote dissociation of the antibody–antigen complex (Figure 3.11). Immunoaffinity chromatography is used for the simple, rapid purification of proteins in the research laboratory both for preparation of the protein itself and as a quantitative measure, i.e. assay. In addition, immunoaffinity chromatography is used commercially for the purification of a number of pharmaceutical proteins, particularly high value proteins which are difficult to purify by other means. Because the interaction between antibody and antigen is very specific, a high degree of purification is achieved in a single step. Therefore, the use of immunoaffinity chromatography negates the requirement to develop protein-specific, multi-step purification protocols and allows rapid isolation of protein for research purposes.

Elution from the immunoaffinity chromatography column is achieved by disruption of the antibody–antigen interaction using mild denaturation such that both antigen and antibody are not irreversibly denatured. Commonly used eluents include extremes of pH and chaotropic agents such as thiocyanate, which often allow active material to be obtained. However, if the affinity of the antibody for antigen is very high it may be difficult to elute bound antigen without irreversible denaturation. Relatively low affinity MAbs are therefore ideal for successful immunopurification. Antibodies with affinity constants in the range 10^{-4}–10^{-8} M have been suggested to be most suitable (Phillips, 1989). However, the best way to select an antibody for immunopurification is by testing a panel of antibodies, to see which allow elution of the antigen under mild conditions. This can be achieved conveniently using ELISA or BIAcore format assays in which several eluent solutions are also screened, or antibody coupled to solid-phase beads can be used in small-scale experiments. When suitable antibodies have been identified, the antibody is coupled to a solid phase, usually a cross-linked beaded agarose material such as Sepharose, suitable for use in immunoaffinity chromatography. Many such matrices are commercially available in pre-activated form, which allows simple attachment of the antibody through amine groups with matrices derivatised with linkers such as *N*-hydroxysuccinimide or cyanogen bromide (see Section 2.6.2). Matrices are also available which allow site-specific attachment of antibodies, such that more of the antigen-binding ability is retained.

Random coupling through lysine residues often results in only 10–15% of the antibody retaining its antigen-binding activity due to a combination of chemical modification of

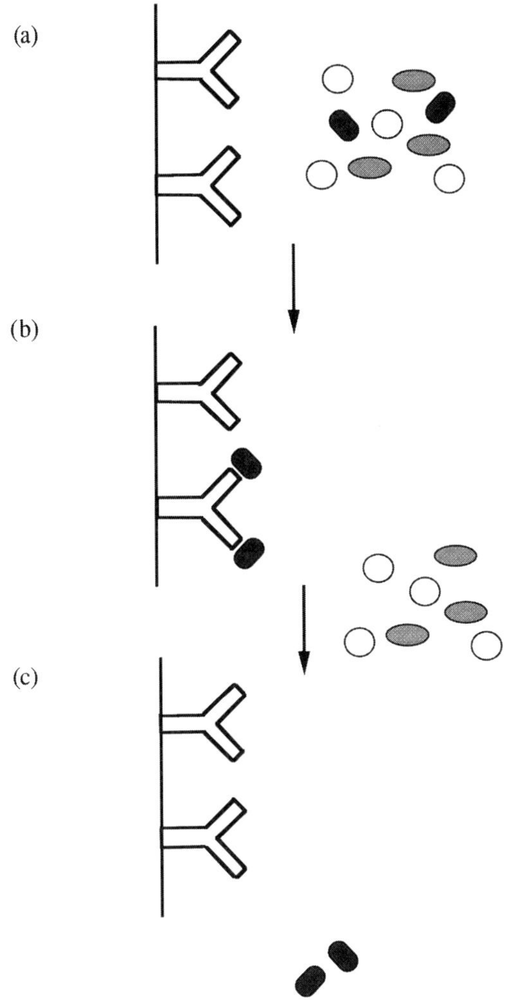

Figure 3.11 Immunoaffinity chromatography: immobilised antibody is used to selectively bind the ligand of interest (a); after washing to remove non-bound material (b), bound ligand is eluted in purified form (c), regenerating the immobilised antibody for re-use

important amino acid residues in the binding site of the antibody, and lack of orientation of antibody on the solid phase, such that many binding sites may be occluded (Fowell and Chase, 1986). Similar methodology is used for site-specific attachment as for site-specific protein–protein conjugates (see Section 2.6.2). Site-specific attachment can be achieved through attachment to Fc carbohydrate by periodate oxidation followed by reaction with hydrazide activated matrix (O'Shanessy, 1990). By attaching the antibody via the Fc this method orients the antibody such that the antibody-binding sites are less likely to be occluded by the matrix as well as avoiding modification of important amino acids in the CDR regions. Similarly, Fc region binding proteins have been used to orient the antibody correctly such that the antigen-binding site is still available. Staphylococcal protein A and Streptococcal protein G are widely used in antibody purification as they bind IgG specifically through Fc region binding sites between the CH2 and CH3 domains (see Chapter 5). Protein A and protein G linked to solid phases are therefore widely available reagents.

Binding of antibody for immunopurification to the protein A matrix followed by chemical cross-linking can therefore be used to immobilise antibody in oriented fashion, retaining a larger proportion of active antigen-binding sites (Sisson and Castor, 1990). Fab′ fragments can also be used for site-specific attachment to matrices through the hinge region free thiol groups. Maleimide or iodoacetyl activated matrices can be used to generate Fab′ solid phases in which most of the antigen-binding ability of the Fab′ is retained (Prisyazhnoy *et al.*, 1988). Immobilised Fab′ fragments may have additional advantages in that the lower avidity of the monovalent Fab′ fragment for antigen may allow elution of bound antigen under less harsh conditions than for intact IgG, particularly when low immobilised density on the solid phase is used.

Immunoaffinity chromatography is widely used for the purification of both natural and recombinant proteins for research purposes, particularly when the source of the protein contains only small amounts among many other contaminating proteins. With recombinant proteins it is not always necessary to have a MAb to the protein, as specific ‘tags’ can be added to the gene before expression such that the resulting fusion protein can be easily purified. The commercially available FLAG system is one such ‘tag’ (Hopp *et al.*, 1988). The FLAG epitope consists of the 8 amino acid sequence Asp-Tyr-Lys-$(Asp)_4$-Lys, and vectors are available which allow fusion of the gene for this sequence to the recombinant gene of interest. The resulting fusion protein can then be purified using an immobilised MAb which recognises a calcium complex of the FLAG peptide. The MAb used is conformation-specific and in the absence of calcium the antibody dissociates from the peptide tag. This system therefore allows for very mild elution of the tagged recombinant protein using EDTA to remove calcium or, alternatively, a competitive peptide can be used (Hopp *et al.*, 1996). The FLAG system has proved of wide applicability and has been used to purify a wide range of protein targets including interleukin-3 (Park *et al.*, 1989), β2-adrenergic receptors (Guan *et al.*, 1992), tumor necrosis factor (Su *et al.*, 1992), transcription factors (Chiang and Roeder, 1993), antibody fragments (Knappik and Pluckthun, 1994) and interleukin-5 receptors (Brown *et al.*, 1995).

Immunoaffinity chromatography is also increasingly used as a form of immunoassay. Immobilised antibody is used to capture the analyte of interest, which is then eluted and quantitated by UV absorbance or other means. The development of useful assays has been facilitated by the development of new solid-phase materials which can be operated at high flow rates allowing very fast assays to be performed and automated (Afeyan *et al.*, 1992). Assays are not restricted to proteins, and for other analytes immunoaffinity chromatography is often coupled with detection techniques such as high-performance liquid chromatography (HPLC), gas chromatography (GC) or mass spectrometry (MS). For example, the β2-agonists salbutamol and clenbuterol have been determined in tissue samples (Pou *et al.*, 1994), and both aflatoxin contamination in cheeses and cannabis metabolites in saliva have been determined by combining immunoaffinity chromatography with HPLC (Dragacci *et al.*, 1995; Kircher and Parlar, 1996).

The high affinity of the antibody:antigen interaction also allows immunoaffinity chromatography to be used for the specific removal of trace contaminants which may be present at low levels. This includes the removal of protein contaminants from pharmaceutical proteins for *in vivo* use, and applications such as the removal of trace bacterial contamination from foodstuffs. Molloy *et al.* (1995) have demonstrated that an immobilised scFv which recognises a lipoprotein surface component of *Pseudomonas aeruginosa* can remove cells of this bacterium from mixed cultures and from milk. This study demonstrated that immobilised antibody fragments could be effective and potentially economic in this type of application and suggests that scale-up of this technology may be possible to allow the use of immunoaffinity columns for removal of toxic materials from the environment.

The use of immunoaffinity columns for specific removal of materials may also find clinical application. Several groups have attempted to develop systems in which plasma is removed from the patient, passed through an immunoaffinity column and then returned to the patient. Sato *et al.* (1989) have suggested that specific removal of IgE is useful in the treatment of patients with allergy or other conditions in which the level of IgE is abnormally high. Another application is the use of immunoaffinity techniques for the removal of radiolabelled antibodies from the blood following radioimmunotherapy of tumours (Norrgren *et al.*, 1991). Radioimmunotherapy is described in Section 4.2.5. Briefly, antibody to a tumour-associated antigen is used to target a radioisotope to tumour cells in an attempt to kill them. However, to deliver a toxic dose to the tumour requires injection of a large amount of radiolabelled antibody which is often toxic to normal tissues such as bone marrow. Therefore attempts have been made to allow antibody to localise to tumour cells and then remove excess antibody from the blood using immunoaffinity techniques, hence reducing normal tissue toxicity.

Many recombinant proteins are now used as pharmaceuticals. Immunopurification would be an attractive method for purification of many of these pharmaceutical proteins. However, the production of MAbs in large amounts is relatively expensive compared to other chromatographic ligands, and therefore the use of immunopurification techniques in the production of proteins for pharmaceutical use has been limited. Also, when intended for use in the manufacture of a pharmaceutical protein the MAb must be produced to a standard which is as high as that demanded of the pharmaceutical protein itself. Nevertheless, immunopurification has been used for some high-value proteins which are difficult to prepare by other means, for example interferons, factor VII, factor VIII and factor IX (Bailon and Roy, 1990; Kim *et al.*, 1992).

Antibody fragments which can be produced in bacterial expression systems may be manufactured much more economically than intact IgG (see Chapter 5) and thus there has been interest in the development of immunoaffinity columns using Fv and scFv fragments. Engineered antibody fragments can also be produced with sites for specific attachment to the solid phase designed such that minimal antigen binding activity is lost on immobilisation. For example a hinge region containing a single thiol group can be engineered onto the C-terminus of an scFv fragment to allow site-specific conjugation (King *et al.*, 1994). Both Fv and scFv fragments have been immobilised and used for immunoaffinity chromatography (Berry *et al.*, 1991; Spitznagel and Clark, 1993; Molloy *et al.*, 1995). scFv may be a better reagent to use as the covalent linkage between VH and VL results in less leakage from the column during elution (Berry and Pierce, 1993). In a comparison of IgG, Fab′ and Fv fragments it was found that immobilisation of Fv resulted in the highest binding capacity, presumably due to the small size of the fragment (Spitznagel and Clark, 1993). However, antibody fragments with exposed hydrophobic patches on the surface of the protein must not be used, as these may result in high levels of non-specific binding to the column. Use of a single VH domain which could bind antigen (dAb) was not successful as an immunopurification reagent for this reason (Berry and Davies, 1992).

The development of these low-cost affinity materials will no doubt open up a range of applications for immunopurification which have been limited to date by cost. This could include the design of reagents for the simple purification of pharmaceutical proteins, and applications such as removal of environmental contaminants. A further example of this is the generation of an scFv to the herbicide paraquat specifically for use as an immunopurification reagent (Graham *et al.*, 1995). The use of phage technology will allow the selection of the most suitable antibody for a particular application, which can then be reconstructed in the most suitable format for immobilisation.

3.8 Antibodies in structural biology

MAb fragments have also found a role in the determination of protein structures by X-ray crystallography. Many proteins of biological interest are unable to form the high-quality crystals required for structural determination due to heterogeneity, insolubility, flexibility or polydispersity in solution. Antibody fragments such as Fab and Fv are soluble, bind tightly to antigen and can thus effectively transform aggregated protein into a soluble monodisperse sample suitable for crystallisation (Kovari *et al.*, 1995). The HIV capsid protein p24 was able to be crystallised only as a Fab–p24 complex which prevented oligomerisation of the p24 protein (Prongay *et al.*, 1990). In the case of HIV reverse transcriptase, a Fab fragment was used to decrease the mobility of a region of the protein allowing rapid determination of the structure (Jacobo-Molina *et al.*, 1993). Fv fragments are also particularly useful and have been used to allow determination of the membrane protein bacterial cytochrome c oxidase by enlargement of the polar surface to allow formation of crystal lattice (Ostermeier *et al.*, 1995). It is usually necessary to screen a large number of antibody fragments to find a suitable one for crystallisation which not only stabilises the protein of interest but also allows suitable crystal contacts to form. Ideal antibody fragments will not alter the conformation of the protein of interest, although it should be remembered that it is often the case that the conformation of the antibody itself may change on binding antigen (see Chapter 1). Selection of appropriate antibody fragments from phage display may allow rapid identification of suitable reagents.

3.9 *In vivo* diagnostics

The use of antibodies for imaging disease has been developed to the point where there are now several products marketed for clinical use (Table 3.4) and many more under development. Antibodies to suitable antigens can be used to confirm the presence and extent of disease in the body, identify its location and monitor the progress of any therapy. To achieve this the antibody is labelled with a suitable radioisotope, usually a short half-life gamma emitter, and the emissions from the isotope detected with a gamma camera. Antibodies can also be labelled with positron emitting isotopes for use in positron emission tomography scanning. This technique is variously known as radioimmunodetection (RAID) or radioimmunoscintigraphy (RIS). Advances in instrumentation, developing such techniques as single photon emission computerised tomography (SPECT), have contributed to the popularity of development of RIS by allowing improved accuracy and detection of lesions not seen using other techniques. An alternative form of radioimmunodetection is used to aid surgical procedures, known as radioimmunoguided surgery (RIGS). In this technique radiolabelled antibody localised to the site of disease is detected by a hand-held monitor to allow simple identification and resection of diseased tissue during surgery.

Table 3.4 Some commercially marketed monoclonal antibody imaging agents

Imaging agent	Antibody	Isotope	Disease	Company
OncoScint CR/OV	B72.3 IgG (anti-TAG72)	^{111}In	Colorectal and ovarian cancer	Cytogen
CEA-Scan	Immu-4 Fab′ (anti-CEA)	^{99m}Tc	Colorectal cancer	Immunomedics
LeukoScan	anti-NCA-90 Fab′	^{99m}Tc	Infection/inflammation	Immunomedics
Myoscint	anti-myosin Fab	^{111}In	Heart disease	Centocor
ProstaScint	7E11-C5.3 IgG	^{111}In	Prostate cancer	Cytogen
Verluma	NR-LU-10 Fab′	^{99m}Tc	Small-cell lung cancer	NeoRx

3.9.1 ***Radioimmunodetection of human tumours***

A major impetus to the development of RAID has been the desire to develop techniques to allow detection of human tumours and determine the extent and location of disease, particularly monitoring metastatic spread. In many cases accurate, early diagnosis using RAID can be a contributor to effective treatment, with the major use of RAID being to monitor disease in patients following treatment by surgery, radiotherapy or chemotherapy (Larson, 1995). Many tumour-associated antigens are now known and characterised, and antibodies to these allow targeting to different tumour types (Table 3.5). However, obtaining high-quality tumour images from radiolabelled antibodies is not straightforward and the immunoconjugate to be used requires careful optimisation of not only the antibody specificity and the radioisotope, but also the form of the antibody used and the method of attachment of the isotope to the antibody. Some of the many factors affecting RAID are given in Table 3.6.

Tumour-associated antigens

Useful tumour-associated antigens are absent on normal tissues and present at high levels on tumour cells, preferably homogeneously on all cells of the tumour. Antigen should also not be shed from the tumour into the blood. In reality there are no perfect, tumour-specific antigens and the choice of a suitable target requires compromise over one or more of these characteristics. Nevertheless, high-quality tumour imaging can be achieved through optimisation of the immunoconjugate and the use of alternative strategies to overcome the individual problems of particular targets. For example, antibodies to antigens which are present on normal tissues, but inaccessible to the administered antibody, have been used successfully in imaging studies, and the presence of circulating antigen can often be overcome through increasing the dose of antibody administered to saturate circulating antigen, or the use of an unlabelled antibody pre-dose (van Hof *et al.*, 1996).

Table 3.5 Some commonly used tumour-associated antigens, with examples of antibodies raised against them

Antigen	Tumour type	Representative antibody
Tumour-associated glycoprotein72 (TAG72), 72 kDa glycoprotein	Pancarcinoma	B72.3, CC49
Carcinoembryonic antigen (CEA), 180 kDa glycoprotein	Pancarcinoma	NP-4, A5B7
Polymorphic epithelial mucin (PEM), >100 kDa glycoprotein	Ovarian, breast, lung	HMFG1
Epithelial membrane antigen (EMA), 40 kDa glycoprotein	Colorectal (and other epithelial tumours)	17-1A
Epidermal growth factor receptor (EGFR), 175 kDa glycoprotein	Breast, lung	425
$p185^{HER2}$/ c-*erb*-B2 (185 kDa glycoprotein)	Breast, lung	4D5
Prostate-specific membrane antigen (PSMA), 100 kDa glycoprotein	Prostate	7E11-C5.3
CD33 67 kDa glycoprotein	Myeloid leukemia	P67.6, M195
CD20 35 kDa glycoprotein	Lymphoma	C2B8
GD2 ganglioside	Melanoma, neuroblastoma	14-18

The homogeneity of tumour-associated antigen expression is less important for RAID than for targeted therapy (see Chapter 4), as good imaging can be achieved as long as enough labelled antibody can reach the tumour site to generate a high tumour : backgound ratio. Hence tumours with as few as 15% of the cells expressing the relevant antigen have

Table 3.6 Factors affecting radioimmunodetection of cancer

Tumour-associated antigen	Specificity (presence in normal tissues) Antigen density and homogeneity in tumour Presence of shed antigen in blood
Antibody	Specificity Affinity and avidity Molecular size – penetration into tumour – pharmacokinetics Internalisation Immunogenicity
Radioisotope	Method of attachment to antibody (stability of conjugate and retention of antigen binding) Half-life Suitability of emission energy for imaging
Tumour	Location in the body Size Vascularisation and vascular permeability

been successfully imaged (Doerr *et al.*, 1991). When developing an antibody for RAID it is desirable that it can be used for several different types of tumour. Such pan-reactive antibodies have been identified using widely distributed tumour-associated antigens such as carcinoembryonic antigen (CEA) which is present on cells of colorectal, breast, ovarian and lung tumours, and polymorphic epithelial mucin which is present on ovarian, lung and breast tumour cells (Table 3.5).

Form of antibody

The generation of a high tumour : background ratio is also the key to the selection of the best form of antibody to use for tumour imaging. Intact antibody circulates in the blood for long periods and builds up relatively high levels of activity at the tumour site. However, the long residence time of labelled antibody in the blood means that relatively poor tumour : blood ratios are generated at early time points, and tumour imaging is only possible days later when sufficient antibody has cleared from the blood. Antibody fragments clear much more rapidly (see Section 2.7) and although lower levels of activity are accumulated at the tumour site, higher tumour : blood ratios are generated at early time points, allowing imaging to take place on the same day. Many studies have been carried out with $F(ab')_2$ and Fab′ fragments generated by digestion from IgG and, more recently, the generation of recombinant fragments of anti-tumour antibodies has allowed smaller fragments such as Fv and scFv to be tested. Studies of Fab′ and $F(ab')_2$ compared to IgG demonstrate that faster imaging is possible with antibody fragments leading to the detection of small tumour deposits with higher sensitivity (Lane *et al.*, 1994; Behr *et al.*, 1995). In comparative studies of IgG, $F(ab')_2$, Fab′, Fv and scFv, higher ratios were generated with smaller fragments, the best being the Fv and scFv fragments (Milenic *et al.*, 1991; King *et al.*, 1992b). There is also strong evidence that smaller fragments penetrate further into the tumour mass, again with scFv more effective than Fab′ and $F(ab')_2$ (Yokota *et al.*, 1992). Clinical imaging with scFv has also been found to be very effective. Using an anti-CEA scFv derived from phage display, Begent *et al.* (1996) have demonstrated that high-quality tumour images could be obtained with improved sensitivity over conventional imaging using X-ray computerised tomography (CT). Several liver metastases of colorectal tumours not seen at all by conventional CT were detected using scFv imaging. However, good tumour imaging is also dependent on the biology of the system under investigation. Hence although for most tumours antibody fragments are preferable, for some tumours which are less accessible or less well vascularised, early imaging with antibody fragments may be difficult and better results may be observed by imaging later with an intact IgG (Behr *et al.*, 1995).

Further attempts to optimise the form of antibody for RAID have led to the testing of divalent and trivalent Fab′ and scFv fragments. Fab′ and scFv fragments are monovalent with respect to antigen binding which leads to low avidity of binding (see Chapter 1). Production of multivalent forms of scFv may enable the benefits of small size to be retained while enabling better binding ability leading to increased tumour uptake. scFv′s expressed with a hinge region attached have been used to prepare di-scFv′s which have shown increased tumour uptake leading to superior tumour : blood ratios (Adams *et al.*, 1993; King *et al.*, 1994). Tri-scFv′s have improved ratios further due to increased antigen-binding ability with only a small increase in blood retention (King *et al.*, 1994). Cross-linked Fab′ fragments also show improved tumour targeting which may lead to improved agents for both RAID and targeted therapy (see Chapter 4).

Table 3.7 Commonly used radioisotopes for development of RAID

Radionuclide	Emission	Half-life (hours)	Energy (keV)	Comments
^{131}I	$\gamma + \beta$	193	364	Reactor produced, also used as therapeutic isotope (β emission)
^{123}I	γ	13	159	Cyclotron produced
^{111}In	γ	68	171	Cyclotron produced, matched to ^{90}Y for therapy, used clinically
^{99m}Tc	γ	6	141	Generator produced, most suited to available cameras, readily available and widely used clinically
^{67}Ga	γ	80	184	Cyclotron produced, in clinical use as citrate
^{64}Cu	$\beta+$	13	511	For use in PET scanning

Radioisotopes

Suitable radioisotopes for RAID are usually short half-life gamma emitters, although positron emitting isotopes are also under development for positron emission tomography (PET) applications (Table 3.7). For use in RAID, radioisotopes must be of sufficient energy to reach the detector outside the body whilst having a low linear energy transfer (LET, the energy deposited by the isotope over its pathway) resulting in the minimum damage to cells within the body. Optimal tumour imaging requires a good match between the energy of the gamma emission and the gamma camera detector. Most current gamma cameras are best used with gamma emissions in the range of 100–200 keV. Isotopes also need to be available in carrier-free form and must have suitable chemical properties to allow attachment to antibody. They must not have hazardous daughter isotopes, and should be of suitable half-life to allow preparation of the radiolabelled antibody, localisation and clearance of blood activity. As such the half-life of the isotope used needs to be matched to the biological half-life of the form of antibody used. For example, indium-111 is well suited to RAID with intact IgG, while technetium-99m is well suited to rapidly clearing antibody fragments such as Fab′ and scFv.

Iodine isotopes have been widely investigated due to their ready availability and well-developed methods for radioiodination of antibodies at either tyrosine or lysine residues (see Section 3.2.2). ^{131}I is not an ideal isotope for RAID, and is used primarily in experimental studies of tumour targeting for therapy. However, the ability to image tumours using the same labelled immunoconjugate used for therapy is valuable as it allows tumour dose estimates to be determined (DeNardo *et al.*, 1996a). ^{123}I is a potentially useful diagnostic nuclide although it is relatively poorly available and expensive. Its short half-life means it is best suited to imaging with antibody fragments and has been used successfully to image colorectal tumours with anti-CEA $F(ab')_2$, Fab and scFv (Goldenberg *et al.*, 1990; Begent *et al.*, 1996). One problem common to all radioiodinated antibodies is their relative instability *in vivo*. Antibodies are deiodinated *in vivo* which leads to iodine leaking out from the tumour and often to the accumulation of free iodine in the thyroid and stomach. The accumulation of free iodine can be blocked by pre-treatment with cold iodine, although deiodination cannot be prevented. Internalising antibodies may also be rapidly metabolised leading to the expulsion from the cell of small molecular weight iodinated metabolites (Press *et al.*, 1996).

Technetium-99m is probably the isotope of choice for most RAID applications. It is cheap, very readily available from generators in all hospital nuclear medicine departments, leads to low radiation exposure to the patient and has ideal physical properties for detection by gamma cameras. ^{99m}Tc can be attached to antibody either directly or via a chelating agent. ^{99m}Tc is prepared from a generator in the +7 state as pertechnetate and requires reduction, usually achieved with Sn^{2+} or ascorbate, to the +5 state for antibody labelling. Thiol groups are particularly good ligands for binding ^{99m}Tc and direct labelling of IgG is achieved through reduction of disulphide bonds in the antibody molecule, presumably in the hinge region, which are then used to bind ^{99m}Tc. Sn^{2+} can be used to reduce both pertechnetate and the antibody itself, and often an 'intermediate chelator' is used such as glucarate. Several kit formulations have been developed to allow ^{99m}Tc labelling via such direct labelling methods (e.g. Pak *et al.*, 1992, Alauddin *et al.*, 1992). Similarly, Fab′ fragments can be labelled via free thiol groups in the hinge region, and scFv fragments have been engineered specifically to contain C-terminal cysteine residues to allow ^{99m}Tc labelling (George *et al.*, 1995; Verhaar *et al.*, 1996).

Bifunctional chelating agents have been developed for radiolabelling with different metallic radionuclides. These are termed bifunctional due to having a reactive group for antibody attachment and a chelating group for binding the radiometal. Thus the bifunctional reagent can be used to form an antibody–chelator conjugate which can then bind radiometal under mild conditions. Alternatively, the chelator can be pre-labelled with radiometal before attachment to the antibody, although this method has the disadvantage that more handling of radiolabelled materials is required and manipulations need to be carried out rapidly to avoid extensive radioactive decay. Several chelators for ^{99m}Tc labelling have been developed, many of which use thiol ligands. Diamide dimercaptide ligands, also known as N_2S_2 ligands, form a stable tetradentate complex with technetium (Figure 3.12), although problems with non-specific binding of technetium to antibody protein have resulted in a pre-labelling method being most successful (Fritzberg *et al.*, 1988). Nevertheless methods have been developed to reduce this to a kit form for simple radiolabelling of antibody fragments (Kasina *et al.*, 1991). N_3S ligands (Figure 3.12), alternative N_2S_2 chelates based on bis-aminoethanethiols, and other chelating groups such as hydrazino nicotinamides have also been developed which allow formation of the antibody–chelator conjugate before labelling (Weber *et al.*, 1990; Eisenhut *et al.*, 1996; Ultee *et al.*, 1997). Macrocyclic ligands based on cyclam which form very stable technetium complexes have also been examined (Morphy *et al.*, 1988). Recombinant proteins with specific groups capable of binding technetium are also under development, for example fusion of the metal-binding protein metallothionein to antibody fragments can be used to allow subsequent binding of ^{99m}Tc (Das *et al.*, 1992).

Comparisons of direct labelling with chelation have demonstrated that the use of defined chelation gives a more stable complex and results in higher retention of technetium in tumour tissue and less non-specific uptake in normal tissues, resulting in improved tumour images (Hnatowich *et al.*, 1993; Ultee *et al.*, 1997). Nevertheless, clinical imaging of tumours has been successful with both types of procedure (Behr *et al.*, 1995; Eary *et al.*, 1989), and commercially available preparations include technetium-based tumour imaging agents using Fab′ fragments labelled with both methodologies. Immunomedics' CEA-scan™ for imaging colorectal tumours uses a direct labelling procedure, whereas Verluma™ developed by NeoRx for imaging lung tumours is based on chelation methodology.

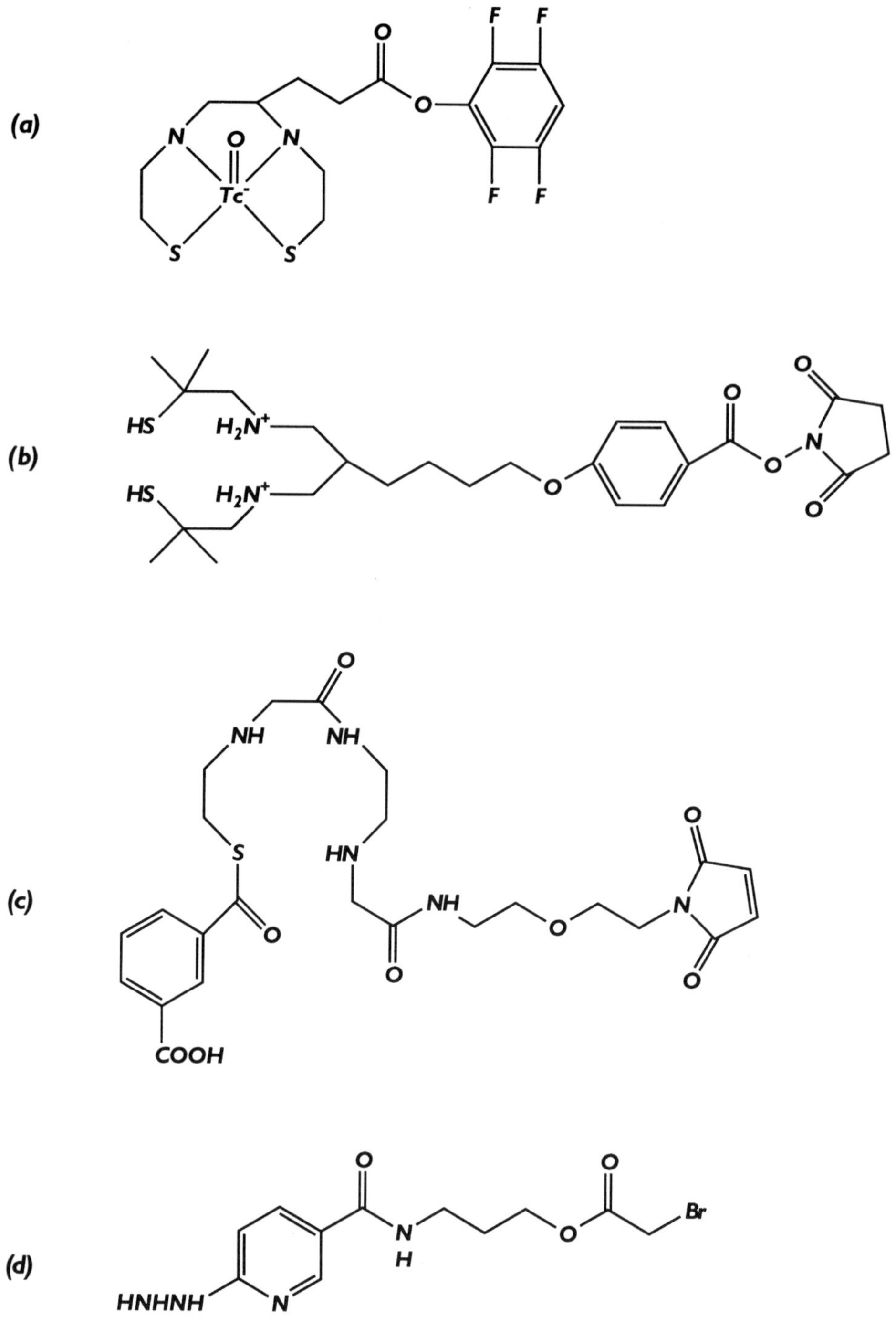

Figure 3.12 Examples of ligands for attachment to MAb which can be used to chelate technetium-99m: (a) active ester of ^{99m}Tc-4,5-bis-(thioacetamide)pentanoate for attachment to amine groups (Fritzberg *et al.*, 1988); (b) 6-(4′-(4″-carboxyphenoxy)butyl) -2-10-dimercapto-2,10-dimethyl-4,8-diazaundecane for attachment to amine groups (Eisenhut *et al.*, 1996); (c) *N*-(5-maleimido-3-oxapentyl)[2-[(3-carboxybenzoyl)-thio]acetyl]glycylglycylglycinamide (N_3S ligand) for attachment to thiol groups (Weber *et al.*, 1990); (d) bromoacetyl hydrazinonicotinamide hydrobromide for attachment to thiol groups (Ultee *et al.*, 1977)

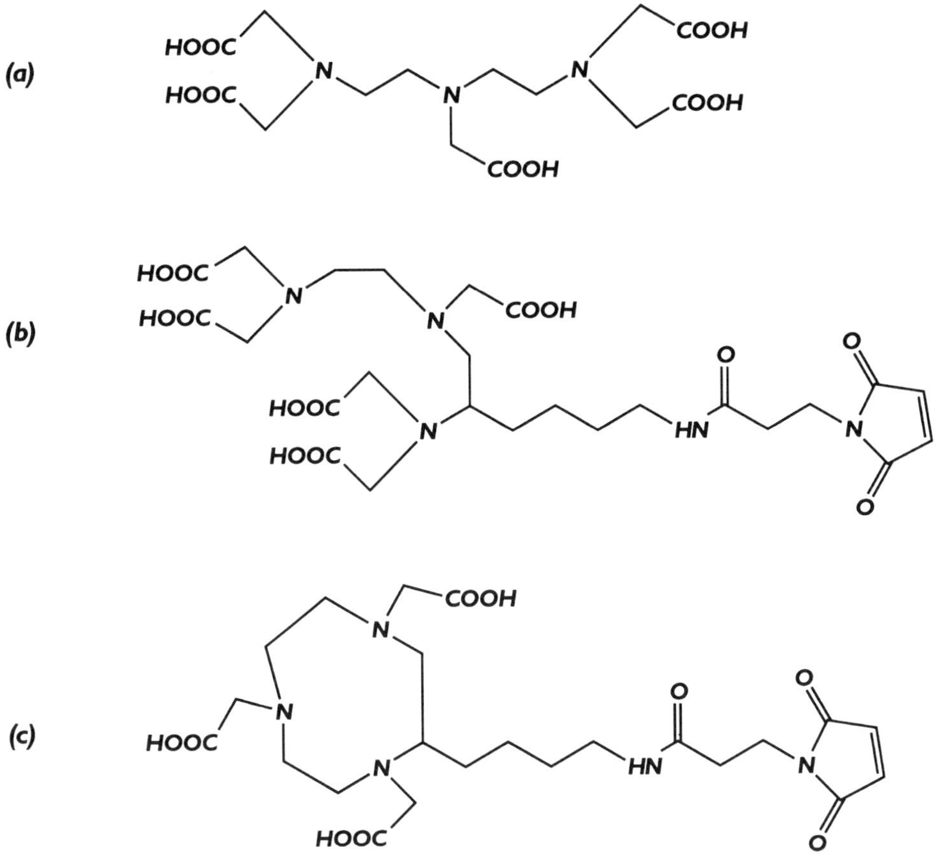

Figure 3.13 Examples of ligands for attachment of indium-111 to MAb: (a) DTPA; (b) derivatised DTPA (Harrison *et al.*, 1991); (c) 9N3 macrocycle (Turner *et al.*, 1994)

Indium-111 is also widely used for RAID applications and requires the use of a bifunctional chelator to allow labelling. Early studies used the cyclic anhydride of diethylenetriaminepentaacetic acid (DTPA) for attachment to antibody and subsequent chelation of ^{111}In (Hnatowich *et al.*, 1983). Although efficient radiolabelling can be achieved, the stability of the ^{111}In–DTPA complex is relatively low as one of the chelation arms is used for attachment to the antibody. *In vivo* dissociation of the complex can take place, leading to the formation of ^{111}In–transferrin which is subsequently deposited in the liver (Schumacher *et al.*, 1990). Subsequently, new derivatives of DTPA were produced such that a separate antibody attachment site could be used, preserving all eight coordination sites for ^{111}In (Figure 3.13), and resulting in more stable immunoconjugates (Brechbiel *et al.*, 1986). However, the most stable complexes of ^{111}In known are formed with macrocyclic ligands, and maleimide derivatives for attachment to antibodies have been developed which retain all of their stability toward binding ^{111}In (Figure 3.13; Craig *et al.*, 1989). Comparative biodistributions of the 9N3 macrocyclic ligand with DTPA and a bifunctional DTPA derivative revealed that the macrocyclic ligand resulted in improved tumour localisation with higher levels of activity in tumour and less in normal tissues (Turner *et al.*, 1994). The macrocyclic ligand DOTA (12N4), best known for labelling with ^{90}Y for radioimmunotherapy (see Section 4.2.5), can also be used to form a stable complex with ^{111}In. Comparative biodistributions demonstrate little difference between the 9N3 and 12N4 macrocycles, although labelling of 12N4 with ^{111}In is more difficult (Turner *et al.*, 1994). Labelling procedures for 9N3 macrocyle immunoconjugates have been developed to allow reproducible ^{111}In incorporation efficiencies of >95% (Haines, A.M.R. and King, D.J., unpublished data), and clinical studies with such conjugates show good tumour imaging characteristics (van Hof *et al.*, 1996).

Fewer studies have been performed with alternative radionuclides, although gallium-67 is a potentially useful reagent which forms stable complexes with macrocyclic ligands (Craig *et al.*, 1989). Positron emission tomography (PET) imaging can be used with suitable radionuclides such as copper-64, bromine-76 and zirconium-89, allowing high resolution imaging to be carried out. Suitable macrocyclic ligands for stable attachment of ^{64}Cu have also been developed (Moi *et al.*, 1985). However, the application of antibodies to PET has been slow to develop due at least partly to the expensive equipment required for imaging studies to be performed. Experimental PET imaging studies have demonstrated good tumour imaging in animal models with a variety of isotopes (Anderson *et al.*, 1992; Lovquist *et al.*, 1997; Meijs *et al.*, 1997). In clinical studies of colorectal carcinoma, ^{64}Cu-labelled antibody demonstrated impressive radioimmunodetection of small tumours in the abdomen and pelvis (Philpott *et al.*, 1995).

Two- and three-step targeting approaches

An alternative approach to improving tumour : blood ratios for RIS is the use of two-step targeting strategies. The simplest form of two-step tumour imaging is to use a second antibody reagent to clear blood background activity, hence improving the signal : noise ratio and the quality of the image. Immune complexes formed by a second antibody are rapidly removed from the circulation by the reticuloendothelial system, particularly in the liver. This can be viewed as an alternative to the use of rapidly clearing antibody fragments. Antibodies to the antibody itself or the radiolabelled chelator can be used (Reardan *et al.*, 1985; Pedley *et al.*, 1989). The use of antibodies to the chelator ensures that any radiolabelled metabolites in the circulation are also rapidly removed. Biotinylated antibodies have also been investigated and shown to be rapidly cleared by the administration of streptavidin (Marshall *et al.*, 1994).

In alternative approaches the administration of antibody and radiolabel are separated. Antibody is allowed to localise to tumour and sufficient time is allowed for antibody clearance from the blood and non-target tissues. Radioisotope is then injected separately in a form which can be readily captured by the tumour-bound antibody (Figure 3.14). Bispecific antibodies have been developed with specificity for tumour and a radiometal chelator such as DTPA (Goodwin *et al.*, 1988; LeDoussal *et al.*, 1989). After localisation of the bispecific antibody, and clearance from the blood, radiolabelled metal chelate is added which is then bound by the antibody localised at the tumour site and rapidly cleared from the rest of the body through renal excretion. Early studies revealed that divalent metal chelators which were able to bind two antibody binding sites simultaneously resulted in enhanced affinity for the tumour site and improved imaging (Goodwin *et al.*, 1988; LeDoussal *et al.*, 1989). A successful strategy may be the use of a di-Fab or tri-Fab construct, with one or two Fab arms binding to the tumour antigen and one available for binding the radiolabelled chelator. Bivalent chelator may then cross-link two antibodies at the tumour surface, resulting in its increased affinity. Such bispecific antibodies have been successfully used to image medullary thyroid carcinoma, colorectal

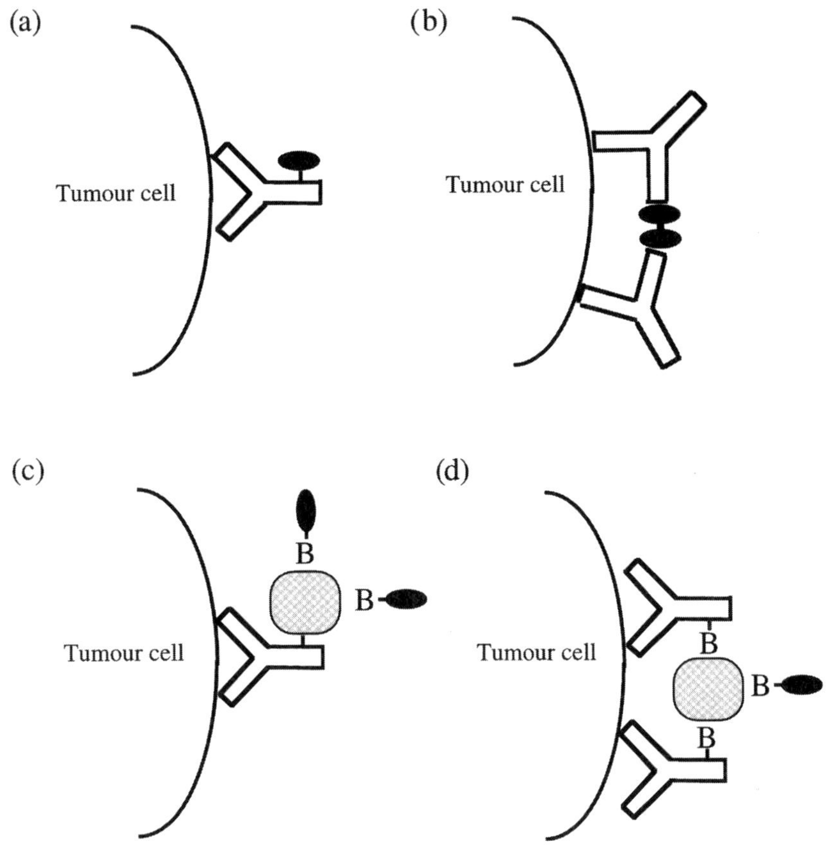

Figure 3.14 Strategies for tumour imaging: (a) one-step, directly labelled antibody; (b) two-step, bispecific anti-tumour antibody and anti-labelled hapten; (c) two-step, anti-tumour antibody–avidin conjugate and labelled biotin; (d) three-step, anti-tumour antibody–biotin conjugate, avidin, then labelled biotin (see text for details)

tumours and non-small cell lung tumours in patients (Peltier *et al.*, 1993; LeDoussal *et al.*, 1993; Vuillez *et al.*, 1997).

Alternative strategies have also been developed using the high affinity of the avidin:biotin system to capture radiolabelled small molecules from the blood as a two- or three-step imaging system. In the two-step version, antibody–avidin conjugate is injected and allowed to localise to tumour and clear from the blood. This is followed by injection of a low molecular weight biotinylated radiolabelled ligand which is captured by the tumour-bound avidin complex (Hnatowich *et al.*, 1987). Further development of this system has resulted in a three-step system in which biotinylated MAbs are injected and allowed to bind to the tumour site. This is followed by injection of avidin (or streptavidin) which binds to the tumour-bound antibody and has the additional advantage of forming a complex with antibody remaining in the blood which is rapidly cleared. Radiolabelled biotin is then added which is bound at the tumour site and cleared from the rest of the body (Paganelli *et al.*, 1991). Such three-step targeting allows the use of cocktails of biotinylated antibodies if desired which improves sensitivity through simultaneous detection of several tumour antigens, and has been used to detect small tumour deposits in patients, not visible by other techniques (Magnani *et al.*, 1996). The disadvantages are their relative complexity, and the immunogenicity of some of the components used, such as streptavidin. Whether such methods will find long-term application remains to be seen.

3.9.2 Radioimmunoguided surgery

Radioimmunoguided surgery (RIGS) may prove to be a useful aid to surgery for the removal of tumour tissue. Patients are pre-administered radiolabelled antibody, and a hand-held gamma detecting probe is used intraoperatively to locate occult tumour which may otherwise escape detection. Some studies have suggested that the sensitivity obtained with RIGS may be better than with radioimmunoscintigraphy (Hinkle *et al.*, 1991), and RIGS has been shown to improve assessment of tumour spread and improve surgical intervention (Arnold *et al.*, 1992). For example, in patients with primary colorectal tumours RIGS with the anti-TAG72 antibody CC49 allowed detection of 86% of tumours, whereas in patients undergoing second-look surgery for recurrent disease 97% of tumour deposits were detected (Arnold *et al.*, 1992), the sensitivity of detection apparently being linked to antigen expression. Most studies have used antibody labelled with the weak gamma emitter ^{125}I, although ^{99m}Tc-labelled antibody has also been successfully tested (Ind *et al.*, 1994). Many of the parameters involved in effective RIGS are the same as those described above. For example, improvements are offered by the use of rapidly clearing scFv fragments allowing RIGS to be performed more rapidly after administration of the labelled targeting molecule (Nieroda *et al.*, 1995). Additionally, limitations due to antigen expression can be addressed by appropriate selection of antibodies or, in the case of CC49, by up-regulation of TAG72 antigen expression with γ-interferon treatment (Nieroda *et al.*, 1995).

3.9.3 Non-tumour radioimmunodetection

MAb-based radioimmunodetection has also been applied in several other disease situations, where precise imaging may offer useful clinical information. These include detection of myocardial necrosis, imaging of blood clots and the detection of infection and inflammation.

Myocardial necrosis takes place in a range of ischaemic, inflammatory and toxic heart diseases and results in breakdown of cell membranes, exposing cardiac myosin which is then accessible to a MAb administered systemically. Fab to the heavy chain of cardiac myosin labelled with ^{111}In has been used to image damage as a result of myocardial infarction (Khaw *et al.*, 1987), myocarditis and cardiac transplant rejection (Carrio *et al.*, 1988). It may also be beneficial in the identification of cardiac involvement in the autoimmune disease systemic lupus erythematosus (Morguet *et al.*, 1995), and in the assessment of cardiac toxicity following cancer chemotherapy with doxorubicin (Carrio *et al.*, 1995). Localisation of labelled antibody to the site of damage is usually very rapid, even when blood flow is reduced as a result of that damage, and the quality of the images obtained is dependent on rapid clearance of blood activity. Therefore most studies have used Fab fragments with rapid clearance times, usually labelled with ^{111}In via the chelator DTPA. However, with ^{111}In-Fab it normally takes 24–48 hours for blood pool activity in the heart to decrease sufficiently to allow good interpretation of images, which has limited the use of imaging in conditions such as myocardial infarction where rapid imaging is required. New agents are now being designed to allow more rapid imaging. In an animal model system, an anti-myosin scFv has been shown to retain the same accumulation at sites of infarction as the Fab fragment, although clearing more rapidly from the blood (Nedelman *et al.*, 1993). In addition the scFv was labelled with ^{99m}Tc via a bifunctional chelator, which allows earlier imaging than with ^{111}In as discussed above (Section 3.9.1). Another approach has been to chemically modify anti-myosin Fab′ so as to impart a negative charge (Khaw *et al.*, 1991). Positively charged Fab′ may interact non-specifically through ionic interactions with negatively charged cell surfaces. Chemical modification of the Fab through attachment of a negatively charged polymer decreased background activity and allowed earlier imaging.

There has been much interest in the development of anti-fibrin antibodies for imaging blood clots, in both deep vein thrombosis and arterial thrombosis. Anti-fibrin antibodies have been identified which bind to fibrin but not circulating fibrinogen and thus can be used to visualise thrombosis. Both ^{111}In and ^{99m}Tc have been used in clinical studies, with high specificity and sensitivity of thrombus detection in deep vein thrombosis (DeFaucal *et al.*, 1991; Schiable *et al.*, 1992). Results in imaging arterial disease have been less impressive and ^{99m}Tc-antifibrin Fab′ was less effective in thrombus detection than labelled platelets (Stratton *et al.*, 1994). Venous thrombi comprise mostly thrombin and red blood cells whereas acute arterial thrombi contain more platelets with smaller amounts of fibrin and red blood cells. Hence a smaller antigen pool is present in arterial disease, which may be responsible for poorer imaging results. Anti-platelet MAbs have also been investigated as agents for detection of thrombi (Oster *et al.*, 1985), and may be potentially useful in the detection of active arterial thrombi, although further clinical evaluation is required. Again attempts to use more rapidly targeting and clearing molecules have been made. An scFv has been expressed with specificity for fibrin fragment D-dimer (Laroche *et al.*, 1991), and a peptide designed from an antibody CDR loop to mimic an anti-platelet antibody binding site has been produced (Knight *et al.*, 1994). Although some specificity was seen for ^{99m}Tc-labelled peptide, the levels accumulated at the thrombus were too low for useful imaging.

RAID is also under development in several other disease states to enable localisation of disease and to monitor therapy. Antibodies specific for polymorphonuclear leukocyte (PMN) surface antigens have been developed for imaging inflammatory processes at sites of infection. A ^{99m}Tc-Fab′ to the cell surface antigen NCA-90 has shown rapid targeting in clinical studies detecting soft-tissue infections and osteomyelitis, and is the basis of the agent LeukoScan™ (Becker *et al.*, 1994). Similarly, ^{99m}Tc-labelled anti-CD15 antibody has been used to image inflammatory disease, labelled either directly or via a DTPA-based chelator with no significant difference in image quality (Thakur *et al.*, 1996). Labelled polyclonal human IgG is also capable of imaging sites of inflammation, probably due to increased vascular permeability at the site or to binding to Fc receptors up-regulated at the site (Rubin *et al.*, 1989). The use of specific reagents may, however, increase confidence in interpretation of the scans as non-specific reagents may lead to scans which are difficult to interpret with often low or diffuse uptake (Thakur *et al.*, 1991). An interesting follow-up to non-specific imaging uses a two-step strategy (Kranenborg *et al.*, 1997). An unlabelled, monoclonal anti-DTPA antibody is used to localise to the site of infection non-specifically. After localisation and clearance from the blood has taken place, DTPA labelled with ^{99m}Tc is administered which binds to the sequestered antibody with the remainder rapidly cleared from the circulation, leading to low background levels and clearer imaging.

An alternative to imaging via antibodies to PMNs is to image the inflamed endothelium. Certain adhesion molecules are up-regulated during inflammation as part of the mechanism of PMN recruitment to the site. Up-regulation of the selectins allows rolling of the PMN along the endothelial cells and up-regulation of intercellular adhesion molecule-1 (ICAM-1) allows the PMN to become firmly attached to the endothelial cell and then to migrate into the interstitium. Imaging sites of inflammation by antibodies to these endothelial cell adhesion molecules may therefore give additional selectivity. ^{111}In-labelled antibodies to both E-selectin and ICAM-1 have been investigated in animal models with promising results (Keelan *et al.*, 1994; Sasso *et al.*, 1996).

The abundance of CD4 on inflammatory cells present in arthritic joints has led to interest in targeting CD4 to image arthritic joints. Antibodies to CD4 have been used to assess targeting in animal models of arthritis. Design of the reagent is crucial for success, as intact IgG demonstrated no benefit over a non-specific antibody, whereas a ^{99m}Tc-labelled Fab′ led to improved imaging (Kinne *et al.*, 1995). Anti-CD4 reagents may also be useful in monitoring the distribution of CD4 positive lymphocytes, which could be useful in assessing patients with HIV infection (Rubin *et al.*, 1996). RAID has also been investigated for detection of atherosclerotic plaques using an antibody towards proliferating smooth muscle cells (Narula *et al.*, 1995) and lesions in Alzheimer's disease using a cationised antibody towards beta A4 protein (Bickel *et al.*, 1994). Cationisation through chemical modification of the antibody with charged groups is believed to aid transport of antibodies through the blood:brain barrier, which is obviously helpful to image brain lesions in Alzheimer's disease.

4

Monoclonal Antibodies in Therapeutic Applications

4.1 Introduction

The idea of targeted therapy using antibodies dates from the beginning of the twentieth century when it was proposed by Paul Ehrlich, and indeed antibodies have been used in humans for many years as polyclonal antisera, particularly for passive immunisation. Since the discovery of the technology for production of MAbs there has been intense interest in their development as therapeutic agents for human disease. However, until recently progression of many MAb-based therapeutics was limited by the human immune response generated by the administration of murine or rat antibodies. Since the introduction of recombinant chimeric and humanised antibodies, and ultimately human antibodies, interest in developing therapeutics has been revived. There are now five antibody-based products licensed for human therapeutic use, and many more under clinical investigation (Table 4.1).

MAbs can be used in several different modes, depending on the required therapeutic effect. In some cases a simple blocking or neutralising effect may be required, for example in the neutralisation of an inflammatory cytokine or the blocking of a specific receptor. In other applications therapy may require an active role for the antibody by the targeting of an effector function. Therapeutics can utilise natural antibody effector functions such as complement activation, phagocytosis or ADCC, or entirely novel functions can be introduced, as in the case of targeting radioisotopes, drugs or toxins to kill tumour cells. In some cases antibodies may also find a role through their ability to mediate signal transduction from binding to cell surface receptors (Vitetta and Uhr, 1994). Available mechanisms for generation of therapeutic effects can be considered as a range of antibody effector functions (Table 4.2). Many of these effector functions can be designed into, or out of, antibody molecules for specific therapeutic purposes as described in Chapter 2. In this chapter the basis of targeted therapy is illustrated using examples of the various modes of antibody-based therapy in different disease states. The emphasis is placed on the design of antibody-based therapeutics rather than a catalogue of antibodies used in therapeutic studies.

Table 4.1 Commercially available monoclonal antibody therapeutics

Therapeutic agent	Antibody	Disease	Company
Orthoclone OKT3	Murine OKT3	Acute transplant rejection (kidney, liver and heart)	Ortho Biotech
ReoPro	Chimeric 7E3 Fab	Complications of post-coronary angioplasty	Centocor
Panorex	Murine 17-1A	Colorectal cancer	Centocor
Rituxan	Chimeric 2B8	Lymphoma	IDEC/Genentech
Zenapax	Humanised anti-tac	Acute kidney transplant rejection	Protein Design Labs/Roche

Table 4.2 Effector functions for antibody targeted therapy

Blocking/neutralising	Antigen binding
Natural – Fc mediated effects	Complement fixation ADCC Phagocytosis
Cell signalling	Receptor cross-linking
Natural immune responses	Generation of anti-idiotype response Other 'vaccination' approaches
Artificial effectors	Radioisotopes Toxins – bacterial – plant Cytotoxic drugs Cytokines Enzymes – prodrug activation – direct toxicity
Bifunctional	Cross-linking cytotoxic effector cells Two-step targeting strategies for radioisotopes, toxins, etc.

4.2 Cancer

One of the major targets of antibody-based therapeutics has been in the development of anti-cancer agents. Initially MAbs to tumour-associated antigens were raised and investigated without modification, in attempts to target natural antibody effector mechanisms to tumour cells. Results of such studies were initially disappointing in many cases, particularly with the common solid tumour types. However, beneficial effects were seen when treating patients with minimal disease and in treating lymphomas and leukemias, suggesting that further research to investigate the role of antibodies as anti-cancer agents may lead to improved results. In addition, the use of antibodies to target traditional anti-cancer agents such as cytotoxic drugs and radioisotopes has made steady progress, particularly since the introduction of recombinant antibodies and improved methods of coupling cytotoxic agents (Chapter 2). As described in Chapter 3, many tumour-associated antigens representative of different tumour types are now characterised (see Table 3.5) which allow therapeutic molecules to be designed.

4.2.1 Cancer therapy with unmodifed (naked) antibodies

The use of unmodified, or naked, antibodies for tumour therapy initially resulted from attempts to harness the immune system through the natural antibody effector functions, such as ADCC or complement-mediated lysis. More recently it has become apparent that cell signalling mechanisms may be involved in many of the anti-tumour effects observed, through arrest of the cell cycle or inducing apoptosis. Design of antibodies to elicit natural effector functions has largely been a process of choosing the optimal isotype to harness human ADCC and complement effects, although as described in Section 2.6.1 the selection of a suitable isotype alone is not sufficient to ensure good cell killing, and suitable antigen targets must be individually tested. Mouse IgG2a is the best of the murine isotypes for eliciting human ADCC, with rat IgG2b also potent. Of the human isotypes, IgG1 and IgG3 are the most potent in cell killing studies. Of these IgG1 is most commonly used for construction of humanised antibodies, partly because IgG3 antibodies are more difficult to purify and to handle *in vitro* due to a tendency to aggregate. The preparation of chimeric or humanised antibodies from rodent antibodies thus has the dual benefit of reduced immunogenicity and the ability to use a constant region best suited to recruitment of human effector functions. Similarly human antibodies isolated from phage display, or other means, can be reconstructed and expressed with the desired isotype constant regions. Expression of the antibody in mammalian cells is required as the glycosylation of the CH2 region is required for maintainence of the ability to elicit effector functions. However, cell killing effects are dependent not only on the constant region, but also on the disposition of antigenic sites and other poorly understood mechanisms. Many cells also have protective mechanisms against attack from the immune system, and thus the *in vivo* effects of antibodies capable of eliciting cell killing *in vitro* are often difficult or impossible to predict.

Several alternative mechanisms have also been suggested for tumour cell killing by unmodified antibodies. Antibodies against cell surface markers on many types of tumour cell can act as ligands eliciting anti-tumour effects by signal transduction (Vitetta and Uhr, 1994). Signal transduction can result in arrest of the cell cycle, hence preventing tumour cell growth, or in some cases the induction of programmed cell death known as apoptosis. In addition, antibodies to growth factors or their receptors may exert anti-tumour effects through blocking binding of growth factors needed for tumour cell growth. Growth factors such as EGF (epidermal growth factor) and IL-6 have been implicated in the growth of a number of tumour types, and target tumour cells overexpress large numbers of molecules of the growth factor receptors. Receptor blocking antibodies prevent interaction with the ligand, and can lead to down-regulation of the number of receptor molecules present on the tumour cell such that growth is inhibited. However, unravelling the contribution of each of these factors to effective tumour cell killing is not straightforward and often several mechanisms may operate concurrently. Future research may allow the selection of antibodies based on an improved understanding of the importance of eliciting each of these anti-tumour effects. For example, it may be possible to select combinations of MAbs capable of different mechanisms of cell killing.

As described in Chapter 2, dimeric or polymeric antibodies have been produced either by chemical cross-linking or by recombinant means which have increased activity for both complement activation and ADCC. Chemically cross-linked constructs comprising Fab fragments linked to two Fc regions $FabFc_2$ have also been produced in attempts to increase therapeutic efficacy through effector functions. However, studies comparing such constructs with a humanised antibody for immunotherapy of multiple myeloma showed no advantage in complement activation or ADCC (Ellis *et al.*, 1995).

Homodimerisation of IgG has also been reported to be able to confer signalling activity, leading to cell cycle arrest or apoptosis, to antibodies which are unable to do so in monomeric form (Ghetie *et al.*, 1997a). Chemically cross-linked IgG homodimers were produced for several antibodies and shown to be capable of specific anti-tumour effects, presumably due to cross-linking the antigen and/or a lower dissociation rate. In the case of an anti-CD19 antibody the Fc region was not required for anti-tumour activity, leading to the possibility that small multimeric fragments may be capable of potent anti-tumour effects.

Although many studies with unmodified antibodies for cancer therapy have been disappointing, with no evidence of anti-tumour effects, several antibodies have shown promising effects in the clinic. It is only in the past few years that recombinant chimeric and humanised antibodies have reached clinical evaluation. In these studies it has been apparent that the ability of the human constant regions to interact with the human immune system plus the ability to re-treat without the generation of a prohibitive immune response is leading to more effective therapeutic agents.

The first humanised (CDR-grafted) antibody to be used clinically, CAMPATH-1H, recognises the CAMPATH–1 antigen, also known as CDw52, which is present on most human lymphocytes and monocytes but not stem cells. CAMPATH-1H has been investigated as a potential therapeutic agent in non-Hodgkin lymphoma (B-cell lymphoma) as well as in several inflammatory diseases (Section 4.5.3). Although only limited studies have been reported, use of CAMPATH–1H for the treatment of B-cell lymphoma has led to tumour remissions in some patients (Hale *et al.*, 1988). Studies with different isotypes of the rat parent antibody suggest that human effector cells are involved, as the IgM and rat IgG2a versions, which can activate complement but not ADCC, led to only transient falls in blood cell counts. In contrast, the rat IgG2b version which could bind to and activate human effector cells led to much more efficient depletion of tumour cells *in vivo* (Dyer *et al.*, 1989).

Another antibody, which has shown promising effects in B-cell lymphoma, recognises CD20, a phosphoprotein present on the surface of B cells. A chimeric IgG1 version of the antibody 2B8, termed C2B8, has been shown in *in vitro* assays to be an effective mediator of both complement-mediated effects and ADCC using human effector cells (Reff *et al.*, 1994). In addition, this antibody has been shown to induce transmembrane signalling leading to cell cycle arrest and occasionally apoptosis of CD20 positive cells, and thus several mechanisms may be important in the anti-tumour effects observed (Demiden *et al.*, 1995). Phase II and III clinical studies for relapsed B-cell lymphoma have shown tumour shrinkage of 50% or more in approximately half of the patients studied (Maloney *et al.*, 1997). In addition, C2B8 was found to sensitise resistant lymphoma cells to certain cytotoxic drugs, which may lead to synergistic effects in combination therapy. In a phase II clinical study of C2B8 with standard chemotherapeutic treatment, responses were seen in all patients studied (Czuczman *et al.*, 1995). Although as a single agent, tumour responses were not as impressive as those observed with radiolabelled antibodies to CD20 (Section 4.2.5), treatment with this unmodified antibody was much less toxic and is more straightforward than handling large amounts of radioisotope. Therefore, it is likely that this agent will find clinical application alongside chemotherapy for B-cell lymphoma treatment.

The HER2 gene (also known as *neu* or *c-erb-B2*) is a proto-oncogene which encodes a transmembrane growth factor receptor ($p185^{HER2}$) which is overexpressed in 25–30% of patients with breast cancer, particularly those with a poor prognosis. A humanised antibody to $p185^{HER2}$ has been developed which has shown significant growth inhibition of cells overexpressing this receptor (Carter *et al.*, 1992b). The humanised antibody was also shown to be more efficient at mediating ADCC with human effector cells than the parent murine antibody. In a phase II clinical study in patients with metastatic breast cancer that overexpressed HER2, objective responses were seen in 5 of 43 patients including one complete remission and 4 partial remissions (Baselga *et al.*, 1996). The mechanisms of action of this agent are also not clear at present. The antibody induced a clear down-regulation of the growth factor receptor which may reverse the malignant phenotype. This antibody is also known to be capable of activation of a signal transduction pathway that leads to inhibition of tumour cell proliferation and possibly cell death, and also elicits ADCC.

4.2.2 Anti-idiotype antibodies

MAbs may also be used as vaccines to generate an anti-idiotype response. The idotype of an antibody is made up of a cluster of epitopes at the antigen-binding site. As each MAb has a different antigen-binding site, the idiotype of each MAb is unique. Antibodies can be raised through an immune response to the idiotype which have a conformation that resembles the antigen of the original antibody (Figure 4.1). These anti-idiotype antibodies are also known as anti-id or ab2 antibodies. Immunisation with an ab2 antibody can in turn lead to the elicitation of an anti-idiotype to the ab2 antibody, an anti-anti-id, known as ab3, a proportion of which will have the same or overlapping specificity as the original ab1 antibody (Figure 4.1). Structural studies suggest that anti-idiotype antibodies carry an 'internal image' of the original antigen which allows the development of ab3 responses that recognise the original antigen. The crystallographic structure of an anti-idiotype antibody in complex to an antibody raised against lysozyme showed that the same antigen-binding residues used to bind lysozyme were used for binding to the ab2, and that the ab2 mimicked lysozyme (Fields *et al.*, 1995).

Vaccination with ab2 has been investigated in attempts to raise a human ab3 response to target antigens which may be poorly immunogenic or difficult to prepare or use in immunisation. ab2 may also be able to raise antibodies not raised to the original antigen through breaking immunological tolerance and also give rise to T cell responses which may have therapeutic significance. Many studies, both preclinical and clinical, have been conducted in attempts to achieve therapeutic effects with ab2 vaccines for cancer (reviewed by Herlyn *et al.*, 1996). In several cases antigen-specific immune responses have been achieved, including both antibody and cellular responses. For example, the antibody 105AD7 is an anti-idiotype antibody that mimics a colorectal tumour-associated antigen. Phase I clinical studies demonstrated that a T cell response to tumour was induced on treatment with 105AD7 and this resulted in delayed tumour growth and an increase in survival time compared with patients at the same stage of disease that did not receive the antibody (Buckley *et al.*, 1995).

It is, of course, also possible for antibodies to tumour-associated antigens administered for tumour therapy (ab1) to elicit anti-idiotype responses. The murine MAb 17-1A has

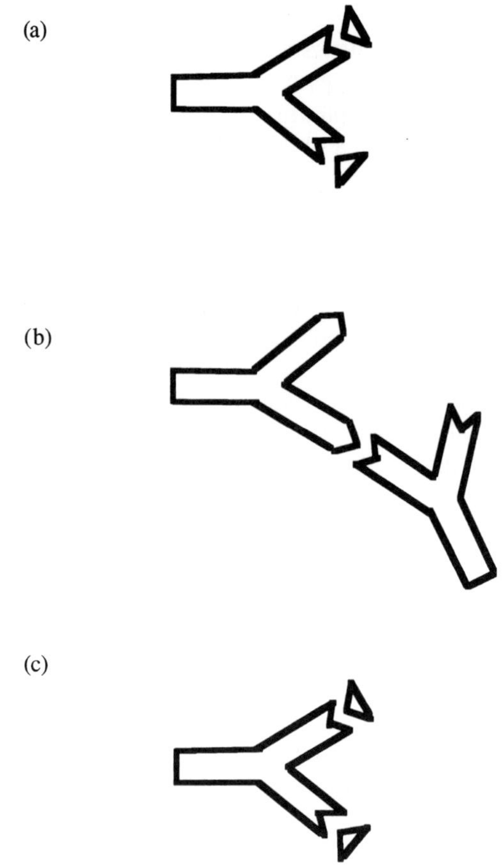

Figure 4.1 Representation of anti-idiotype antibodies: (a) initial antibodies (ab1) recognise the antigen to which they were raised; (b) anti-idiotype antibodies (ab2) recognise the antigen-binding region of ab1 and thus can be considered as containing an 'internal image' of the antigen; (c) anti-anti-idiotype antibodies (ab3) raised against ab2 have similar specificity to the original antibody (ab1) and can thus recognise the same antigen

been extensively studied. It recognises the epithelial membrane antigen (EMA) present on the majority of colorectal tumour cells. Originally developed for its ability as a murine IgG2a to elicit ADCC in humans, early clinical studies revealed occasional responses in patients with advanced disease which were delayed following antibody administration, implying an active immune response. Most patients developed an anti-idiotype response to 17-1A which may have led to the observed therapeutic effects (Fagerberg *et al.*, 1996). In a phase III trial of 17-1A for the treatment of minimal residual disease in colorectal cancer patients following surgical removal of the primary tumour, 17-1A treatment resulted in a statistically significant improval in survival (Riethmuller *et al.*, 1994). After following patients for five years, 17-1A therapy reduced the overall death rate by 30% and decreased the recurrence rate by 27%. These results led to the approval of 17-1A as a therapeutic agent, sold under the name 'Panorex' (Table 4.1). Analysis has identified patients with high levels of ab3 as those that live longer, suggesting that the induction of an anti-idiotype response is important in the mechanism of the therapeutic effect of this antibody (Fagerberg *et al.*, 1996).

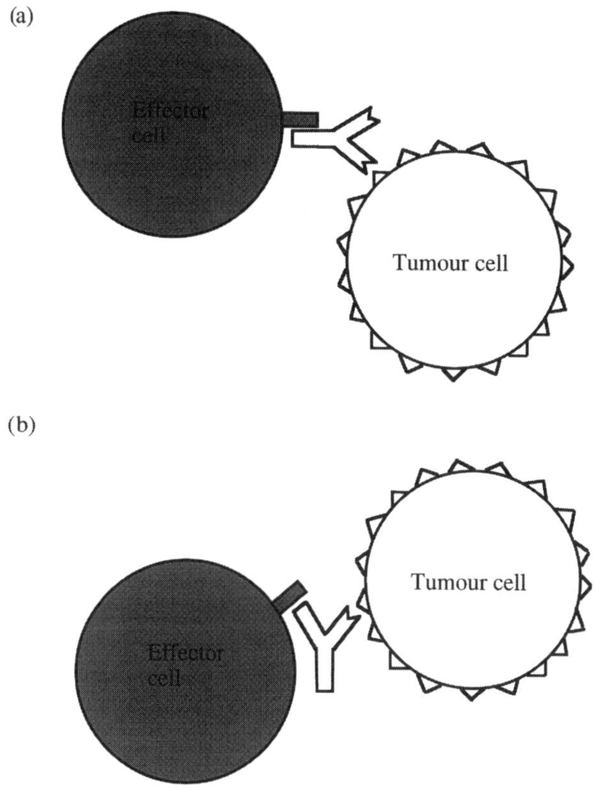

Figure 4.2 Mechanisms of targeting effector cells to tumours: (a) active isotypes which can bind to Fc receptors on the effector cell can be used to target the cell to a tumour-associated antigen; (b) alternatively, bispecific antibodies may be used which bind to both the tumour cell and the effector cell through the antigen-binding arms of the antibody (for further details see text)

4.2.3 *Bispecific antibody-mediated effector cell targeting*

The limited success that has been achieved in using antibodies to elicit ADCC responses has prompted the search for other approaches to recruit effector cells, such as cytotoxic T cells, to the tumour site. One such approach is to use bispecific antibodies with one specificity for a tumour-associated antigen and one for the effector cell such that the effector cell is linked directly to the tumour cell (Figure 4.2). A surface 'trigger molecule' is used as the antigen on the effector cell such that the cell is activated to lead to a cytotoxic response. Several different types of effector cell and trigger molecule have been investigated, including Fcγ receptors and molecules of the CD3/T cell receptor complex (Table 4.3). The advantages and disadvantages of each of these have been reviewed (Renner and Pfreundschuh, 1995). In addition to requiring triggering, cytotoxic effector mechanisms may also require the involvement of accessory molecules and activation via cytokine release.

On monocytes, macrophages, eosinophils and polymorphonuclear neutrophils (PMNs), activation is achieved through linking to Fcγ receptors. Natural killer cells are activated through FcγRIII and CD2, whereas activation of T cells is usually achieved through the

Table 4.3 Cytotoxic effector cells and trigger molecules

Effector cell	Trigger molecules
T cells	CD3 T cell receptor (γ/δ) CD2 CD28
Natural killer cells	Fcγ receptor III (CD16) CD2
Monocytes/macrophages	Fcγ receptor I (CD64) Fcγ receptor II (CD32) Fcγ receptor III (CD16)
Granulocytes (eosinophils, PMNs)	Fcγ receptor II (CD32)
PMNs activated with interferon-γ, GM-CSF or G-CSF	Fcγ receptor I (CD64)

CD3/T-cell receptor complex. The CD3/T-cell receptor complex is normally responsible for antigen-specific T cell responses, but the use of bispecific antibodies allows activation through this complex which can be redirected to the antigen recognised by the other arm of the bispecific antibody. Activation of effector cells is greatly increased by cytokines such as interleukin-2 or interleukin-7, or by co-stimulatory signals. CD2 and CD28 are co-stimulatory molecules which when stimulated increase the activation of T cells. For example, combinations of triggering CD3 and CD28 allow particularly potent anti-tumour effects. Cell killing by a bispecific antibody cross-linking the Hodgkin lymphoma antigen CD30 to CD3 resulted in relatively little tumour cell killing unless co-administered with CF30–CD28 bispecific antibody (Renner *et al.*, 1994). The combination was also far more potent in treatment of xenografted human tumour in immunodeficient mice repopulated with human T cells. Cures of established tumours could only be achieved with the combination of both bispecific molecules (Renner *et al.*, 1994). Similar effects can be achieved by co-administration of anti-CD28 IgG with anti-CD3/anti-tumour bispecific antibodies (Demanet *et al.*, 1996). An alternative approach has been the use of a trispecific antibody, made by cross-linking three Fab′ fragments, with specificity for tumour antigen as well as two different T cell antigens (Tutt *et al.*, 1991b).

Several phase I clinical studies have been carried out with bispecific antibodies with interesting results, although these have been limited by the need to activate T cells *ex vivo* and re-infuse to the patient at the time of bispecific antibody therapy. IL-2 activated cells targeted by an anti-CD3/anti-tumour bispecific antibody administered intracranially to glioma patients resulted in survival of 76% of patients over two years, compared to 33% of those treated with activated cells alone (Nitta *et al.*, 1990). Anti-FcγRIII/anti-tumour antibodies designed to activate natural killer cells and mononuclear phagocytic cells have also been used clinically, with minor responses observed (Weiner *et al.*, 1995).

The production of bispecific antibodies can be achieved using a variety of approaches, including the production of hybrid hybridomas, chemical cross-linking and recombinant approaches (Section 2.5). The affinity of the antibody for the triggering molecule may also be important. Humanised Fab′ fragments to the breast tumour antigen $p185^{HER2}$ and to CD3 have been expressed and cross-linked to form a bispecific antibody capable of efficient lysis of tumour cells (Zhu *et al.*, 1995). In these studies a high affinity version of the humanised anti-CD3 Fab′ was found to be more effective at tumour cell lysis than a lower affinity version. The same antibody specificities have been used to produce a bispecific diabody which was equally active in tumour cell lysis but was much simpler to produce as it was directly expressed by secretion from *E. coli* at high yield (Zhu *et al.*, 1996). Under some circumstances diabodies may be more effective than bispecific IgG at achieving tumour cell lysis, possibly due to the smaller size of the diabody bringing target and effector cell closer together (Holliger *et al.*, 1996).

4.2.4 Other approaches to recruit the immune system using MAbs

Superantigens are bacterial or viral proteins, such as the exotoxin produced by *Staphylococcus aureus*, that stimulate T cells by cross-linking the T cell receptor to MHC class II molecules. Superantigens bind to the T cell receptor outside the antigen recognition site and to MHC class II molecules outside the peptide binding groove normally involved in antigenic recognition, bypassing normal immunological specificity. Superantigens are thus the most potent known activators of T cells. Fab′–superantigen fusion proteins have been produced which stimulate T cells and target to tumour cells. The fusion proteins are intended to replace MHC class II binding with tumour cell binding through the Fab′. Such fusion proteins have been shown to be capable of inhibiting tumour growth in animal models through causing activated T cells to infiltrate and attack the tumour (Dohlstein *et al.*, 1995). One problem with the use of such agents, however, is residual MHC class II binding which leads to relatively high toxicity through systemic T cell activation and accumulation of inflammatory cytokines in serum. Attempts to improve Fab′–superantigen fusion proteins for tumour therapy are thus being made by mutations in the superantigen binding site for MHC class II which prevent binding and reduce systemic toxicity (Hansson *et al.*, 1997).

Other approaches in attempts to improve the recruitment of the immune system by antibodies have included the use of antibody–cytokine fusion proteins. Certain cytokines, such as interleukin-2 (IL-2) are able to activate multiple immune mechanisms, although attempts at tumour therapy with cytokine alone have led to severe systemic toxicity. Fusion proteins have thus been designed to target a high local concentration of cytokine to the tumour site, minimising systemic toxicity. Several different cytokines have been investigated. IL-2 stimulates T cell proliferation and T cell mediated killing and has thus been widely investigated, but tumour necrosis factor α (TNFα), TNFβ, GM-CSF, IL-5, IL-8 and interferon α have also been tested. Fusion proteins have been produced with several antibodies and encouraging pre-clinical results obtained. Chimeric 14.18, an antibody that recognises the GD2 ganglioside expressed on neuroblastoma, melanoma and certain other tumours, has been used to produce fusions with several cytokines, the most promising of which appears to be IL-2 (Hank *et al.*, 1996). The fusion protein maintained antigen binding and IL-2 activity and was able to exert anti-tumour effects in mice bearing human tumour xenografts of neuroblastoma and melanoma. Activation of human effector cells has also been demonstrated (Hank *et al.*, 1996). Cytokine fusion proteins have also been produced with small antibody fragments such as scFv, which may allow more effective tumour penetration (Dorai *et al.*, 1994). However, whether such fusion proteins will be sufficiently less toxic than free IL-2 when administered to patients remains to be seen.

4.2.5 Radioimmunotherapy

As with diagnostic tumour imaging (Section 3.9.1), the design of antibodies for radioimmunotherapy (RIT) requires careful consideration of the individual components of the immunoconjugate, including the form of the antibody, the radioisotope and the means of attachment of the radioisotope to the antibody. In addition, the choice of a suitable tumour-associated antigen becomes more critical with therapeutic doses of isotope, as the presence of antigen on normal tissues will result in the deposition of activity, which may lead to unacceptable toxicity. Many factors influence the degree of tumour localisation in patients. These include heterogeneous expression of antigen, presence of circulating antigen, tumour vascularisation and penetration of antibody into tumour tissue (Boxer *et al.*, 1992). However, there may not be a need to achieve localisation to each individual tumour cell, as a radioisotope can be chosen with a pathlength of many cell diameters, allowing for 'bystander cell' killing at the tumour site. Nevertheless, a high proportion of tumour cells expressing the antigen is desirable. This is because the major limitation to RIT is the dose of isotope which can be delivered at the tumour site without resulting in unacceptable toxicity. In general, only a small proportion of the dose administered localises to the tumour site in patients. Typically tumour-localisation levels of 10% injected dose per kg of tumour (0.01% injected dose per gram of tumour) are observed in human studies where good targeting is achieved (van Hof *et al.*, 1996). The total administered dose is usually dependent on the toxicity to normal body tissues. Bone marrow is the most sensitive normal tissue to radiation and therefore doses are usually limited by the dose to bone marrow from radiolabelled antibody circulating in the blood (Badger, 1990). Several strategies have been investigated in attempts to reduce bone marrow toxicity, including the use of short-range radioisotopes which require cell internalisation for cytotoxicity (Auger emitters), the use of specific clearing mechanisms to remove circulating activity, two-step targeting strategies and the use of rapidly clearing antibody fragments.

Careful optimisation of an immunoconjugate for RIT, through antibody engineering and development of suitable chemistry for radioisotope attachment, can lead to improved properties both *in vitro* and *in vivo* (King *et al.*, 1994), and it is likely that such optimised immunoconjugates will provide the next generation of molecules of clinical utility for RIT. Some of the parameters which need to be considered in the design of radioimmunoconjugates for therapy are given in Table 4.4.

Table 4.4 Parameters for design of radioimmunoconjugates for therapy

Radioimmunoconjugate component	Properties to consider
Antibody	Specificity, affinity for tumour antigen Lack of cross-reactivity with accessible normal tissues
Form of engineered antibody	Immunogenicity Pharmacokinetics Accessibility/penetration into normal tissues Incorporation of residues for site-specific labelling
Radioisotope	Physical properties (type of emission, energy, half-life) Stability of attachment to antibody Ease of radiolabelling/suitability for clinical protocol Immunoreactivity of final conjugate

Form of antibody

In contrast to tumour imaging *in vivo*, generation of a high tumour : blood ratio is not the major consideration for the optimal form of the antibody for RIT. An important consideration is the absolute amount of antibody localised to the tumour site. Therefore, the ideal molecule would localise to the tumour in large amounts, delivering a high dose of radiation while clearing rapidly from the circulation and the rest of the body, minimising non-specific toxicity. In practice this ideal is not reached and thus there has been much debate over the ideal form of antibody to use. Intact IgG circulates for a long period of time and accumulates high levels of activity at the tumour site, whereas antibody fragments clear more rapidly, sparing the dose to normal tissues but accumulating lower levels of activity at the tumour site.

Comparisons of IgG with $F(ab')_2$ and Fab′ fragments in both animal models and patients have suggested that there may be a significant advantage for $F(ab')_2$ fragments. Studies with radioiodinated $F(ab')_2$ in animal models revealed more effective tumour therapy with less toxicity than IgG (Buchegger *et al.*, 1990; Pedley *et al.*, 1993). More efficient early uptake into tumours has also been found for $^{131}I\text{-}F(ab')_2$ fragments in clinical studies, suggesting a significant advantage in tumour penetration (Lane *et al.*, 1994). Also, theoretical predictions of the optimal molecule for tumour therapy based on pharmacokinetic and tumour accumulation data suggest that $F(ab')_2$ may be the best targeting molecule (Yorke *et al.*, 1991). Smaller fragments such as Fab′- and Fv-based molecules result in higher tumour : blood ratios but significantly less accumulated at the tumour (e.g. Brown *et al.*, 1987; Milenic *et al.*, 1991; King *et al.*, 1994). Although such fragments are useful for imaging applications, the tumour levels reached are unlikely to be useful for achieving therapy. These findings in animal models have been confirmed in clinical studies with scFv in which high tumour : blood ratios were generated although absolute uptake into the tumour was low (Begent *et al.*, 1996).

Further development of antibody fragments for delivery of ^{131}I has suggested that the use of chemically cross-linked $F(ab')_2$ fragments may be particularly useful. This was investigated due to the relative instability of some $F(ab')_2$ fragments where the exposed hinge region may lead to proteolytic and/or reductive breakdown. In addition, different $F(ab')_2$ fragments have different *in vivo* stabilities. In a study comparing human IgG1, 2 and 4 $F(ab')_2$, it was found that IgG4 $F(ab')_2$ was relatively unstable compared to IgG2 $F(ab')_2$, with the stability of IgG1 $F(ab')_2$ intermediate between them (Buchegger *et al.*, 1992). This leads to differences in the efficiency of different $F(ab')_2$ fragments in tumour localisation. The differences may be due to disulphide bond formation in the hinge region. IgG2 has four hinge disulphide bonds whereas IgG1 has two. The least stable $F(ab')_2$, IgG4, also has the potential for two disulphide bonds, although it is known that due to the structure of the hinge region in human IgG4 molecules, a proportion have only one disulphide bond formed, and some none at all (Angal *et al.*, 1993). This conclusion is verified by the finding that the introduction of more disulphide bonds into the hinge region of a $F(ab')_2$ fragment results in increased *in vivo* stability (Rodrigues *et al.*, 1993).

The introduction of a chemical cross-link between two cysteine residues in the hinge region also leads to increased *in vivo* stability (Quadri *et al.*, 1993; King *et al.*, 1994). Chemical cross-linked $F(ab')_2$ can be produced through the use of bis-maleimide linkers (Section 2.4.2, Figure 2.4). The hinge region of recombinant Fab′ fragments can be engineered to contain a single hinge thiol group to allow simple and high-yield production of chemically cross-linked molecules (King *et al.*, 1994). Such chemically cross-linked molecules have been termed DFM, for a di-Fab prepared with a bis-maleimide linker. DFMs of the antibody chimeric B72.3 which recognises a tumour-associated glycoprotein have been shown to improve targeting to human tumours in a mouse xenograft model, resulting in both higher tumour accumulation and improved tumour : blood ratios compared to $F(ab')_2$ (King *et al.*, 1994). Also, DFM may have higher association rates for antigen than $F(ab')_2$, which may be a consequence of the increased spacing or flexibility of the linker. This was the case for DFM of the anti-carcinoembryonic antigen antibody, A5B7, which led to this being the preferred molecule for RIT applications with ^{131}I as the radioisotope (Casey *et al.*, 1996).

The best form of antibody for tumour targeting also depends on the radioisotope used. Several metallic isotopes, such as ^{90}Y and ^{67}Cu, are potentially attractive for RIT (see below), but the biodistribution of antibody fragments labelled with metallic isotopes is very different. Once taken into cells, metallic isotopes are less readily excreted than halogens such as iodine, which leads to intracellular accumulation (Press *et al.*, 1996). This can be an advantage for metallic isotopes when internalised into tumour cells, as the isotope is retained at the tumour site for longer. However, when taken into normal cells the build-up of isotope can lead to high levels in normal organs. Most antibody fragments are metabolised and excreted through the kidneys. During excretion re-adsorption of protein takes place in the kidney tubules. This can lead to unacceptably high levels of metallic isotopes deposited in the kidney tubule cells. High kidney levels of activity have thus been observed with many antibody fragments labelled with metallic isotopes, including Fab-, $F(ab')_2$- and Fv-based fragments (Brown *et al.*, 1987; Schott *et al.*, 1992). Similar high kidney uptake takes place with DFM (King *et al.*, 1994). However, very interesting data have been obtained with chemically cross-linked tri-Fab conjugates.

Chemically cross-linked tri-Fabs (TFM) have been prepared with several different antibodies using tri-maleimide linkers (Figure 4.3(b)). These molecules localise to tumour cells at high levels, clear from the blood relatively rapidly and yet do not give rise to unacceptably high kidney levels when labelled with ^{90}Y (Figure 4.3(c)). In addition, linkers have been developed for the attachment of ^{90}Y to the cross-linker, which allows site-specific labelling to high specific activity, suitable for RIT, with no loss of antigen-binding activity (King *et al.*, 1994; Antoniw *et al.*, 1996). Pharmacokinetic studies of chimeric B72.3 TFM revealed a longer alpha phase but a similar beta-phase half life to $F(ab')_2$ and DFM (King *et al.*, 1994). The rapid blood clearance of TFM is surprising considering that this molecule has a molecular weight similar to that of intact IgG. However, the absence of an Fc region means that FcRn-mediated recycling does not operate, as it does for IgG, to maintain blood levels (see Section 2.7) and it also suggests that clearance through the kidney is not solely dependent on molecular weight but that shape and possibly charge of the molecule may also be important. The longer pharmacokinetic alpha phase probably contributes to the increased tumour levels obtained relative to other fragments, but TFM molecules also show increased avidity for antigen due to the presence of an extra binding site (King *et al.*, 1994, 1995). Comparative radioimmunotherapy studies in nude mice bearing human tumour xenografts have been carried out with IgG and TFM of the anti-tumour antibody A33 (Antoniw *et al.*, 1996). Results demonstrated that complete cures of established tumours could be achieved with both A33 IgG and TFM when labelled with ^{90}Y. Higher doses were required for TFM to achieve the same therapeutic effect, but these resulted in less toxicity. Humanised versions of A33 IgG and TFM have also been prepared to allow repeat dose human studies (King *et al.*, 1995). Similar results have also been observed for a TFM prepared from the antibody A5B7, suggesting that in this case also ^{90}Y-TFM is an attractive agent for RIT (Casey *et al.*, 1996).

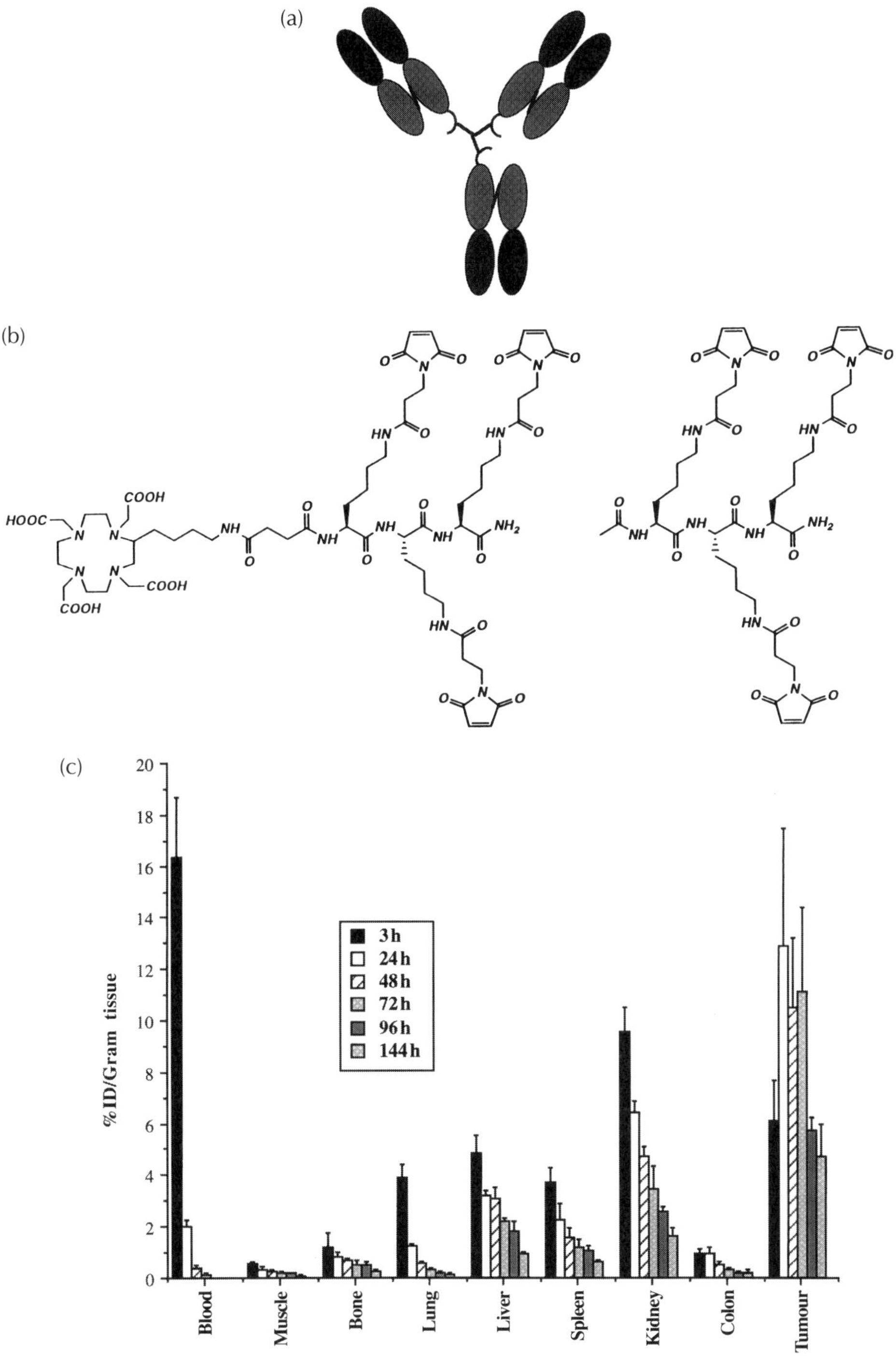

Figure 4.3 Tri-Fab′ production allows improved tumour targeting compared to IgG: (a) tri-Fab′ produced through chemical cross-linking of hinge region cysteines; (b) linkers used to produce tri-Fab′ including one containing a macrocycle for site-specific attachment of therapeutic radioisotopes; (c) biodistribution of yttrium-90-labelled chemically cross-linked tri-Fab′ from the anti-tumour antibody A33 in nude mice bearing SW1222 human tumour xenografts (tissue distribution was determined at 3, 24, 48, 72, 96 and 144 hours post-injection, n = 4, error bars indicate standard deviations)

Table 4.5 Some potentially useful radioisotopes for radioimmunotherapy

Radionuclide	Emission	Half-life	E (MeV)	γ (keV)
^{131}I	β	193 h	0.61	364
^{90}Y	β	64 h	2.25	—
^{67}Cu	β	62 h	0.4, 0.48, 0.58	185
^{186}Re	β	89.2 h	1.07, 0.94	137
^{188}Re	β	17 h	2.12, 1.96	155
^{177}Lu	β	162 h	0.50	210
^{111}Ag	β	179 h	1.04	342
^{199}Au	β	75 h	0.29, 0.25	159
^{221}At	α	7.2 h	5.87, 7.45	—
^{212}Bi	α	60.5 min	6.09–8.79	727
^{125}I	Auger	59 days	—	35
^{195m}Pt	Auger	96 h	—	65–130
^{64}Cu	β Auger positron	12.8 h	0.19	511

Radioisotopes

In contrast to imaging applications, radioisotopes for radioimmunotherapy are required to deliver a sterilising dose of radiation to the tumour. Therefore isotopes with a high linear energy transfer (LET, the energy deposited by the isotope over its pathway) are desirable. Three different types of radioisotopes have been investigated: Auger emitters, alpha emitters and beta emitters. Potentially, Auger emitters offer the ability to deliver the highest dose of radiation to the tumour cell but only over a range of less than one cell diameter. Therapy with Auger emitters therefore requires the isotope to be taken into the cell and localised at the nucleus. Alpha emitters also have the potential to deliver a high radiation dose over a short range. In this case the range is about 50 μm in tissue, so little 'bystander killing' of adjacent cells can be expected. Thus both Auger and α emitters require a very homogeneous distribution of the antibody at the tumour to allow a good therapeutic effect. Such homogeneous distribution may be difficult to achieve in many cases. β emitters deposit less energy per cell, but have a longer range and therefore do not require such a homogeneous distribution in tumour tissue. Ranges of β particles in tissue can be over 1 cm for high-energy β emitters such as ^{90}Y. This is favourable for larger tumours although when targeting small micrometastases, much of the energy may be deposited outside the tumour in the surrounding tissue. Candidate radioisotopes for use in RIT are given in Table 4.5.

In addition to suitable energy, candidate radioisotopes also need to fulfil other criteria to allow their use for RIT. A suitable half-life is required which can be matched to the biological half-life of an antibody delivery vehicle, the isotope must not have any hazardous daughter products, suitable methods for stable attachment to antibody must be available and it must be available carrier-free to allow preparation of high specific activity immunoconjugates. Many useful β emitters also have a gamma emission. This is undesirable as the gamma emission contributes little to therapy although it can play a major role in the total body dose (Langmuir, 1992). Also, the presence of large amounts of penetrating gamma radiation may present a significant hazard to medical personnel, and in waste storage and disposal, often making patient handling difficult. However, the gamma emission does allow for imaging studies to be performed using the same labelled immunoconjugate which allows tumour dose estimates to be determined (DeNardo *et al.*, 1996a).

The majority of radioimmunotherapy studies to date have been carried out using ^{131}I as the therapeutic isotope. Although ^{131}I has a suitable β emission for therapy, it has a relatively long half-life (8 days) and an abundant γ emission, which are less than optimal for an ideal radioimmunotherapeutic. The reasons for the popularity of ^{131}I stem from the long history of its application in the treatment of thyroid malignancies, its ready availability, and the ease of radiolabelling antibodies with ^{131}I without any specialised chemistry. As mentioned above, one problem with radioiodinated antibodies is their relative instability *in vivo*. Several novel methods for radiolabelling antibodies with iodine have been developed in attempts to reduce deiodination or the loss of iodinated metabolites from cells. These include the use of *N*-succinimidyl-3-(tri-*n*-butylstannyl) benzoate (Schuster *et al.*, 1991), dilactitoltyramine, tyramine cellobiose and other non-metabolisable carbohydrate adducts (Ali *et al.*, 1990; Reist *et al.*, 1995) or compounds designed to retain a positive charge to aid intracellular retention (Reist *et al.*, 1996). However, such methods have not yet achieved the high specific activities achievable with conventional iodination methods, and the benefit of such labelling methods in clinical studies remains to be determined. One potential effect of such 'residualising' iodine labelling methods is that the biodistribution of iodinated antibodies will resemble those of radiometal-labelled antibodies, making it difficult to see any advantage for residualising iodine labels over radiometals with more favourable physical properties. Indeed, recent data suggest exactly this, with a similar biodistribution and therapeutic advantage seen to that of radiometals (Stein *et al.*, 1997).

Many alternative isotopes for tumour therapy are now becoming increasingly available together with methods for their attachment to antibodies. Several of these alternative metallic isotopes offer potential advantages, for example in the amount of energy deposited per unit of administered activity, in increased retention in tumour tissue and in the possibility of out-patient treatment with less specialised facilities required.

Apart from ^{131}I the most studied β emitting isotope for RIT has been ^{90}Y. The shorter half-life of ^{90}Y and the higher energy mean that the absorbed dose per unit of activity and the dose rate is higher for ^{90}Y than ^{131}I and the dose is delivered over a shorter time. This may be a particular advantage for RIT as the biological half-life of the antibody delivery agent should also be taken into account. Some useful theoretical examples have been described by Harrison *et al.* (1991). If the labelled antibody remains at the tumour site for infinite time then the percentage of the total dose (182 rad/g/μCi ^{90}Y, 127 rad/g/μCi ^{131}I) delivered over 10 days is 93% for ^{90}Y and 58% for ^{131}I. If a biological half-life of 2.5 days is assumed, the percentage of the dose delivered to tumour in 10 days by ^{90}Y is reduced from 93% to 45% and from 58% to 14.8% for ^{131}I. This may be further reduced for ^{131}I if a residualising label is not used, as described above. In addition, ^{90}Y is a pure β emitter with no gamma component, which is potentially advantageous in terms of patient handling.

In early studies with ^{90}Y attachment to antibody was achieved largely using the chelating group DTPA. However, in common with the imaging isotope, ^{111}In, DTPA is a poor chelator for *in vivo* use and significant leakage of ^{90}Y occurs resulting in deposition, in this case in the bone, leading to unacceptable toxicity at high doses. In clinical studies, administration of EDTA has been investigated to mop up free yttrium which has leaked from the DTPA based immunoconjugate (Stewart *et al.*, 1990). However, subsequently improved chelators have been developed for ^{90}Y, including DTPA derivatives with all eight coordination sites for ^{90}Y retained (Brechbiel *et al.*, 1986). These ligands demonstrated improved stability over DTPA in *in vivo* studies (Washburn *et al.*, 1991). A

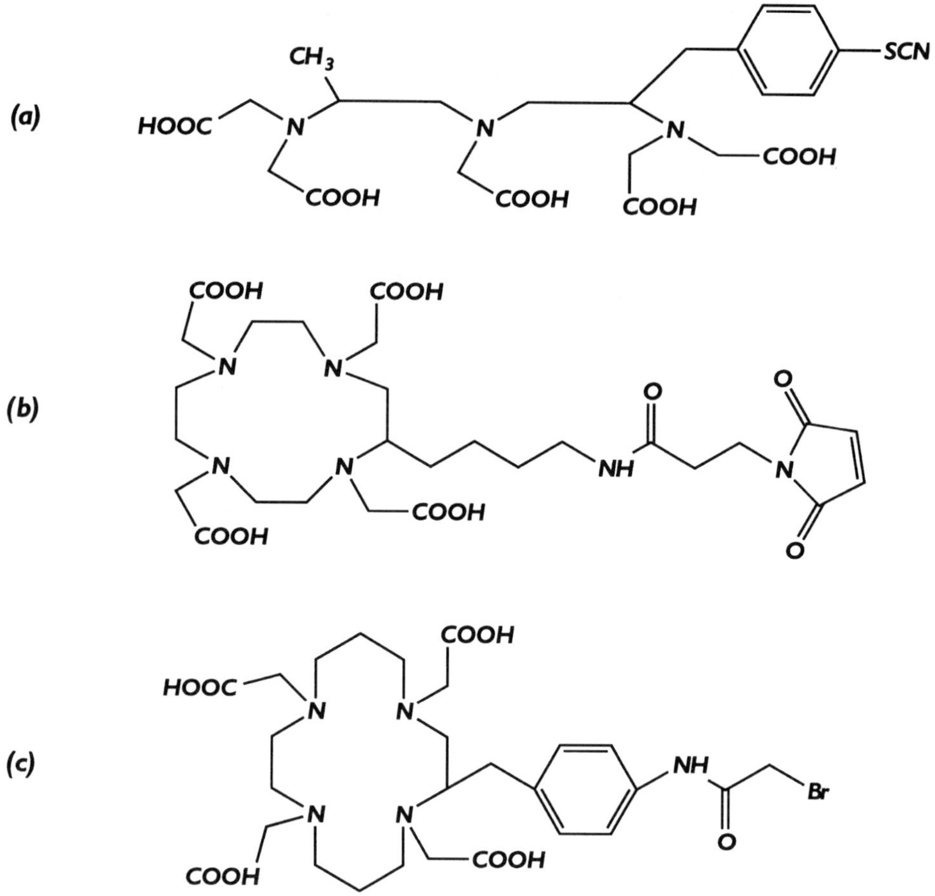

Figure 4.4 Structures of ligands used for attachment of therapeutic radioisotopes: (a) 1B4M-DTPA (Camera *et al.*, 1944); (b) 12N4-maleimide for labelling with yttrium-90 (Harrison *et al.*, 1991); (c) 14N4 derivative for labelling with copper isotopes (Morphy *et al.*, 1989)

further advance came with the development of stable macrocyclic ligands such as tetra-azacyclododecane tetraacetate (DOTA or 12N4) (Figure 4.4; Cox *et al.*, 1989; Deshpande *et al.*, 1990). Comparative biodistribution studies have demonstrated improved targeting with ^{90}Y bound through DOTA, and lower levels of activity in normal tissues, especially bone, which is of course important to prevent unnecessary bone marrow toxicity (Harrison *et al.*, 1991). Initial clinical studies using DOTA to chelate yttrium resulted in an immune response against the macrocyclic structure (Kosmas *et al.*, 1992) which led to renewed interest in less stable chelators such as the DTPA derivative CHX-DTPA (Figure 4.4). Studies on the stability and biodistribution of CHX-DTPA and other DTPA derivatives revealed that all were less stable than DOTA (Camera *et al.*, 1994). However, DTPA derivatives have been used to chelate ^{90}Y in clinical studies of RIT using an antibody to polymorphic epithelial mucin in breast carcinoma (Schrier *et al.*, 1995). Partial responses to treatment were observed, although it was necessary to provide autologous stem cell support in some patients to overcome bone marrow toxicity. More recent data suggest that immune responses against DOTA may be less of a problem than first thought: in a study in lymphoma patients immune responses to macrocycles were no more frequent than those seen with other chelating groups, and were less than those to the antibody (DeNardo *et al.*, 1996b).

The use of DOTA for chelation of yttrium has been combined with a humanised tri-Fab to produce an effective immunoconjugate for tumour therapy (Antoniw *et al.*, 1996). This conjugate was produced using a Fab′ cross-linker which incorporates the DOTA macrocycle, allowing site-specific labelling with ^{90}Y. Versions of cross-linkers have been produced with both DOTA and phosphinate DOTA derivatives, which are equally stable chelators of ^{90}Y (Norman *et al.*, 1995a). Methods for high-specific activity labelling have been developed which operate at >95% labelling efficiency, allowing realistic clinical use of the conjugate (Haines and King, unpublished data).

Stable macrocyclic chelators have also been developed for ^{67}Cu labelling, based on tetraazacyclotetradecane tetraacetate (TETA or 14N4, see Figure 4.4) (Cole *et al.*, 1987; Morphy *et al.*, 1989). ^{67}Cu is potentially an attractive radioisotope for RIT but its wide-spread application has been limited by poor availability. Nevertheless, antibodies labelled with ^{67}Cu via a macrocyclic ligand have been shown to be effective in therapy of tumours in nude mouse models (Connett *et al.*, 1996), and have shown improved biodistribution characteristics in colorectal cancer patients compared to iodinated antibody (Bischoff-Delaloye *et al.*, 1997). The DOTA macrocycle used for ^{90}Y can also be used for chelation of lutetium-177, and this complex has been investigated attached to the antibody CC49, which recognises a tumour-associated glycoprotein, both in animal models and for the treatment of ovarian cancer in patients (Meredith *et al.*, 1996). In this phase I study in patients, ^{177}Lu-CC49 was well tolerated and showed some evidence of anti-tumour activity, particularly in patients with microscopic disease.

Two isotopes of rhenium, ^{188}Re and ^{186}Re, are of interest for RIT (Table 4.4). Rhenium chemistry is very similar to that of the imaging isotope technetium (see Section 3.9.1), and therefore similar methods for attachment to antibody are used. These include direct labelling through reduced hinge thiol groups, and the use of chelators such as N_2S_2 or N_3S (see Figure 3.15). Encouraging results have been obtained in preclinical studies with ^{186}Re-antibody conjugates. Treatment of small cell lung cancer xenografts with antibody labelled with ^{186}Re via an N_3S chelator resulted in long-term tumour regressions in some mice (Beaumier *et al.*, 1991), and clinical studies with rhenium-labelled antibodies are in progress.

As an alternative to β emitting radioisotopes, α emitters emit their energy over a short range and are therefore attractive for treatment of tumours only if homogeneous distribution of antibody throughout the tumour can be achieved. However, the other implication of short-range isotopes is that relatively little normal tissue damage may result. Application of α emitters has been limited by the need to achieve a homogeneous distribution, and also by their short half-life and the unstable daughter products produced which make working with these isotopes a technical challenge. Nevertheless, α emitters have been examined in a number of animal model studies. α emitters may be well suited to RIT for therapy of easily penetrated tumours such as leukemias, micrometastases and malignancies with sheet-like geometry and free floating tumour cells, such as ovarian cancer or neoplastic meningitis where antibody can be administered directly into the body compartment. Astatine-211 labelled antibody has been shown to be effective in therapy of neoplastic meningitis in a rat model (Zalutsky *et al.*, 1994). Bismuth-212 is also a potentially useful isotope, which has shown efficacy in an adjuvant setting for i.p. tumours (Hartmann *et al.*, 1994). It has been suggested that ^{212}Bi-labelled anti-Tac antibody may be useful for therapy of T cell leukemias. Similarly to ^{90}Y, the biodistribution of bismuth-labelled antibody can be significantly improved by use of DOTA to chelate the isotope, compared to DTPA or DTPA derivatives such as CHX-DTPA (Junghans *et al.*, 1993). Studies so far have used whole antibody; however, because of the short half-life of ^{212}Bi, it is likely that the use of a rapidly clearing antibody fragment would be significantly better. Studies with Fab′ and F(ab′)$_2$ have been limited by their accumulation in the kidney, and so far experiments with improved antibody forms such as those mentioned above have not been reported.

Auger emitters which are potentially useful as therapeutic radionuclides for RIT include ^{125}I, ^{64}Cu, ^{195m}Pt and some isotopes also used for imaging at low doses such as ^{111}In , ^{99m}Tc and ^{67}Ga. As with alpha emitters, applications are likely to be restricted to conditions in which good tumour distribution is achieved. In this case, internalisation into the cell is required for efficacy and preferably location to the nucleus such that there is a good chance of this intense, short-range energy damaging cellular DNA. Therefore Auger emitters have also been investigated as potential therapeutic isotopes in an adjuvant setting, in leukemias and well-vascularised tumours. Because of the requirement for internalisation for cytotoxic effects, Auger emitters are potentially less toxic in the circulation than beta emitting isotopes. Modelling analysis has suggested that ^{195m}Pt-labelled antibody would be effective at delivering a sterilising dose to blood-borne tumour cells without delivering a toxic dose to bone marrow (Willins and Sgouros, 1995). Clinical studies have been carried out with several ^{125}I-labelled antibodies. Studies with ^{125}I-chimeric 17-1A and ^{125}I–murine A33 in colorectal cancer have demonstrated that high doses of this Auger emitting radionuclide can be given without significant toxicity, although no, or very modest, anti-tumour effects were observed in either case (Meredith *et al.*, 1995; Welt *et al.*, 1996).

^{64}Cu is an attractive isotope as a combined Auger and beta emitter which can be used to radiolabel antibodies using the TETA macrocycle, as for ^{67}Cu. Comparisons of ^{64}Cu and ^{67}Cu-labelled antibody for RIT showed that both had similar anti-tumour effects with good therapeutic effects against small tumours (Connett *et al.*, 1996). ^{64}Cu is also under investigation as an isotope for radioimmunodetection using PET (see Section 3.9). The half-life of ^{64}Cu suggests that improved RIT results might be expected using an antibody fragment. Initial biodistribution experiments have used F(ab′)$_2$, although kidney accumulation of this fragment has precluded RIT studies (Anderson *et al.*, 1995).

An approach to improve therapeutic effects with Auger emitters is to target the isotope to the cell nucleus. One attempt at achieving this has been to prepare a synthetic linker comprising a macrocycle for radioisotope chelation with a known DNA intercalator, such as an acridine (Norman *et al.*, 1995b). When internalised into cells it is likely that many antibodies will be degraded, but macrocyclic ligands are known to be retained in the cell. A DNA intercalator may then bind to DNA in the cell nucleus and hence retain the isotope where the Auger emitter may result in more potent cytotoxic effects.

Clearance mechanisms and two-step targeting

Clearance mechanisms and two-step tumour targeting strategies for RAID have been described in Chapter 3. Many of these strategies have also been suggested as potential approaches to improving the therapeutic ratio achievable in RIT. Clearance mechanisms which clear circulating isotope *in vivo* by the administration of second antibody may reduce bone marrow toxicity but are likely to deliver a large dose to the organ of clearance, usually the liver. Hence immunoadsorption devices have been developed (Norrgren *et al.*, 1991). In these systems, radiolabelled antibody is allowed to bind to tumour antigen and then circulating antibody is removed by passing blood through an extracorporeal immunoaffinity chromatography column. For example, an anti-mouse antibody can be immobilised as an immunoaffinity material and packed into a column format (see Section 3.7 for more details of immunoaffinity chromatography). Blood can then be pumped outside the body and through the column which binds radiolabelled murine antibody before the blood is returned to the patient. An alternative system of potential application to human and humanised antibodies is to use biotinylated, radiolabelled antibody for RIT. This can then be removed through an extracorporeal column of immobilised avidin, sparing the radiation dose to bone marrow from circulating activity (Garkavij *et al.*, 1997).

Two-step targeting protocols are also under development for RIT. Antibodies to chelate–isotope complexes, e.g. DOTA–^{90}Y, have been developed which may allow production of bispecific reagents which can be localised to the tumour and allowed to clear from circulation (Goodwin *et al.*, 1994). Radiolabelled chelate can then be injected which is bound to the anti-chelate antibody at the tumour site and readily excreted from the rest of the body. Potential problems with this approach are that a high-affinity interaction is required to bind the radiolabelled ligand from a relatively low concentration in blood as it passes the tumour site. The high-affinity interaction of the avidin:biotin system has therefore been employed in attempts to improve tumour loading levels of isotope. Streptavidin–antibody conjugate is localised to tumour tissue and circulating conjugate cleared using a biotin–protein conjugate. ^{90}Y–DOTA–biotin is then administered which localises to the tumour site and clears from the rest of the body, generating high tumour : blood ratios. Promising results demonstrating cures of human tumour xenografts in nude mice have been presented, and clinical evaluation of this approach is underway (Axworthy *et al.*, 1994).

Clinical results with RIT

The best clinical results with RIT to date have been obtained in leukemias and lymphomas. These hematopoietic tumours have the advantage of being readily accessible to intravenously administered antibody and are often relatively sensitive to radiation. Particularly encouraging results have been obtained in trials for the therapy of non-Hodgkin lymphoma. A high proportion of patients have been found to respond to treatment in phase II studies of ^{131}I–B1, a MAb which recognises CD20 (Press *et al.*, 1995; Kaminski *et al.*, 1996). In these studies patients were screened with a tracer dose of ^{131}I–B1, and those with a favourable biodistribution progressed on to therapy. Therapy with non-myeloablative doses resulted in long lasting, complete responses in 50% of patients (Kaminski *et al.*, 1996). Higher doses could also be used together with autologous bone marrow transplantation to cope with the toxicity of the radiolabelled IgG. In this case responses in a phase II trial were seen in 18 of 21 patients, including 16 complete remissions (Press *et al.*, 1995). Responses have also been observed with ^{90}Y-labelled anti-CD20 antibody therapy of lymphoma. In a dose-escalation study, a dose of up to 40 mCi was found not to be myeloablative and resulted in an overall response rate of 78% even though in this study patients were not pre-screened for a favourable biodistribution (Knox *et al.*, 1996).

Promising results have also been obtained in leukemia studies. An early trial of ^{90}Y-labelled anti-Tac antibody in adult T cell leukemia resulted in responses in 9 of 16 patients treated although a relatively low dose was used (Waldmann *et al.*, 1995). Several studies have been carried out with antibodies to CD33 in acute myeloid leukemia. Both murine and humanised versions of M195 have shown significant anti-tumour effects when labelled with ^{131}I, including complete remissions in several patients. This has led to the incorporation of these agents into preparative regimens for bone marrow transplantation in which complete remissions were observed in most patients (Jurcic *et al.*, 1995). The drawback of this approach is the myelosuppression seen with high doses of radiolabelled antibody. Complete responses have also been obtained in studies with another anti-CD33 antibody, P67, when labelled with ^{131}I (Applebaum *et al.*, 1992). As this antibody internalises into leukemic cells it is also under investigation for targeting the cytotoxic drug calicheamicin, in an approach to reduce myelosuppression (Section 4.2.7).

Results in the major solid tumour types (colorectal, lung, breast and ovarian cancer) have been less impressive, with few tumour responses. Occasional complete responses to treatment have been reported (e.g. Juweid *et al.*, 1997), but these are the exception rather than the rule. Several trials in colorectal cancer have resulted in only minor responses. In a phase II study of the anti-TAG72 antibody CC49 labelled with ^{131}I, no significant tumour responses were observed (Murray *et al.*, 1994). Another anti-TAG72 antibody, chimeric B72.3, resulted in stable disease in 4 of 12 patients treated with ^{131}I-labelled antibody, with one minor response, approx. 40% reduction in tumour size, seen in a repeat dose study (Meredith *et al.*, 1992b). Antibodies to CEA have also been extensively studied in colorectal cancer. Treatment with ^{131}I–A5B7 resulted in one partial response in a lung metastasis and one complete resolution of liver metastasis (Lane *et al.*, 1994). In a study of ^{131}I-labelled chimeric L6 in 10 women with metastatic breast cancer, partial responses of up to 5 months' duration were seen in four patients, with two other minor responses (DeNardo *et al.*, 1994). Following on from successful results for lymphoma when combining RIT with bone marrow support to overcome the toxicity of the circulating labelled IgG, a study has been carried out in breast cancer with ^{90}Y-labelled BrE3 antibody to polymorphic epithelial mucin (Schrier *et al.*, 1995). In this study of 9 patients, 4 required bone marrow support and partial responses were observed in approximately 50% of patients treated.

There is now increasing evidence that although doses delivered to large tumour deposits are inadequate for therapy, significant tumour doses can be achieved to small tumours. Investigations have been made into the use of RIT as an adjuvant therapy, for the treatment of minimal residual disease following surgical excision of primary tumour, or for treatment after tumour reduction by conventional chemotherapy. A clear correlation between tumour response and size of the tumour has been demonstrated in both animal models and patients, presumably as a result of better tumour penetration into smaller tumours. With ^{131}I-antibody administered intraperitoneally for ovarian cancer, no responses were seen in patients with tumours of diameters greater than 2 cm, two responses were seen from 15 patients with smaller tumours (13% response) and three of six (50% response) patients with malignant cytology but no measurable disease had their malignant cells cleared from the peritoneum (Stewart *et al.*, 1989). Encouraging results have also been observed in a study of ^{131}I-anti-CEA $F(ab')_2$ in patients with tumours less than 3 cm in diameter (Juweid *et al.*, 1996). Disease stabilisation was observed in approximately half of the patients treated with some tumour shrinkage. However, strong HAMA responses to this murine antibody fragment prevented re-treatment. In another ovarian cancer study with intraperitoneally administered ^{131}I-labelled antibody, treatment of minimal residual disease resulted in complete responses in 5 of 16 patients with a mean disease-free period of over 10 months (Crippa *et al.*, 1995). Also in ovarian cancer, a study with ^{90}Y-antibody to polymorphic epithelial mucin resulted in increased survival in patients treated with the antibody and who had no measurable disease at the time of treatment (Hird *et al.*, 1993). These patients were viewed as having received adjuvant therapy for the minute micrometastases present, and were compared to historical control groups of patients. Over four and a half years, more than 90% of patients treated survived compared to less than 40% in a large group of historical controls. It is not known how much of this effect is due to cytotoxicity from the radioisotope or to biological effects of the antibody administered, such as the induction of an anti-idiotype response (see Section 4.2.2), and these promising results are thus the subject of further study.

Although initially disappointing, RIT is now beginning to prove clinically useful in some disease situations, particularly the liquid tumour types and as an adjuvant therapy or for treatment of minimal residual disease. The majority of agents tested to date have been murine antibodies, often labelled with ^{131}I, which have not taken advantage of the ability to develop improved targeting agents through the developments in molecular biology and improved chemistry mentioned above. As these developments allow improved agents for RIT to be brought forward for clinical testing, it is likely that clinical benefit will be seen in an increased range of diseases. Combinations of RIT with conventional cancer therapy by chemotherapy or external beam radiation are already beginning to be explored in experimental systems, with synergistic effects observed, allowing better therapy than either approach alone (Tschmelitsch *et al.*, 1997; Vogel *et al.*, 1997).

4.2.6 ***Immunotoxins***

Protein toxins are extremely potent cell killing agents and have been widely investigated as potential cytotoxic agents for antibody targeted cancer therapy. Antibody–toxin conjugates, or immunotoxins, are being evaluated with plant toxins such as ricin and abrin and bacterial toxins such as diptheria toxin (DT) and *Pseudomonas* exotoxin (PE) (Table 4.6). The target of these toxins is the inhibition of protein synthesis. In the case of the plant toxins this is achieved through inactivation of ribosomes by modification of ribosomal RNA, whereas the bacterial toxins inactivate the elongation factor, EF-2, which normally interacts with the ribosomal RNA.

Immunotoxins require internalisation into the cell to exert their cytotoxic effect; however, protein toxins are enzymes, and therefore catalytic, and small numbers of toxin molecules may be sufficient for cytotoxicity. The mode of action of toxins (Figure 4.5) requires not only internalisation but translocation of the toxin across a membrane of the endosome or the trans-golgi network (depending on the toxin used) to reach the site of action in the cytosol. Therefore, although under some conditions a single toxin molecule may be sufficient to kill a cell (Yamaizumi *et al.*, 1978), many more must be delivered to the cell surface as internalisation and translocation to the site of action are inefficient. Many toxins have a subunit or domain of their own which is responsible for cell binding and internalisation and this must be removed or blocked to reduce non-specific toxicity. PE has three domains, one of which is responsible for cell binding. Truncated versions of

Table 4.6 Toxins under evaluation for the preparation of immunotoxins

Plant toxins	Bacterial toxins
Ricin	*Pseudomonas* exotoxin (PE)
Abrin	Diptheria toxin (DT)
Gelonin	
Saporin	
Pokeweed antiviral protein (PAP)	

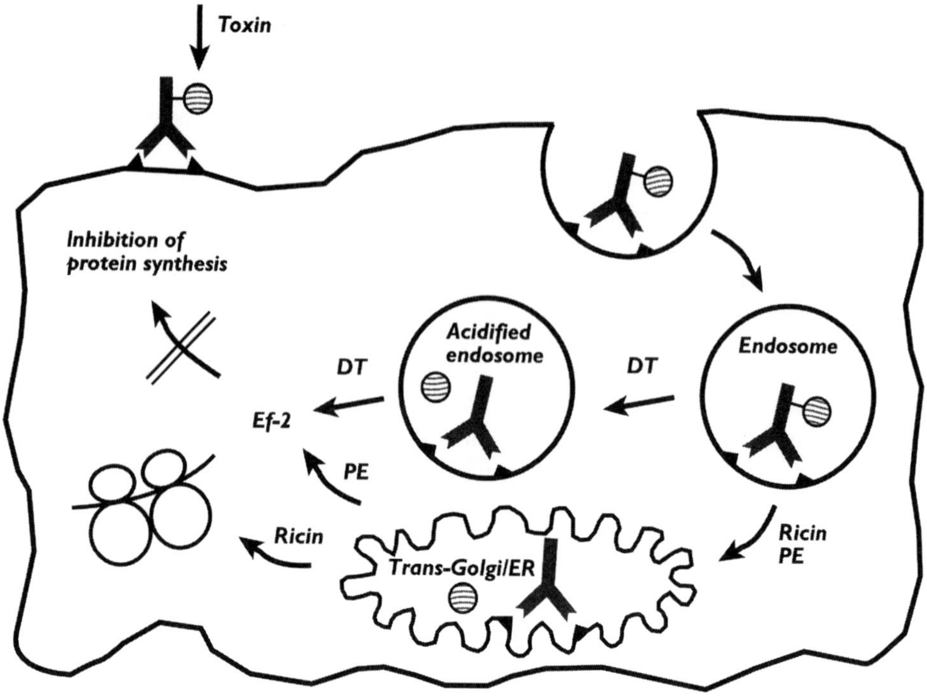

Figure 4.5 Illustration of the mechanism of cytotoxicity of immunotoxins. Immunotoxins bind to cell surface antigen and are internalised by endocytosis. Toxins such as ricin A-chain and *Pseudomonas* exotoxin are routed to the trans-golgi/endoplasmic reticulum network whereas diptheria toxin is routed to an acidified endosome. In these compartments release of toxin is achieved which can be translocated to the cytoplasm and inhibits protein synthesis.

PE have been expressed comprising the catalytic and translocation domains of the toxin, but not the cell binding domain. Ricin comprises an A- and B-chain, the A-chain being responsible for catalytic inactivation of ribosomes whereas the B-chain is responsible for cell binding and internalisation. Immunotoxins have therefore been constructed using ricin A-chain alone, which must be stringently purified away from B-chain and intact toxin. An alternative approach is to block the cell binding site using an affinity ligand. Blocked ricin conjugates have been prepared by covalent linkage of an affinity ligand (Lambert *et al.*, 1991). Blocked ricin conjugates have the advantage that the intact toxin is retained with its full ability to translocate into the cytosol, a property which is reduced in A-chain conjugates.

Immunotoxins can be prepared either by chemical linkage or by expression of recombinant fusions of the toxin gene to the antibody gene. Linkage of toxin to antibody must be stable enough to remain intact in the circulation, but be able to release free toxin inside the cell to allow cytotoxicity. Toxins such as PE contain an internal disulphide bond which is reduced as part of its mechanism of release inside the cell. Stable linkages of PE to antibody therefore result in effective immunotoxins, and these have been prepared through thioether linkages or as recombinant fusions (Pai *et al.*, 1996; Benhar and Pastan, 1995). Ricin A-chain is normally linked to B-chain through a disulphide linkage which is also reduced intracellularly. This cysteine on the A-chain has thus been the site of conjugation to antibody, in the place of B-chain. Initial work with disulphide linked ricin A conjugates resulted in constructs which were unstable in the circulation, therefore linkers containing 'hindered' disulphide linkages were developed with a methyl or phenyl group adjacent to the disulphide. These are more stable in circulation but still effectively reduced inside the cell (Thorpe *et al.*, 1988). An additional problem with plant toxins such as ricin is their glycosylation which promotes binding to cells of the reticuloendothelial cell system, particularly the liver. Therefore chemically deglycosylated ricin, or A-chain produced by recombinant means in *E. coli* has been used to avoid non-specific toxicity through carbohydrate binding (Thorpe *et al.*, 1988). Alternative linkage of ricin A-chain has been achieved through proteolytically sensitive peptides in a recombinant ricin A-chain fusion protein. When internalised, the proteolytically sensitive site is cleaved to generate free ricin A-chain which is then cytotoxic (O'Hare *et al.*, 1990).

The optimal form of antibody for use in immunotoxins has been less well studied. Intact IgG conjugates can be efficiently internalised and have the highest degree of accumulation at the tumour due to their long circulating half-life. One drawback is their large size which reduces penetration into tumour tissue. Antibody fragments may therefore be advantageous in achieving more effective penetration into tumour tissue. In animal model experiments, antibody fragment toxin conjugates have proved to be superior to intact IgG. Conjugates of BR96, an antibody which recognises a variant of the Le^y antigen on human carcinomas, have been prepared with the intact IgG and PE40 (truncated *Pseudomonas* exotoxin) and as a recombinant scFv fusion protein (Friedman *et al.*, 1993). The IgG conjugate was able to retard the growth of human tumour xenografts in mice, whereas the scFv fusion protein was capable of inducing complete regressions. Similar results have been obtained with immunotoxins prepared with other antibodies. An scFv–PE immunotoxin of the antibody B3, which recognises the same antigen, was shown to be more effective than an IgG conjugate at producing cures of carcinomatous meningitis in a rat model (Pastan *et al.*, 1995). Similarly, a Fab′–PE fusion protein was also shown to be more effective than the equivalent IgG based immunotoxin (Debinski and Pastan, 1995). Both scFv–PE and dsFv–PE immunotoxins have been prepared and shown to be equipotent in their ability to cause long-term regressions of tumour xenografts (Benhar and Pastan, 1995). The scFv–PE conjugate was less stable yet retained more of the binding activity of the IgG, whereas the dsFv was more stable but less immunoreactive, with the overall result that they were approximately equipotent *in vivo*.

Bispecific antibodies have also been suggested as means of delivery of toxins to tumour cells. Antibodies in which one arm recognises a lymphoma-associated antigen and one the toxin saporin have been produced, and complexes of the bispecific antibody with saporin shown to be capable of cell killing (Bonardi *et al.*, 1993). A potential advantage of this approach is that the toxin is held via non-covalent forces in the antibody-binding site and is therefore not damaged through chemical modification. Also, the toxin may be released more easily on internalisation into the cell, and non-specific binding sites on the toxin may be blocked through the antibody interaction. A small clinical study using this aproach for lymphoma therapy has been carried out using an anti-CD22, anti-saporin bispecific di-Fab (French *et al.*, 1995). In this study all of the four patients treated appeared to show rapid and beneficial responses to treatment with only mild toxicity.

Clinical studies have been carried out with a number of different immunotoxins produced by direct linkage of the toxin to antibody (reviewed in Ghetie and Vitetta, 1994). Anti-tumour activity has been reported in lymphomas, and recently in breast and colorectal tumours (Pai *et al.*, 1996). In this phase I study of 38 patients objective anti-tumour activity was observed in 5 patients, 18 had stable disease and 15 progressed. Two major problems have limited efficacy in this and other studies: immunogenicity and toxicity.

The major toxicities seen in immunotoxin trials have been hepatotoxicity, neurotoxicity and often the most limiting, vascular leak syndrome. Vascular leak syndrome is probably caused through effects on endothelial cells, and research is ongoing to attempt to overcome this problem through engineering or inhibition with anti-inflammatory drugs (Siegall *et al.*, 1997).

Protein toxins used to date have been derived from plant or bacterial sources and are highly immunogenic. This immunogenicity has limited the number of doses which can be given in patients and therefore limits anti-tumour effects. Attempts to overcome this through engineering, immunosuppressive agents or PEGylation are being explored. One approach is to use a human enzyme as a toxin which would hopefully be less immunogenic in man. Human ribonucleases which kill cells only after internalisation have been explored as potential immunotoxins both by chemical conjugation to antibody and as recombinant fusion proteins (Gadina *et al.*, 1994). However, the cytotoxicity of such 'humanised immunotoxins' is poor compared to those produced with ricin or PE. Attempts to design improved immunotoxins using scFv–RNAse fusion proteins are a subject of current research (Newton *et al.*, 1996).

4.2.7 Drug conjugates

Chemotherapy using cytostatic or cytotoxic drugs has been the mainstay of cancer therapy over the past 50 years. However, there is little selectivity of these drugs for tumour cells over rapidly dividing normal cells, such as those in the bone marrow, and consequently chemotherapy is limited by normal tissue toxicity, and usually results in severe toxic side-effects. The ability to target chemotherapeutic drugs such that they are inactive until taken into tumour cells is thus an attractive concept with the potential to allow higher doses to tumour cells without exposing normal cells to high concentrations of drug. In common with immunotoxins, drug conjugates require internalisation to be able to kill cells, but they have the potential advantage over toxins of being small molecules which are less likely to raise an immune response. Attempts to address this problem have suggested that choice of suitable chemical linkers, and the use of humanised antibodies, may result in non-immunogenic or only weakly immunogenic conjugates (Johnson *et al.*, 1991a). The other major problem of immunotoxins, non-specific toxicity, may also be overcome with the use of drug conjugates, as many drug derivatives may be screened to identify optimal characteristics.

Initial attempts to produce conjugates of cytotoxic drugs with anti-cancer MAbs focused on drugs used in the clinic. Conjugates have been prepared with many of these, including methotrexate, doxorubicin, mitomycin C, neocarzinostatin, daunorubicin and vinca alkaloids. Early attempts suffered from loss of activity of both the antibody and drug on conjugation, and relatively poorly active immunoconjugates resulted. Several problems were identified, including identifying drugs of sufficient potency given the inefficiency of antibody uptake by solid tumours, and identifying suitable chemistry to generate linkage of the antibody to the drug without a major loss of activity. Since these early experiments, improved linkage methods have been developed (reviewed by Pietersz *et al.*, 1994). The best linkage chemistry varies from drug to drug, but a key feature is to allow stable conjugates in circulation which can also release active drug during antibody degradation inside the cell. In addition, site-specific linkages to antibody have been investigated, for example, to antibody carbohydrate (see Section 2.6.2) which allow retention of antibody activity (Schrappe *et al.*, 1992). However, the low potency of many of these drugs has required many molecules of drug to be coupled per antibody to achieve sufficient doses for therapeutic effects. Nevertheless, some good anti-tumour effects have been observed with optimised conjugates of several of these drugs in animal models. Long-term growth suppression of glioma xenografts have been observed with conjugates of vinblastine (Schrappe *et al.*, 1992) and cure of xenografted lung, breast and ovarian tumours has been observed in some mice with doxorubicin conjugates of the anti-Le^y antibody BR96 (Trail *et al.*, 1993). In general, however, such responses have required treatment before establishment of the tumour or the use of doses close to the maximum tolerated dose, which may lead to limited benefits over non-conjugated drug in the clinical situation.

Attempts to improve potency have included the use of polymeric carriers to conjugate large numbers of molecules of drug to antibody without loss of antigen-binding properties. In this strategy the polymeric carrier is conjugated to many molecules of drug before site-specific attachment of a single polymer to the antibody. Polymers can be synthetic, proteins such as HSA or carbohydrate such as dextran (Shih *et al.*, 1991). However, results with such conjugates have been disappointing, with little real improvement over directly linked conjugates observed either *in vitro* or *in vivo.*

Attempts have also been made to produce immunoliposomes to target large amounts of the drug to the tumour site. Liposomes are synthetic vesicles consisting of one or more concentric phospholipid bilayers which can be used to encapsulate an aqueous compartment, which can be filled with drug. Antibody can be attached to the surface of the liposome, resulting in an immunoliposome. Interaction of the liposome with the target cell either by direct fusion with the cell membrane or by endocytosis should result in the release of large numbers of drug molecules to the tumour cell. However, immunoliposomes are rapidly taken up by the reticuloendothelial system *in vivo*, and thus PEG-modified phospholipid is often incorporated into the liposome, resulting in 'sterically stabilised' or 'stealth' immunoliposomes which are less prone to reticuloendothelial uptake and can be used to target to tumour cells (Allen *et al.*, 1995). Attempts to deliver doxorubicin using immunoliposomes have shown promising tumour targeting, although little improvement over free doxorubicin has been demonstrated in some studies (Park *et al.*, 1995). Immunoliposome mediated targeting of drugs may find a niche, however, in the treatment of newly established micrometastatic disease or metastatic cells migrating in blood or lymph (Allen *et al.*, 1995).

More impressive results have come with the development of more toxic cytotoxic agents. Several classes of extremely potent new drugs have been identified, at least 100-fold more cytotoxic than drugs used for chemotherapy in the clinic. Such drugs are too toxic for use as the free drug, but can give rise to very potent immunoconjugates. These are principally the enediynes (Nicolaou *et al.*, 1993), maytansinoids (Chari *et al.*, 1992), tricothecenes (Liu, 1989) and analogues of CC-1065 (Chari *et al.*, 1995). Immunoconjugates have been produced with these agents which have impressive anti-tumour effects in mouse xenograft experiments. Calicheamicin, a member of the enediynes, has been conjugated to an anti-PEM antibody, CTM01, and shown cures of established breast and ovarian tumour xenografts in nude mice (Hinman *et al.*, 1993). In this case, several derivatives of calicheamicin were compared to identify the optimal drug for conjugate activity without toxicity. A hindered disulphide was used to allow a stable linkage in circulation which was labile inside the cell, a strategy also used for ricin immunotoxins (see Section 4.2.6). This antibody has been humanised (Baker *et al.*, 1994), and humanised CTM01–calicheamicin conjugates prepared which retain the potent anti-tumour effects of the murine antibody. Clinical trials with this conjugate are under way. A similar calicheamicin conjugate has been developed for the anti-CD33 antibody, hP67. Early clinical studies with this molecule for the therapy of acute myeloid leukemia have revealed an excellent safety profile and encouraging anti-tumour effects, with several complete and partial responses seen (Bernstein, 1996). Similarly potent anti-tumour effects have been observed with maytansinoid conjugates for therapy of colorectal cancer (Liu *et al.*, 1996). Eradication of established xenograft tumours was achieved even though antigen distribution was heterogeneous, a model in which immunotoxins with PE linked to the same antibody performed relatively poorly. Such potent drug immunoconjugates thus represent considerable promise for cancer therapy.

4.2.8 Antibody-directed enzyme prodrug therapy (ADEPT)

ADEPT is a two-step targeting method for cytotoxic drug therapy. Similarly to drug conjugates, ADEPT aims to improve the therapeutic ratio of cytotoxic drugs using antibody targeting, but in this case an enzyme is attached to the antibody and targeted to tumour tissue. The enzyme is then used to convert an inactive prodrug to a cytotoxic drug which generates a high local concentration of drug at the tumour site (Figure 4.6). This an attractive concept, as less antibody conjugate may be required at the tumour site due to the enzyme acting catalytically and being able to turn over many molecules of prodrug to drug. Also, there is less need to target every cell as killing of bystander cells should be possible by the high local drug concentration. The disadvantages of this approach include the fact that free drug is generated which may diffuse away from the tumour site, damaging non-target tissues. Also, an enzyme must be chosen which does not occur in humans, or at least not in extracellular locations, to avoid activation of the prodrug in normal tissues.

Several different enzyme systems have been investigated for ADEPT (Table 4.7) (reviewed by Melton and Sherwood, 1996). The first system to be developed was based on carboxypeptidase G2 (CPG2) cleaving a glutamate derivative prodrug to liberate a nitrogen mustard drug (Searle *et al.*, 1986). Subsequently this and other enzyme systems have shown anti-tumour effects in mouse xenograft experiments. Studies with an anti-CEA F(ab′)$_2$ linked to CPG2 revealed only modest anti-tumour effects at first. This was improved by the introduction of a third step to the system, a clearing antibody (Sharma *et al.*, 1990). In this revised three-step system the antibody–enzyme conjugate is administered and allowed to localise to the tumour. After localisation has taken place residual

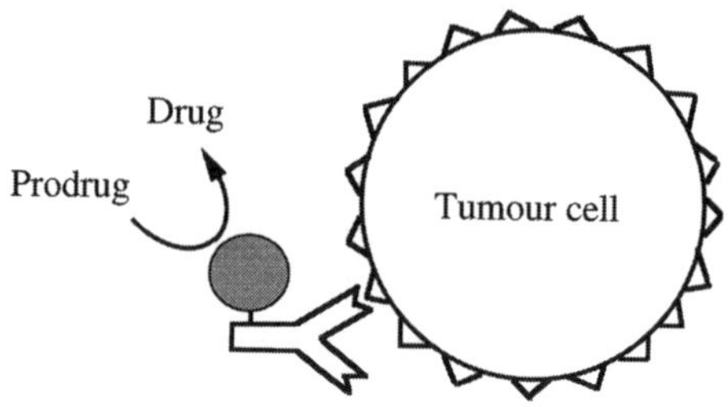

Figure 4.6 Conversion of prodrug to active drug through ADEPT. Antibody–enzyme conjugate is localised to tumour and allowed to clear from the circulation. Inactive drug is then administered which is converted to a cytotoxic drug by the enzyme attached to antibody at the tumour site. This results in a high local concentration of drug with resultant tumour cell cytotoxicity

Table 4.7 Enzymes used for construction of antibody–enzyme conjugates for ADEPT

Enzyme	Origin	Active drug	Substrates
Carboxypeptidase G2	*Pseudomonas*	Benzoic acid mustards	Glutamic acid derivatives
Carboxypeptidase A	Bovine	Methotrexate	Phenylalanine derivatives
Alkaline phosphatase	Bovine	Etoposide, doxorubicin, phenol mustards	Phosphate derivatives
β-lactamase	*E. coli* *Enterobacter*	Nitrogen mustards, vinblastine, taxol	Cephalosporin derivatives
Cytosine deaminase	Yeast	5-fluorouracil	5-fluorocytosine
Penicillin amidase	*Fusarium*	Doxorubicin, melphalan	Hydroxyphenylacetyl derivatives
β-glucuronidase	*E. coli*, human placental	Doxorubicin, epirubicin, aniline mustards	Glucuronide derivatives
β-glucosidase	Sweet almond	Hydrogen cyanide	Amygdalin

circulating conjugate is removed by the administration of a clearing antibody against the CPG2 enzyme. This clearing antibody was galactosylated to allow rapid clearance through galactose receptors in the liver. After clearance of circulating conjugate the prodrug is administered for activation at the tumour site. This approach was found to reduce non-specific activation of prodrug and allowed successful anti-tumour effects in xenografts to be seen.

Early conjugates were prepared by chemical coupling of IgG or $F(ab')_2$ to the enzyme of interest, but subsequently recombinant constructs have been designed and expressed as scFv–enzyme fusions (Goshorn *et al.*, 1993). These have the advantage of smaller size which may assist penetration into tumour tissue and more rapid clearance which may avoid the need for clearing antibody. scFv–β-lactamase conjugates were expressed in *E. coli* in an active form and were therefore relatively simple to produce, and as defined molecules, simpler to characterise compared to chemical conjugates (Goshorn *et al.*, 1993). Disulphide-linked Fv–β-lactamase conjugates have also been produced which are more stable in circulation than the scFv fusion proteins, although a direct comparison has not yet been reported (Rodrigues *et al.*, 1995).

The CPG2 system has been examined in a pilot clinical study which confirmed the suspected major drawback of ADEPT, i.e. immunogenicity of the enzyme in humans (Bagshawe *et al.*, 1995). In this study repeat dosing was limited by immune responses to both the murine $F(ab')_2$ used and the bacterial enzyme, and attempts to suppress the immune reponse with cyclosporin were unsuccessful. Overcoming the immune response to the antibody may be achieved by use of a human or humanised antibody, although overcoming responses to non-human enzymes may be more difficult. A humanised antibody–human enzyme conjugate has been prepared using β-glucuronidase, which has shown anti-tumour effects in mice (Bosslet *et al.*, 1994).

An alternative approach which may be non-immunogenic is the use of catalytic antibodies (Miyashita *et al.*, 1993). Antibodies can be produced to transition state analogues which are capable of catalysing a variety of chemical reactions (see Chapter 6). These can be humanised to reduce immunogenicity, and used to produce a bispecific antibody with binding specificity for a tumour-associated antigen as well as catalytic activity. Use of such bispecific antibodies may be a feasible approach, and a catalytic antibody capable of activating a carbamate prodrug to a nitrogen mustard has been produced (Wentworth *et al.*, 1996). The catalytic activity of this antibody was too weak to be of use *in vivo*, and therefore it has been used to set up assays to screen phage libraries for improved catalytic activity.

4.2.9 Vascular targeting

The continued growth and viability of a tumour is critically dependent on maintaining an effective blood supply which can supply the oxygen and nutrients required. An alternative strategy for tumour therapy is thus to shut off tumour blood vessels, cutting off nutrient supply which rapidly leads to tumour cell death. Two approaches have been investigated, targeting existing blood vessels or preventing the formation of the new blood vessels required to support growth. The process of blood vessel formation is known as angiogenesis, and thus these two approaches can be termed anti-vascular and anti-angiogenic. As these targets are vascular the problem of accessibility of antibodies to solid tumours is overcome, and thus antibody approaches have played a major part in attempts to develop such therapy.

Several growth factors secreted by tumours are known to be involved in the process of tumour angiogenesis, such as basic fibroblast growth factor (bFGF) and vascular endothelial growth factor (VEGF). Secreted factors cause migration and proliferation of vascular endothelial cells from existing blood vessels to form new tumour vasculature. Blocking this process has been attempted using antibodies which bind to bFGF or VEGF and block receptor interaction. Alternatively, antibodies to the growth factor receptors could be used. Both antibodies to bFGF and VEGF have shown inhibition of tumour growth (Hori *et al.*, 1991; Kim *et al.*, 1993). Particularly exciting data have arisen from inhibition of VEGF-induced angiogenesis which appears to be applicable to several types of tumour. Antibodies to VEGF are capable of inhibiting tumour growth and metastasis in experimental systems (Asano *et al.*, 1995). Complete inhibition of angiogenesis and growth of micro-tumours has been observed following treatment with anti-VEGF, changing the characteristics of a rapidly growing malignancy to a dormant microcolony (Borgstrom *et al.*, 1996).

Cell adhesion interactions, mediated by integrins, are also crucial in the process of angiogenesis. The integrin $\alpha_v\beta_3$ (CD51/CD61) is required on vascular endothelial cells during angiogenesis for their differentiation, migration and division. Blocking $\alpha_v\beta_3$ with a MAb can inhibit the growth of human tumours by causing apoptosis of the proliferating cells while leaving pre-existing blood vessels unaffected (Brooks *et al.*, 1994).

Early anti-vascular therapies were directed at killing tumour endothelial cells. Destruction of tumour blood vessels could lead to the death of many tumour cells, as many cells are supplied by each blood vessel. Animal model studies in which an antibody–toxin conjugate was targeted to tumour vasculature resulted in impressive anti-tumour effects, with eradication of large solid tumours (Burrows and Thorpe, 1993). Another approach is to block tumour blood vessels through inducing blood clots. Targeting of tissue factor, a protein that normally triggers blood clotting at sites of injury, has been shown to be capable of inducing blood clots specifically in the tumour vasculature leading to regression of large tumours (Huang *et al.*, 1997). However, the utility of these approaches may be limited by the availability of markers which are selectively expressed on human tumour vasculature and not normal vasculature. Although several markers which are at least partially selective for tumour vasculature, such as endosialin and endoglin, have been described (Rettig *et al.*, 1992; Burrows *et al.*, 1995). An anti-endoglin antibody has been raised that cross-reacts with mouse endothelial cells, facilitating studies in mouse tumour models (Seon *et al.*, 1997). The antibody reacted strongly with human endothelial cells and more weakly with mouse vasculature. Nevertheless, anti-tumour effects were seen with an immunotoxin made from this antibody and ricin A chain, with prevention of tumour growth in immunodeficient mice (Seon *et al.*, 1997).

4.3 Infectious disease

4.3.1 Antiviral antibodies

Although most bacterial infections can be well controlled by the use of antibiotics, the control of some viral diseases is far more difficult. Vaccination has allowed the control or even eradication of some viral diseases such as smallpox, poliomyelitis and measles, yet others have been more difficult to control. Antibodies have long been considered as potential protective or therapeutic agents for such conditions and MAbs may have a role to play in some cases. MAbs are under development for therapy of a variety of viral diseases, including respiratory syncytial virus (RSV), rabies, hepatitis B and C (HBV and HCV), herpes simplex viruses (HSV-1 and HSV-2), cytomegalovirus (CMV) and human immunodeficiency virus (HIV).

RSV is the leading cause of pneumonia and bronchitis in infants, and immunity appears to be only partially effective since re-infection can occur within a few weeks. In infants with cardiopulmonary disease or with immunodeficiency RSV can be fatal. Antibodies can protect against serious disease and current therapy uses polyclonal sera enriched with antibodies to RSV. There has been considerable interest in developing MAbs for RSV therapy. MAb preparations allow a higher concentration of specific antibody and are hence more potent therapeutic agents which are easier to use as they can be administered intramuscularly (i.m.) rather than intravenously (i.v.). The major surface antigens of RSV are the F and G proteins, and animal studies suggest that MAbs to certain epitopes on the F protein efficiently neutralise virus. Also, the F protein is conserved between strains and subgroups of RSV, making it an attractive target for a therapeutic MAb. Several anti-RSV antibodies have been humanised and shown to be effective in neutralising virus in animal models (Tempest *et al.*, 1991). The long half-life of humanised IgG is helpful in providing protection against infection for a prolonged period, and humanised IgG is therefore currently under clinical evaluation.

Anti-RSV antibody specificities have also been obtained from phage libraries (Barbas *et al.*, 1992). In some cases isolation of neutralising antibodies from phage libraries may be difficult due to the dominance of a non-neutralising epitope, such that only antibodies to this epitope are isolated. This has been the case for RSV F protein, and has been overcome by use of selection against antigen which is pre-blocked with an antibody to the dominant epitope. An antibody isolated from initial screening has been used to block the F protein dominant epitope such that antibodies to other less dominant epitopes, including neutralising antibodies, can be isolated (Tsui *et al.*, 1996).

Phage libraries have also been used for the isolation of human antibodies to other viral targets including HIV-1 and herpes simplex viruses (Barbas and Burton, 1996). Fab fragments to the HIV-1 envelope protein gp120 have been isolated. The majority of Fabs were raised to the dominant epitope, the CD4 binding site, although as with RSV it has proved possible to block this epitope to isolate Fabs against other epitopes (Ditzel *et al.*, 1995). Several neutralising Fabs have been isolated, one of which, to the CD4 binding site, has been used to generate a recombinant human IgG and shown to be capable of neutralisation of primary isolates of HIV-1 *in vitro*, and of protecting immunodeficient mice containing human lymphocytes from infection. Attempts to improve binding affinity using mutagenesis have proved successful in some cases. Anti-HIV-1 Fab was improved in affinity by 'CDR walking', in which CDRs are mutated either in turn or in parallel, and favourable mutations combined to develop a high affinity variant (Yang *et al.*, 1995). Antibodies with affinities in the picomolar range have been isolated, and it has been suggested that such high affinity antibodies may have improved *in vivo* potency, possibly extending the time an individual could be protected from disease (Barbas and Burton, 1996). However, the correlation between affinity and potency *in vivo* is not straightforward. Studies with a panel of antibodies to vesicular stomatitis virus (VSV) demonstrated that although there was a clear relationship between avidity and neutralisation *in vitro*, protection *in vivo* was not dependent on avidity provided a minimum threshold value (K_d 2×10^8–5×10^8) was reached (Bachmann *et al.*, 1997). In this study protection depended simply on a minimum serum concentration, suggesting that the isolation of very high affinity antibodies for viral protection may not be beneficial.

Anti-HIV antibodies such as those isolated by phage display, against gp120, may block infection through preventing the gp120/CD4 binding event which is involved in the entry of the virus into T cells. Fc mediated phagocytosis of the virus should also be enhanced. Another approach is to use immunotoxins directed towards the HIV envelope proteins (Pincus and McClure, 1993). Infected cells in which viral replication is occurring express the envelope proteins on the surface of the cell, such that as new virion cores are produced and bud through the cell membrane they become coated with the envelope proteins. Anti-gp120 immunotoxins can therefore be targeted to infected cells in an attempt to kill them before release of further virus. As described in Section 4.2.6, ricin A-chain conjugates can be potent cell killing agents, and anti-gp120–ricin A-chain conjugates can be produced with activity against HIV infected cells (Pincus and McClure, 1993). Antibody conjugates which are capable of virus neutralisation as well as cell killing activity may be additionally beneficial.

4.3.2 Bacterial sepsis

Bacterial sepsis is a clinical syndrome associated with severe bacterial infection from a wide variety of conditions ranging from acute diseases such as meningococcal meningitis to chronic conditions such as sepsis associated with cancer or surgical trauma. Once in the circulation, bacterial products such as the cell wall component lipopolysaccharide (LPS or endotoxin) elicit an inflammatory response that results in activation of an array of host defence systems and the release of a complex series of inflammatory mediators, including a variety of cytokines such as TNF and IL-1. In conjuction with the bacterial products over-activation of such systems can lead to shock, organ failure and death. In fact mortality rates in patients with sepsis are high, 25–70%, despite the availability of potent antibiotics for both Gram-positive and Gram-negative infections. There has been much interest in developing MAb based therapy for sepsis, based on antibodies to endotoxin or to mediators such as TNF. Encouraging results have been seen in preclinical and small-scale clinical studies yet, to date, extensive clinical studies have all resulted in the failure to demonstrate efficacy (reviewed by Quezado *et al.*, 1995).

Endotoxin is a component common to all Gram-negative bacteria, and the lipid A component of endotoxin is conserved across different species. Several antibodies to lipid A have been developed including the murine IgM, E5 and a human IgM HA-1A. Both of these antibodies have been extensively studied and been progressed to large-scale clinical trials. After protective effects were demonstrated in animal models, clinical trials with E5 were believed to show some survival benefit in a subgroup of patients with sepsis who had not gone into refractory shock. However, this finding could not be backed up in a subsequent trial (Bone *et al.*, 1995). HA-1A also demonstrated improved survival in animal models and, in contrast to E5, was believed to have benefit in patients with Gram-negative sepsis accompanied by shock. Again, this finding was not confirmed in subsequent trials (Natanson *et al.*, 1994), and further studies revealed a high degree of non-specific binding of HA-1A to a variety of different antigens and variable results in animal models (Quezado *et al.*, 1995). In addition, mortality was apparently increased in patients without Gram-negative infection.

Antibodies to TNF are under investigation for a range of inflammatory diseases (see also Section 4.5.2) and, as a key mediator in sepsis, inhibition of TNF has also been studied as a potential therapeutic for septic shock. Inhibition of a key mediator may have advantages over antibodies to endotoxin, as both Gram-positive and Gram-negative sepsis should be affected. Selection of TNF as a target to inhibit was based on the evidence that TNF can be measured in the plasma of patients with septic shock, and high levels tend to be prognostic of mortality. The symptoms of septic shock can be reproduced by injection of TNF into animals and anti-TNF antibody can reduce mortality in animal models of septic shock (Beutler *et al.*, 1985). In preclinical studies it has been demonstrated that antibodies with isotypes that are active at recruiting effector functions (human γ1 and murine γ2a) are less effective than inactive isotypes such as human γ4 or murine γ1 (Suitters *et al.*, 1994). This effect may be a consequence of activated complement or effector cells causing the release of additional inflammatory cytokines which in turn potentiate the effects of TNF. However, again studies have failed to show efficacy of anti-TNF antibodies in patients, underlining the complexity and diverse pathology of sepsis and septic shock.

4.4 Cardiovascular disease

4.4.1 *Inhibition of platelet aggregation*

Perhaps the most succesful MAb based therapeutic to date is a mouse/human chimeric Fab fragment from the antibody 7E3. This Fab fragment binds to the gpIIb/IIIa receptor on platelets and blocks the interactions of fibrinogen and von Willebrand factor which are crucial steps in platelet aggregation. Chimeric 7E3 Fab, also known under its tradename ReoPro, is approved for prevention of complications following angioplasty. Angioplasty is an effective treatment for blocked blood vessels which is widely used. In this procedure a balloon catheter is introduced into the vessel and inflated such that the blockage is broken up or displaced. However, some damage to the vascular endothelium usually results, and in a significant proportion of patients, reocclusion of the vessel can occur through activation of platelets leading to their aggregation and adhesion to the damaged endothelium. On activation of platelets there is a conformational change in gpIIb/IIIa which leads to binding of fibrinogen and hence platelet aggregation. Blocking with c7E3 prevents platelet cross-linking and thus prevents the formation of a platelet-rich thrombus which might otherwise reocclude the vessel.

c7E3 is used as a Fab fragment as the function required is purely a blocking one, bivalent IgG might cross-link cells and the presence of Fc might cause unwanted cell binding through Fc receptors, which would be counter-productive to preventing cellular aggregation. Fab fragments are also known to be less immunogenic than IgG (see Chapter 2) and chimeric Fab was developed to reduce immunogenicity further (Knight *et al.*, 1995). The chimeric Fab was considerably less immunogenic in humans than the murine version, even though the entire variable domains were still of murine origin. This was the case even though most of the response detected to the murine Fab was to the variable region, suggesting that in this case the constant region is responsible for modulation of the response to the variable domains (Knight *et al.*, 1995). CDR-grafted antibodies to gpIIb/IIIa have also been produced which are able to prevent platelet-induced thrombosis (Kaku *et al.*, 1995), although the immunogenicity of c7E3 is sufficiently low to allow its therapeutic use.

The efficacy of c7E3 was demonstrated in a 2099-patient trial for the prevention of acute cardiac complications following high-risk angioplasty. In this trial c7E3 demonstrated a 35% reduction in acute ischaemic events (death, myocardial infarction and urgent coronary intervention) at 30 days and a 23% decrease in ischaemic events (Califf, 1994). Principal adverse events were major bleeding at sites of vascular puncture or coronary artery bypass grafting and associated transfusions. Bolus injection followed by a prolonged infusion produced the best results, indicating the importance of inhibiting platelets for a sustained period. Subsequent trials have been designed which have reduced the problems of associated bleeding through fine tuning of the dose of heparin which is co-administered with c7E3, and have examined the role of c7E3 in patients with unstable angina and in lower-risk angioplasty patients with positive results.

4.4.2 Thrombolysis

The ability to lyse blood clots (thrombi) using thrombolytic therapy with plasminogen activators has led to its widespread use in the treatment of acute myocardial infarction, resulting in a significant reduction in mortality. However, present agents are not ideal and have limitations in their ability to lyse all thrombi, they often lead to high rates of reocclusion and in some cases they result in severe side-effects such as cerebral haemorrhage. The ability to target thrombolytic agents to blood clots may offer advantages, with the potential for increased potency and reduced side-effects. Antibodies to fibrin have been examined for this purpose. Importantly, antibodies that recognise fibrin in clots but not circulating fibrinogen have been isolated and used to produce chemically linked conjugates with plasminogen activators. Both two-chain and single-chain urokinase (uPA) could be made fibrin-selective by conjugation to anti-fibrin IgG or Fab fragments (Bode *et al.*, 1990). Similarly, the efficiency of tissue plasminogen activator (tPA) could also be enhanced by conjugation to anti-fibrin antibody (Runge *et al.*, 1988). Attempts to make active fusion proteins were disappointing with tPA, resulting in material which was less effective at clot lysis; however, fusion proteins of anti-fibrin Fab with uPA were constructed which retained the ability to lyse clots effectively (Runge *et al.*, 1991). Fusion proteins have also been produced in the format of scFv–uPA using a 32 kDa truncated version of uPA, resulting in a single-chain plasminogen activator. This protein demonstrated increased thrombolytic potency *in vitro* compared to single-chain uPA, but not compared to two-chain uPA (Holvoet *et al.*, 1991).

Bispecific antibodies have also been examined for their ability to target plasminogen activators to clots. Bispecific antibodies have the potential advantage of not causing any damage to the plasminogen activator on conjugation to the antibody or in the construction of fusion proteins. A bispecific antibody with specificity for fibrin and tPA has been produced from a hybrid hybridoma with better thrombolytic activity than tPA alone in a rabbit model of thrombolysis (Kurokawa *et al.*, 1989). The potential to combine thrombolysis with anti-platelet effects in a single agent has also been explored using a bispecific conjugate. The anti-gpIIb/IIIa Fab′, 7E3, described above has been used to prepare a bispecific di-Fab by cross-linking to a Fab′ fragment that recognises tPA (Neblock *et al.*, 1992). The bispecific agent was able to block platelet aggregation as effectively as 7E3 Fab alone, and could recruit tPA to platelets. Such an agent may be useful in preventing reocclusion of vessels as well as hastening reperfusion.

4.5 Disorders of the immune system/inflammatory diseases

4.5.1 *The inflammatory response*

Inflammation is the body's protective reaction to injury or the presence of foreign material from infection. However, over-activation of the inflammatory response is also associated with a number of human diseases including autoimmune diseases such as rheumatoid arthritis and multiple sclerosis, inflammatory bowel disease, psoriasis, transplant rejection and allergy. The inflammatory response is characterised by increased blood flow to the site, dilation of blood vessels and migration of lymphocytes acoss the vessel wall into the tissue. The events required to achieve this form a complex series of interactions including both soluble factors such as pro-inflammatory cytokines and the up-regulation of cell adhesion molecules involved in adherence, rolling and migration of leukocytes (reviewed by Springer, 1994; Figure 4.7).

Injury or infection within the tissue causes the release of pro-inflammatory cytokines from macrophages, such as IL-1 and TNF, which activate vascular endothelial cells and induce the up-regulation of the adhesion molecules P-selectin and E-selectin. These selectins are responsible for the initial attachment of leukocytes to the endothelium. Leukocytes have considerable momentum in the circulation and once captured they continue rolling along the vessel wall. The relatively weak attachment allows the leukocytes to respond to chemoattractant signals mediated by a range of substances including leukotriene B_4, PAF, complement fragment C5a, *N*-formylated peptides, MCP-1, MIP-1 and IL-8. Such factors cause up-regulation and activation of adhesion molecules on the leukocyte such as the integrins CD11a/CD18 (LFA-1), CD11b/CD18 (MAC-1) and CD29/CD49d (VLA-4). Such integrins mediate tighter binding to the endothelium through interaction with their receptors CD54 (ICAM-1) and CD106 (VCAM-1). In some cases other adhesion interactions such as L-selectin with MAdCAM-1 may also be involved (Table 4.8). The strongly bound leukocytes are then seen to flatten against the endothelium and migrate through to the injured tissue.

Autoimmune diseases are a result of the immune system reacting to self antigens either systemically or in tissues, leading to chronic inflammation. At the heart of many autoimmune diseases are inappropriately activated T cells which interact with antigen

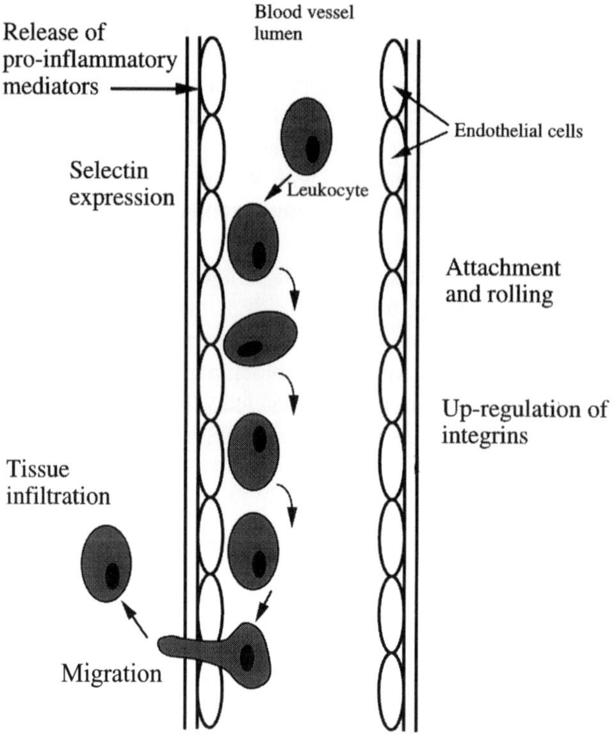

Figure 4.7 Representation of the mechanism of leukocyte-mediated inflammation. The release of pro-inflammatory mediators from tissue macrophages induces the vessel endothelial cells to up-regulate selectins and causes the synthesis of a range of chemokines. Leukocytes are attracted and attach weakly to the endothelium. The flow through the vessel causes rolling of weakly attached leukocytes along the endothelium. As the leukocytes become fully activated, and integrins are up-regulated, the binding becomes stronger and cells flatten and migrate through the endothelial layer to the surrounding tissue. (See text for further details).

Table 4.8 Interactions involved in leukocyte adhesion to the endothelium during inflammation

Endothelium	Leukocyte
P-selectin	PSGL-1, sialyl Lewis^x and others
E-selectin	ESL-1, sialyl Lewis^x and others
CD54 (ICAM-1), CD102 (ICAM-2)	CD11a/CD18 (LFA-1)
CD54 (ICAM-1)	CD11b/CD18 (MAC-1)
CD106 (VCAM-1)	CD49d/CD29 (VLA-4)
GlyCAM-1, CD34	L-selectin
MAdCAM-1	$\alpha4\beta7$, L-selectin

presenting cells (APCs) presenting self antigen leading to a number of direct and indirect effects mediated through cytokine release including the infiltration of activated cells described above. Antibodies to T cells and APCs which could block activation have therefore also been considered as potential therapeutics.

The inflammatory process offers many points of potential therapeutic intervention with MAbs, including antibodies to block T cell activation, antibodies to cytokines such as TNF and adhesion molecules such as selectins and integrins on either the endothelium or the leukocyte. These are all blocking interactions and as such are characterised by the need for antibodies which efficiently block ligand binding. These events are largely taking place in the vasculature and hence there is little need for fragments which would penetrate tissues more effectively. The therapeutic reagent is often required to be present for a long period and thus the long circulating half-life of IgG makes this an ideal form for therapy. Neutral isotypes of IgG are also desirable in most cases as triggering effector functions would exacerbate inflammatory processes.

One problem with developing therapy in such situations is that many antibodies to human proteins do not cross-react with the same protein in animal species. Animal model studies are therefore difficult and usually rely on the use of parallel reagents – antibodies to the equivalent animal protein, to generate information on potential efficacy. Parallel reagents may bind to different epitopes with different affinity to the agent intended for clinical use, and thus great care is needed in extrapolating from animal studies to man.

4.5.2 Blocking inflammatory mediators

Anti-TNF antibodies in rheumatoid arthritis and inflammatory bowel disease

Blocking of TNF with MAbs has been investigated for the therapy of rheumatoid arthritis and inflammatory bowel disease as well as for septic shock (see Section 4.3.2).

Rheumatoid arthritis (RA) is a common autoimmune disease with a prevalance of about 1% of the population, yet it is still poorly understood in molecular terms. RA is characterised by inflammation of synovial joints which can progress to joint damage and eventually destruction. Affected joints are heavily infiltrated with cells, especially T lymphocytes and macrophages, and elevated TNF levels have been found in joint synovial fluid in more than 50% of patients. TNF has many biological effects, including not only endothelial activation and adhesion molecule expression but also granulocyte activation resulting in increased phagocytosis, degranulation and generation of oxygen radicals and prostaglandin E2, stimulation of fibroblast growth, stimulation of cytokine production, and co-stimulation with IL-2 of T cell proliferation. All of these may be important in the RA disease process.

Antibodies to mouse TNF completely prevent collagen-induced arthritis even after the inflammatory process has begun (Williams *et al.*, 1992), and two recombinant anti-human TNF antibodies, a mouse:human chimeric IgG1 (cA2) and an engineered human IgG4 (CDP571) are being examined in clinical trials for RA. CDP571 has been examined at doses up to 10 mg/kg in a double-blind study in patients with active RA (Rankin *et al.*, 1995). The treatment was well tolerated, and patients treated at the top dose demonstrated significant improvements in disease severity measured through reduction in pain, number of tender and swollen joints and serum markers. Similar encouraging results have been seen with cA2 in both open and double-blind clinical studies with rapid improvement in disease symptoms following treatment (Elliott *et al.*, 1994).

Anti-TNF antibodies are also under investigation for the therapy of inflammatory bowel disease (IBD). IBD consists of two histologically distinct disorders, namely ulcerative colitis and Crohn's disease. Crohn's disease is an inflammatory disease that can affect any portion of the gastrointestinal tract whereas ulcerative colitis is restricted to the large bowel. TNF is believed to be pivotal to the disease process, particularly in Crohn's disease and both CDP571 and cA2 have been examined in early clinical trials. In a 31-patient double-blind study, a single dose of 5 mg/kg CDP571 was found to reduce disease activity for two weeks, with some patients experiencing sustained remission (Stack *et al.*, 1997). cA2 treatment also resulted in significant improvements in Crohn's disease. In a 108-patient trial, clinical responses were observed in 81% of patients treated with 5 mg/kg compared to 17% in the placebo group (Feldmann *et al.*, 1997). There was apparently little benefit of higher doses, as response rates of 50% and 64% were seen with 10 and 20 mg/kg doses respectively. Beneficial effects of anti-TNF antibody have also been reported in ulcerative colitis, with a consistent reduction in disease activity determined in a phase II trial.

Anti-C5

Activated components of the complement system are potent mediators of inflammation. Antibodies that block the complement cascade at C5, block the generation of the major chemotactic and pro-inflammatory factors C5a and C5b-9. In animal models neutralisation of C5 can both prevent the establishment of arthritis and alleviate established disease (Wang *et al.*, 1995). Similarly, amelioration of disease has been seen in a mouse model for the autoimmune disease systemic lupus erythematosus (Wang *et al.*, 1996). A humanised antibody for such chronic indications has been developed, but for acute conditions where complement mediated damage is indicated, such as myocardial infarction, stroke or cardiopulmonary bypass, it has been suggested that a single-chain Fv may be more suitable due to rapid tissue penetration. An anti-C5 scFv fragment has been shown to inhibit complement activity both *in vitro* and *in vivo*, and is currently under clinical evaluation (Evans *et al.*, 1995).

4.5.3 **Blocking adhesive interactions**

Blocking adhesion interactions of the vascular endothelium with leukocytes results in anti-inflammatory effects through inhibition of the process of leukocyte adhesion as described above. Antibodies have been described to all of the adhesion molecules known to be involved in the inflammatory process and several of these are under active investigation as anti-inflammatory agents in a number of different diseases.

Binding of leukocytes to E-selectin on the vascular endothelium is an early event in leukocyte recruitment (Springer, 1994). An engineered human antibody to E-selectin has been produced which is capable of blocking leukocyte accumulation into TNF-stimulated baboon skin (Owens and Robinson, 1995). The antibody was designed such that effector functions such as complement binding and Fc receptor binding were not likely to take place, as these might exacerbate the inflammatory process. The relatively neutral isotype, IgG4, was used, and a mutation in the CH2 domain, leucine 235 to alanine, was introduced to make the sequence similar to an IgG2. This mutation has been shown to reduce high affinity Fc receptor binding even further (Lund *et al.*, 1991). This antibody is under clinical investigation for treatment of the inflammatory skin disease psoriasis.

Antibodies to the leukocyte integrins, CD11a/CD18 (LFA-1), CD11b/CD18 (MAC-1) and CD49d/CD29 (VLA-4) as well as to receptors on the endothelium such as CD54 (ICAM-1), CD102 (ICAM-2) and CD106 (VCAM-1) (Table 4.8) have been widely investigated. RA treatment with 5 doses of anti-CD54 was shown to result in clinical improvement in 13 of 23 patients, in 9 of which benefit persisted until day 60 at the end of the study (Kavanaugh *et al.*, 1994). Antibodies to CD18 and CD49d/CD29 have also been shown to improve disease in animal models of arthritis (Issekutz *et al.*, 1996).

The problem of antibodies which do not cross-react with the same protein from animal tissues can be approached in several ways. One approach recently developed for a humanised anti-human CD11a antibody has been the re-engineering of the antibody to allow binding to rhesus monkey CD11a, allowing pre-clinical studies to be carried out (Werther *et al.*, 1996). Alanine scanning mutagenesis was initially carried out and CDR residues important for binding human and rhesus CD11a identified. An engineered variant was then made in which four amino acids in CDR2 of the heavy chain, found to be important in the difference between binding human and rhesus CD11a, were changed. Although it did not prove possible to develop an antibody which bound both human and rhesus CD11a effectively, the resulting antibody was able to bind rhesus CD11a and could be used as a parallel reagent for pre-clinical studies. Anti-CD11a antibodies are currently under development as immunosuppressive therapy of autoimmune diseases and transplant rejection. Several antibodies to CD18 have also been humanised (e.g. Sims *et al.*, 1993) and are currently under evaluation.

Antibodies to CD49d/CD29 (VLA-4) are under development as therapeutic agents for multiple sclerosis (MS). MS is an autoimmune disease characterised by lesions (demyelination) of the central nervous system which lead to the clinical symptoms of weakness of the limbs, paraesthesia and visual disturbances. MS commonly progresses through a series of relapses separated by remissions, to a progressive decline. As disease progresses further symptoms are revealed such as spasticity, fatigue, ataxia, tremor, loss of bladder and bowel control and neuropsychological abnormalities. Lesions in the CNS are caused by activated T cells crossing the blood:brain barrier and initiation of a series of events leading to activation of endothelial cells, recruitment of additional leukocytes, release of pro-inflammatory cytokines and subsequent demyelination of nerve cells. MAb to CD49d has been shown to block binding to CD106 (VCAM-1) and inhibit leukocyte migration. In an animal model of MS, experimental autoimmune encephalomyelitis (EAE), anti-CD49d was able to block onset of the disease (Yednock *et al.*, 1992). In addition, reversal of established EAE has been demonstrated with clearance of leukocytes from the central nervous system (Kent *et al.*, 1995). A humanised form of the antibody has been produced with retention of blocking activity and the ability to reverse active EAE in guinea pigs (Leger *et al.*, 1997). Human IgG4 constant regions were chosen to minimise effector function and maximise half-lfe. A phase I trial demonstrated that the antibody was well tolerated, and phase II studies are currently in progress.

4.5.4 *Antibodies which directly inhibit T cell activation and proliferation*

The murine IgG2a MAb OKT3 was the first to be licensed for human therapy. It recognises the ε-subunit of the CD3 complex (the T cell receptor) on the surface of T cells and is a potent immunosuppressive agent. As such it has found application for the prevention of acute rejection in kidney, liver, heart and lung transplant patients. OKT3 also has T cell activating properties which are thought to be due to cross-linking of T cells via CD3 to Fc receptor bearing cells leading to massive cytokine release. Cytokine release can cause severe side-effects on treatment with OKT3 and these have limited its use in less acute conditions such as autoimmune diseases. Also, an immune response is rapidly generated to OKT3 which can block its effectiveness. Several humanised forms of OKT3 have been produced in attempts to overcome immunogenicity which retain full CD3 binding ability and immunosuppressive properties (Adair *et al.*, 1995). Human IgG4 constant regions were used to minimise Fc receptor binding of the humanised antibody, yet Fc receptor mediated cross-linking still occurs. Mutation of a residue in the Fc receptor binding site, leucine 235 to glutamate, resulted in a 100-fold decrease of the affinity of the antibody for Fc receptor bearing cells (Alegre *et al.*, 1992). This resulted in a marked reduction in T cell activation as measured by proliferation and cytokine release, and may offer an attractive profile for human therapy, maintaining strong immunosuppression while minimising toxicity.

CD4 functions as a co-receptor for stabilising the T cell receptor with antigen presented by the MHC class II molecule and is also involved in signal transduction pathway which leads to the activation of a T helper cell. The binding of anti-CD4 antibodies leads to a down-regulation of T cell activity and suppression of immune responses. Antibodies to CD4 have been widely investigated in a number of inflammatory conditions, including RA, MS, psoriasis, diabetes and systemic lupus erythematosus as well as in prevention of transplant rejection. Antibodies that either deplete all CD4 positive cells or merely block the CD4 molecule have been shown to be effective in preventing and reversing the symptoms of autoimmune disease in animal models. Small open trials with two depleting antibodies, one murine and one chimeric, showed promising results, but large randomised placebo-controlled trials in RA did not show any effectiveness of these antibodies (Wendling *et al.*, 1996; van der Lubbe *et al.*, 1995). Non-depleting anti-CD4 antibodies may have an advantage in their ability to temporarily inhibit T cell activation without major inhibition of other immune functions. Several non-depleting antibodies have been humanised (Pulito *et al.*, 1996) and a monkey–human chimeric antibody, termed a primatised antibody, has also been produced (Newman *et al.*, 1992). The primatised antibody may be less immunogenic than mouse:human chimeric antibodies due to the increased homology of monkey and human variable regions. Clinical results with such antibodies seem to be more promising. For example, a double-blind, placebo-controlled study with the primatised antibody in RA resulted in clinical improvement in 77% of patients at the highest dose compared to 17% in the placebo group (Levy *et al.*, 1996). CAMPATH-1H, which recognises the CDw52 antigen, has been described above (Section 4.2.1) as a potential therapeutic antibody in non-Hodgkin lymphoma. As the antigen is also widely expressed on T cells, there has been interest in its use to deplete T cells in vivo for therapy of autoimmune diseases. This humanised antibody binds to an epitope of CDw52 close to the membrane, which may contribute to its effectiveness in mediating cell lysis. Treatment with this antibody leads to severe lymphocyte depletion from the blood followed by a gradual repopulation. Encouraging effects in early RA trials were seen with some significant clinical benefits (Isaacs *et al.*, 1992). However, subsequent larger studies revealed significant toxicity due, at least in part, to cytokine release, suggesting T cell activation takes place (Brett *et al.*, 1996). In addition, treatment resulted in dose-related susceptibility to infection, suggesting lymphocyte depletion with this antibody may be too severe for use in RA.

Interaction between IL-2 and its receptor is required for the activation of cytotoxic T cells and for their proliferation. The IL-2 receptor is made up of three subunits (α, β and γ), and antibodies to both the α and β chains have been investigated as potential immunosuppressive therapies to prevent transplant rejection. IL-2 binds to all three subunits individually but it binds with much higher affinity to the non-covalently associated combination of all three chains. It was hoped that transplant rejection would be reduced through targeting those T cells involved in immune rejection; antibodies to the IL-2 receptor might be expected to inhibit antigen activated T cell responses without suppressing natural immunity. Initial studies used antibodies to the α chain, also known as the Tac receptor (T activated cell) or CD25. A humanised antibody to the α chain has been produced which retains high antigen-binding affinity and can activate ADCC with human effector cells (Junghans *et al.*, 1990). The humanised antibody was more effective than the murine parent antibody in prolonging survival of monkeys undergoing heart transplants (Brown *et al.*, 1991). However, although graft survival was prolonged without toxic side-effects, antibody treatment was not sufficient alone to prevent eventual rejection. The IL-2 receptor β chain is shared by several cytokine receptors including that for IL-15 which is also involved in T cell proliferation. A humanised antibody to the β chain has also been produced and shown to prolong primate heart transplant survival (Tinubu *et al.*, 1994). Antibody to the β chain showed synergistic effects with antibody to α chain *in vitro*, leading to increased inhibition of T cell proliferation. However, there was no beneficial effect of combined therapy *in vivo* (Tinubu *et al.*, 1994). A humanised bispecific antibody comprising one arm to the α chain and one to the β chain has also been produced (Pilson *et al.*, 1997). This was more active in inhibiting proliferation *in vitro* than the mixture of the two parent antibodies, though *in vivo* studies remain to be carried out. Antibodies to the γ chain may also be attractive as this subunit is also shared by several other cytokine receptors involved in T cell responses, including those for IL-4 and IL-7; however, the effects of blocking the γ chain have not yet been reported.

An attempt to combine the properties of anti-CD3 and anti-IL-2 receptor α chain (CD25) antibodies has been made by the preparation of a bispecific antibody with specificity for both CD3 and CD25 (MacLean *et al.*, 1995). The antibody, produced from a hybrid hybridoma (Section 2.5), was as effective an immunosuppressant as the anti-CD3 alone and more effective than anti-CD25 alone. In addition the toxicity of the anti-CD3 antibody was greatly reduced through the inability to activate T cells via CD3 cross-linking on the surface of the cell. Alternatively, inhibition of IL-2 interaction may be important, although this was less likely as combination of the two original antibodies did not have the same effect. Whether univalent CD3 binding molecules, or $F(ab')_2$ fragments, would be equally effective was not addressed.

Anti-CD25 antibodies have also been investigated as potential therapeutics for graft versus host disease. Graft versus host disease (GvHD) is a life-threatening complication of bone marrow transplantation which is widely used in treatment of leukemia, immunodeficiency and certain other diseases. In GvHD mature T cells carried in the marrow graft recognise the recipient's tissues as foreign and cause immune attack leading to multi-organ damage. GvHD can be reduced by T cell depletion of the marrow before transplantation to the recipient. However, this often leads to increased incidence of graft rejection, relapse of leukemia and delayed reconstitution of the immune system, leading to susceptibility to infection. Antibodies have been investigated as potential treatments to selectively target T cell populations carried in the marrow graft. Treatment with humanised anti-CD25 antibody resulted in improvement in approximately 40% of patients (Anasetti *et al.*, 1994). This relatively poor response may be due to loss of the IL-2 receptor on memory T cells.

CD5 is a marker found on all T cells and a subset of B cells, and antibodies to CD5 have been used in attempts to deplete T cells for therapy of RA and GvHD. Anti-CD5 immunotoxins have been produced in which ricin A-chain was coupled to the antibody as used for anti-cancer immunotoxins (Section 4.2.6). Anti-CD5 immunotoxins were effectively internalised into T cells and depleted cells effectively *in vitro*. However, little therapeutic benefit was seen in clinical trials for RA, and although short-term benefit has been observed in GvHD, long-term benefit did not result (Martin *et al.*, 1996). Anti-CD3 immunotoxins are also under investigation for GvHD, and in animal models anti-CD3 $F(ab')_2$–ricin A-chain conjugates have proved more effective (Vallera *et al.*, 1995). The use of $F(ab')_2$ conjugates rather than intact IgG removes Fc receptor mediated cross-linking which is at least in part responsible for toxicity of anti-CD3 antibodies as described above. Recombinant immunotoxins consisting of anti-CD3 scFv linked to truncated diptheria toxin have also been produced, although attempts at therapy of GvHD resulted in only temporary alleviation of symptoms, probably due to the short half-life of this molecule *in vivo* (Vallera *et al.*, 1996).

4.5.5 Antibody treatment of allergy

The allergic reaction is an excessive immune response to a common substance which is normally harmless. These substances, termed allergens, include pollen, some foods, dust mites, animal fur and some drugs. In some individuals, termed atopic, who have been sensitised by exposure to an allergen, a second exposure can result in a hypersensitivity reaction. Atopic individuals have high levels of circulating IgE and produce large amounts of allergen-specific IgE in response to the allergen. The clinical symptoms depend on the route of exposure and can include atopic dermatitis or eczema (skin contact), allergic rhinitis or asthma (inhalation) or food allergy (ingestion). The mechanism of allergic reactions is shown in Figure 4.8. On first contact with allergen, allergen-specific IgE is produced which is bound to high-affinity receptors present on mast cells in tissues and on basophils in the circulation. On second exposure, allergen can cross-link two receptor bound IgE molecules leading to signal transduction and degranulation – the release of a range of inflammatory mediators including histamine, cytokines and others, which cause the inflammatory reaction.

MAbs to IgE have been raised which bind to IgE at the Fcε receptor binding site and block binding to basophils and mast cells. Importantly, such antibodies do not bind to IgE bound to the receptor as this would result in cross-linking, signalling and subsequent degranulation (Presta *et al.*, 1994). A humanised version has been produced with high affinity for IgE which is effective at inhibition of allergic reactions *in vitro* (Presta *et al.*, 1993). Allergic reactions may be more difficult to block *in vivo*, but recent evidence suggests that treatment with anti-IgE also results in down-regulation of Fcε receptors on basophils in treated patients, presumably as receptor levels are regulated by levels of IgE (MacGlashan *et al.*, 1997). The combination of reduced IgE levels and reduced receptor levels may allow significant therapeutic effects, and clinical studies in allergic rhinitis and asthma are ongoing.

Other approaches to therapy of allergic diseases include the use of antibodies to specific allergens which might neutralise them in the circulation and prevent IgE receptor cross-linking. Human antibodies have been generated to a number of allergens such as pollen, either from immunised donors or by phage display (De Lalla *et al.*, 1996). Alternatively, recombinant IgA molecules can be constructed which can bind the allergen on mucosal surfaces, such as the nasal linings or the lower airways, and inhibit the entry of the allergen across the mucosal epithelium (Sun *et al.*, 1995). Such approaches suffer from the drawback of being specific to one allergen, and thus of limited general utility.

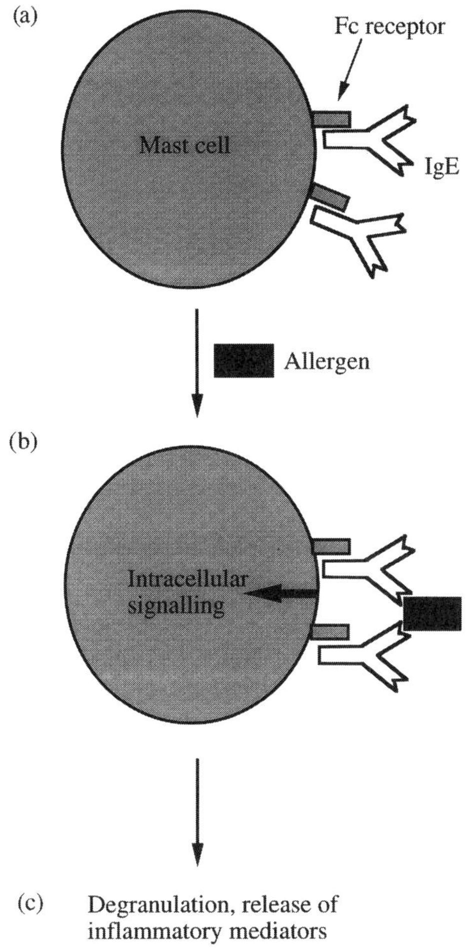

Figure 4.8 Allergen-induced activation of mast cells (see text for details)

A more general approach to asthma therapy may be the neutralisation of key mediators such as IL-5. Eosinophils infiltrate into the lungs during asthma and may cause the damage associated with inflammation of the lung. In animal models antibodies to IL-5 inhibit eosinophil infiltration in the lung and prevent tissue damage and hyperreactivity (Egan *et al.*, 1995). A humanised antibody to IL-5 has been produced which is also effective and is currently under clinical evaluation. Similarly, antibodies to adhesion molecules which are involved in eosinophil migration into the lung may also be effective. Antibodies to CD49d/CD29 (VLA-4) have been shown to inhibit eosinophil accumulation and asthmatic response in a guinea pig model of asthma (Sagara *et al.*, 1997).

5

Production of Monoclonal Antibodies

5.1 Introduction

The ability to apply MAbs in any situation is obviously dependent on the ability to make them! The purpose of this chapter is to review the expression and purification of MAbs, particularly with regard to the ability to manufacture antibodies in an economically viable manner.

The requirements of the production system depend on the form of the antibody required and its intended application. The application will determine the scale of production: for example, MAbs for diagnostic purposes may be required at a scale of tens to hundreds of grams per year, whereas some therapeutic antibodies may be required at tens to hundreds of kilograms per year. In addition, the application will determine the specification for purification of the antibody. The requirement for highly purified antibody is more stringent for antibodies to be used in humans compared to those used for *in vitro* applications. Therefore, it is important at the beginning of a project to develop antibodies with the manufacturing system in mind, including at the design stage where choices can impact on eventual manufacturing processes.

5.2 Expression of antibodies in mammalian cells

Conventional mouse and rat MAbs are expressed from hybridoma cells (see Chapter 1). The selection of the MAb in the first place is often partially dependent on the productivity of the hybridoma for expression and secretion of the antibody. Typical yields of antibody from hybridoma cells are 10–100 μg/10^6 cells/day leading to accumulation of between 50 and 500 mg per litre depending on the culture medium used and the cell biomass available (Brown *et al.*, 1992). Other important factors for antibody production are the stability of the cell line and the absence of undesirable adventitious agents such as viruses or mycoplasma. Adventitious agents can sometimes be removed by a number of methods followed by repeated cloning and selection, but unstable or low productivity hybridomas are often more problematic. The simplest solution to unstable lines is to clone the antibody genes from the hybridoma and re-express them as a recombinant antibody. Although in many cases re-expression of antibody genes in myeloma cell lines has led to lower yields than those seen in hybridomas, high level production systems are now available in both myelomas and nonlymphoid cells (see below), and use of such an improved expression system can lead to higher yields of antibody production, from cells which are suitable for large-scale manufacturing.

The expression of IgG is achieved with high efficency in mammalian cells, and mammalian cells are the only system currently used for large-scale production of intact monoclonal antibodies. Mammalian cells have the most appropriate cellular machinery to allow assembly and secretion of IgG as well as the ability to carry out appropriate post-translational modifications, notably glycosylation. As described in Chapter 1, glycosylation of IgG can be important for maintaining the conformation of the CH2 region of the antibody, and lack of glycosylation leads to a loss of Fc effector functions (Lund *et al.*, 1990). The type of sugar residues attached is also important to allow maintenance of effector function; for example, the attachment of high-mannose carbohydrate by yeast cells in place of the complex carbohydrate attached by mammalian cells leads to loss of the ability to activate complement (Horwitz *et al.*, 1988). Subtle differences in glycosylation between different types of mammalian cells, or between different types of growth conditions, may also be important in some cases (see below).

Two types of expression system need to be considered, transient expression and stable expression. Cloned DNA can be introduced into the nuclei of mammalian cells by several transfection techniques, where it may persist for several days in a high proportion of cells resulting in transient expression of the genes present on the vector. Alternatively, low-frequency integration events can be selected by use of an appropriate marker gene, during which the vector DNA becomes inserted into random sites within the genome. In this case the DNA is replicated along with the host cell genome, and a permanently transfected stable cell line is produced. A substantial investment in time and effort (usually 2–3 months) is required to produce a stable transfected cell line capable of producing large amounts of antibody. Therefore, transient expression systems are very valuable in allowing the rapid generation of small amounts of recombinant antibody, up to a few milligrams, to allow initial analysis of multiple engineered forms of antibody.

5.2.1 *Transient expression systems*

COS cells are the most commonly used cell type for transient expression of antibody genes. COS cells are derived from a monkey kidney cell line, CV1 transformed with an origin-deficient SV-40 virus (CV1 Origin-deficient SV40). COS cells express the SV40 T antigen which is the only protein required to support SV40 replication in these cells. Therefore a plasmid containing the SV40 origin of replication will be replicated to approx. 10^4–10^5 copies per cell within a few days. This high copy number allows high-level transient expression of genes on the vector. Commonly the genes for light and heavy chain are introduced on separate plasmids, each containing the SV40 origin. This facilitates a range of different combinations of heavy and light chains to be tested with each other, to determine the best combination. For example, several CDR-grafted heavy chains can be tested with the same light chain, and vice versa, to select the best combination of chains for further study. A strong promoter/enhancer is required for efficient expression, such as the adenovirus major late promoter in combination with the SV40 enhancer, or the MIE (major immediate early) promoter/enhancer from human cytomegalovirus (hCMV) (Stephens and Cockett, 1989). Yields of antibody achieved in a few days from transient expression in COS cells depend on the particular antibody but are usually in the range 0.1–10 μg/ml (Whittle *et al.*, 1987; Emtage *et al.*, unpublished data).

Alternative transient expression systems have also been developed which use the hCMV-MIE promoter in an adenovirus transformed human embryo cell line, 293 cells, together with a transcriptional activator. One of the early adenovirus proteins, E1a, enhances transcription from a number of promoters including hCMV-MIE when coexpressed in the same cell and allows high levels of transient expression: 7–15 μg/ml has been reported (Carter *et al.*, 1992b). Similarly, a Chinese hamster ovary (CHO) cell line has been developed for transient expression by introducing the E1a gene into the genome, resulting in a system capable of similar expression levels (Cockett *et al.*, 1991).

5.2.2 Stable expression systems

The most successful cell lines for stable expression of antibodies have been the myeloma lines SP2/0 and NS0, and CHO cells. Most recombinant antibodies produced on a large scale have been produced using myeloma cells. As these are the fusion partners used in hybridoma production, they are known to be capable of secreting large amounts of antibody and they grow well in suspension culture in fermentors. CHO cells have also been developed as a system capable of producing large amounts of antibody; they are readily transfectable and grow well in both attached and suspension culture. In addition, as described below, efficient expression systems based on vector amplification have been established for CHO cells (Bebbington, 1991).

Development of stable cell lines requires the use of a selectable marker to select cells in which the relatively rare event of vector integration into the host cell genome has taken place. Bacterial genes are widely used which are incorporated into the vector with mammalian transcription signals. The *gpt* gene encodes the xanthine–guanine phosphoribosyl transferase enzyme from *E. coli* and can be used to confer resistance to mycophenolic acid. This enzyme allows guanosine monophosphate (GMP) synthesis from xanthine via the salvage pathway when *de novo* GMP synthesis is blocked by mycophenolic acid (Mulligan and Berg, 1981). Alternatively the *neo* gene can be used, as this confers resistance to the antibiotic G418, an analogue of the antibiotics neomycin, kanamycin and gentamycin that block protein synthesis (Southern and Berg, 1982). A third, though less commonly used, system is the *hph* gene from *E. coli* which confers resistance to hygromycin (Gorman, 1990).

The use of these markers in combination allows the selection of cells containing two vectors, and therefore the separate transfection of heavy and light chain genes. This can be useful if several variants of an antibody are to be made such as an IgG isotype series and Fab′ variants. In this case a light chain expressing cell line can be made which can be re-transfected with the heavy chain genes as appropriate (e.g. King *et al.*, 1992a). It should be remembered when using this sequential approach that the light chain must be introduced first. This is because heavy chain alone is retained in the endoplasmic reticulum through association with the heavy chain binding protein, grp78, and accumulation of large amounts of heavy chain in the endoplasmic reticulum is usually toxic to the cell (Hendershot *et al.*, 1988). On the other hand, free light chain is readily secreted by the cell.

Vectors which contain both heavy and light chain on the same plasmid are often preferred, as these offer a more balanced production of heavy and light chain (reviewed by Bebbington, 1991). The two genes should be arranged such that there is not interference between them as this might lead to an imbalance of chains. In particular transcription from the first promoter may extend beyond the end of the first gene and interfere with transcription initiation at a second promoter downstream. For this reason the light chain is commonly put upstream of the heavy chain, such that excess heavy chain is less likely to be produced. Also a transcriptional terminator can be introduced between the two genes to minimise interference (Stephens *et al.*, 1995).

Immunoglobulin promoters and enhancers have often been used to direct expression in myeloma cells. Early recombinant antibody expression used V regions obtained from hybridomas by genomic cloning which normally cloned the Ig promoter and enhancer at the same time. Use of these to direct expression of recombinant antibody genes usually results in considerably lower expression levels than seen in the parent hybridoma, with yields often in the range 1–5 $\mu g/10^6$ cells/day accumulating antibody from 1–40 milligrams per litre (Sahagan *et al.*, 1986; Crowe *et al.*, 1992). This is because sequences distant from the V region genes are required for efficient expression. Sequences with additional enhancer activity have been identified for both heavy and light chain genes which are several kilobases downstream of the coding regions (Meyer and Neuberger, 1989; Pettersson *et al.*, 1990). Improved expression levels may be achieved by using strong viral or cellular promoters such as hCMV-MIE or the mouse metallothionein promoter (Gillies *et al.*, 1989a).

Expression levels are also highly variable from one transfectant to another, and many cell lines must be screened to identify the best producers. This is probably dependent on the site of integration into the genome and the level of transcription in that region. For example, the identification of a cell line expressing 5 $\mu g/10^6$ cells/day of the humanised antibody CAMPATH–1H required screeening of 700 transfectant clones (Crowe *et al.*, 1992). Attempts have been made to simplify this process using homologous recombination to target gene integration to a highly transcribed region of host DNA. An extensive region of homology between the vector and a sequence on the host cell genome can lead to homologous recombination resulting in the integration of the vector sequences at a defined site. This is a relatively rare event compared to random integration, but nevertheless this approach has resulted in high-level expression in some cases (Yarnold and Fell, 1994).

The most productive cell lines have been produced using selection for gene amplification to obtain increased copy number of vector sequences integrated into the genome. Gene amplification is a relatively common event in mammalian cells, leading to multiple tandem repeats of large regions of a chromosome. These can be selected for by use of a marker such as an essential enzyme which can be inhibited using a selective inhibitor. Two systems have been widely used, based on dihydrofolate reductase (DHFR) and glutamine synthetase (GS), although others such as adenosine deaminase (ADA) are also available. DHFR is a key enzyme in nucleoside biosynthesis which can be inhibited by methotrexate. Following transfection using a vector containing the DHFR gene, selection is achieved by adding methotrexate and resistant colonies, in which the DHFR gene has been amplified (hopefully along with the antibody genes), are isolated. This is a widely used system for expression in CHO cells. For example, chimeric anti-CD20 antibody has been produced at 30 $\mu g/10^6$ cells/day using a single round of amplification in the presence of methotrexate (Reff *et al.*, 1994) and the humanised antibody CAMPATH-1H has been produced at up to 100 $\mu g/10^6$ cells/day, accumulating 200 mg per litre of antibody (Page and Sydenham, 1991). DHFR mediated amplification can also be used in myeloma cells, although because myeloma cells contain endogenous DHFR activity, a gene containing a methotrexate resistant DHFR has been used, and high levels of methotrexate, up to

500 μM, were required to achieve amplification (Dorai and Moore, 1987). Alternatively DHFR expression can be driven from a stronger promoter, allowing selection against the background of endogenous enzyme.

Glutamine synthetase (GS) catalyses the formation of glutamine from glutamate and ammonia and provides the only pathway for the biosynthesis of glutamine. Therefore, in the absence of endogenous glutamine, GS is an essential enzyme. Methionine sulphoximine (MSX) is a selective inhibitor of GS which can be used to select for gene amplification. CHO cells express sufficient GS to grow without added glutamine, but high-expressing cell lines can be isolated using low levels of MSX. Cell lines producing 15 μg/10^6 cells/day, and accumulating 200 mg/litre of chimeric B72.3, have been produced using GS (Bebbington, 1991), and accumulated yields of antibody greater than 500 mg/litre have now been reached using this system in CHO cells (Brown *et al.*, 1992). Myeloma cells do not express sufficient GS of their own and thus have an absolute requirement for glutamine. Therefore when grown in glutamine-free media, GS selection can be used at lower levels of MSX than required for CHO cells. Expression of chimeric B72.3 in NS0 cells reached 10–15 μg/10^6 cells/day, and could accumulate 560 mg/litre (Bebbington *et al.*, 1992). Yields of other recombinant antibodies have also been very high using GS selection in NS0 cells, for example 700 mg/litre of hA33 (King *et al.*, 1995) and over 1 g/litre for hCTM01 (Baker *et al.*, 1994).

Vectors which contain a selectable, amplifiable marker can also be used to select for integration into particularly active transcriptional sites. For example, the GS expression vector has been set up with GS under the control of the relatively weak SV40 promoter. If a low concentration of MSX is added to suppress GS activity during the selection process, then clones in which the GS gene is integrated at a particularly active site are selected. This reduces the number of transfectants isolated, yet increases the proportion which are high-level producers. Characterisation of such a transcriptionally active region of the chromosome in a clone of a high-yielding NS0 cell line selected using GS identified a site in the IgG2a genomic region as the integration site (Hollis and Mark, 1995). Vectors which allowed targeting to this region via homologous recombination were then produced which increased the frequency of integration at this site and thus the frequency of high-level producers.

NS0 cells are particularly well suited to growth in fermentation due to their ability to grow to high cell density in suspension (approx. 10^7 cells/ml). NS0 cells have also been shown to be capable of maintaining stable expression over many generations following transfection and selection using the GS vector system (Brown *et al.*, 1992). The combination of these factors has allowed the development of fed-batch fermentation systems capable of producing titres of antibody routinely over 500 mg/litre and often between 1 and 2 grams per litre (Figure 5.1) (Brown *et al.*, 1992; Bibila and Robinson, 1995). Although more difficult to scale up to fermentation growth, after fermentation development DHFR-amplified CHO lines have also been shown to be capable of producing antibody titres in excess of a gram per litre (Trill *et al.*, 1995).

Comparisons of NS0 and CHO cells for production of antibodies have been carried out by several groups. Expression of cB72.3 IgG4 from both CHO and NS0 cells using the GS system resulted in CHO cells being approximately half as productive as NS0 cells in fed-batch fermentation (Adair *et al.*, 1993). There was no apparent difference in antigen binding or physical properties of the antibodies produced, and pharmacokinetic and biodistribution studies of the antibodies in mice showed identical results. A detailed comparison of the production of an anti-CD4 antibody has been carried out using DHFR amplified CHO cells compared to GS amplified NS0 cells (Peakman *et al.*, 1994). In cell

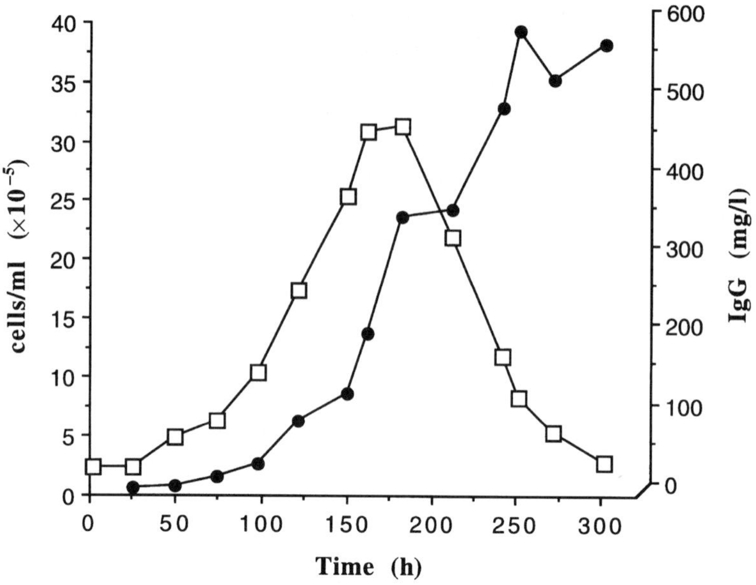

Figure 5.1 Production of chimeric IgG4 in NSO cells using GS system (see text for details): fed-batch 5 litre fermentations carried out in serum-free air-lift fermenters (adapted from Bebbington *et al.*, 1992)

lines producing equivalent amounts of antibody, the copy number of antibody genes in the CHO cells was significantly higher than the NS0 cells, though mRNA levels produced were similar, suggesting that more active transcriptional sites were targeted using GS. Antibody from both types of cell line was identical in antigen binding and effector function.

One potential difference between cell types is the nature of the glycosylation of the antibody molecule. Subtle differences in the structure of the carbohydrate attached by different cell types, and even between different growth conditions for one cell type, have been observed which may affect the ability of the antibody to elicit effector functions (Lund *et al.*, 1996). In addition, it has been shown in one case for a human IgM that alteration in culture conditions could affect the glycosylation of the antibody leading to alteration of pharmacokinetic properties (Maiorella *et al.*, 1993).

Although there is considerable microheterogeneity in the carbohydrate structure of antibodies produced in mouse myeloma or CHO cells, this is also reflected in the analysis of normal human IgG, and overall the structures are rather similar. One difference is the presence of a bisecting *N*-acetylglucosamine residue between the two 'antennae' of the complex carbohydrate chain which is seen in human IgG and not in mouse or hamster antibodies. As expected, humanised antibodies produced in myeloma cells have sugar groups characteristic of the mouse cell, i.e. without the bisecting *N*-acetylglucosamine (Ip *et al.*, 1994). In addition, myeloma cells have been shown to attach a terminal α(1,3)-galactose residue in some cases, which is not present in human IgG or those produced from CHO cells (Lund *et al.*, 1993). For IgG the carbohydrate is buried within the structure of the paired CH2 domains, and alterations may affect biological properties, probably through changes in the conformation and disposition of the domains relative to each other and to the rest of the molecule. At present it is not known whether subtle differences will lead to differences in performance of antibodies in their intended applications, particularly *in vivo*. Concern has been raised over the presence of terminal α(1,3)-galactose residues in antibodies for human therapy because humans often have pre-existing antibodies which recognise this structure (probably as a result of cross-reacting *E. coli* antigens) (Borrebaeck *et al.*, 1993). However, during studies of the many chimeric and humanised antibodies produced in myeloma cells and administered to humans, an immune response to the constant region has not been detected (see Chapter 2).

5.2.3 *Expression of antibody fragments in mammalian cells*

Co-expression of light chain with Fd or Fd′ results in the secretion of Fab or Fab′ fragments. High yields of Fabs (greater than 100 mg/litre) can generally be achieved by mammalian cell expression in much the same way as IgG expression is achieved. For example, chimeric B72.3 Fab′ has been expressed to 200 mg/litre in CHO cells (King *et al.*, 1994), and humanised A33 Fab′ to 500 mg/litre in NS0 cells (King *et al.*, 1995). In common with bacterial expression (Section 5.3), there is generally little material expressed as $F(ab')_2$ unless hinge modification is undertaken (see Chapter 2). However, oxidation of Fab′ produced in CHO cells to $F(ab')_2$ has been achieved with high yield, demonstrating that the hinge cysteine residues are present (King *et al.*, 1992a).

Smaller antibody fragments such as Fv and scFv have also been expressed in mammalian cells, though in general this has been more difficult, and yields have been lower. Expression of Fv fragments in both myeloma and CHO cells has been achieved (Riechmann *et al.*, 1988b; King *et al.*, 1993). A direct comparison of the same Fv fragment expressed in CHO cells and *E. coli* (see Section 5.3 below) revealed that only 4 mg/litre could be achieved from CHO cells compared to 40 mg/litre in *E. coli* shake-flask culture, rising to 450 mg/litre in fermentation. Similarly, expression of scFv is far less efficient in mammalian cells than in *E. coli*, with yields of up to 10 mg/litre reported in myeloma cells after amplification (Dorai *et al.*, 1994). Mammalian cells may be more useful for scFv-based fusion proteins, particularly if the protein being fused to the scFv is structurally complex or requires glycosylation. Several fusion proteins have been expressed and shown to be active, including scFv-IL2 and scFv-B domain from Staphylococcal protein A (Dorai *et al.*, 1994).

5.3 Expression in *Escherichia coli*

The most widely used and convenient host for gene expression remains the simple prokaryote *Escherichia coli. E. coli* has several advantages for gene expression, including the availability of a wide range of expression vectors which can be readily introduced into the cell and tested. *E. coli* is also easy and quick to grow, allowing protein expression to be evaluated rapidly. *E. coli* does, however, suffer two major disadvantages for gene expression; its inability to carry out post-translational modification such as glycosylation, and the tendency of proteins expressed in the cell to accumulate as insoluble aggregates known as inclusion bodies. In many cases, soluble active protein can be recovered from inclusion bodies by *in vitro* protein solubilisation and refolding, although this is a time-consuming and often inefficient process which is not always successful. An alternative approach is to attach a bacterial signal sequence to the gene of interest in an attempt to direct secretion of the expressed protein into the cell periplasm or surrounding medium. In some cases, including those of antibody fragments, this allows the recovery of soluble active protein directly without the need for complicated refolding protocols.

Intact IgG is not readily produced in soluble form in bacterial expression systems. Although full-length heavy and light chain have been co-expressed intracellularly in *E. coli* and refolded material could be associated to form IgG, yields of assembled IgG were low (Boss *et al.*, 1984; Cabilly *et al.*, 1984). Also, as *E. coli* does not contain the cellular machinery to produce glycosylated proteins, the heavy chain is produced in unglycosylated form. Therefore, research has switched to the investigation of *E. coli* as an expression system for antibody fragments. Rapid progress in this field has taken place since the development of *E. coli* secretion systems which allow soluble, functional antibody fragments to be produced in high yields (Skerra & Pluckthun, 1988; Better *et al.*, 1988), so that now *E. coli* expression has become the system of choice for producing many antibody fragments.

In contrast to expression in mammalian cells, expression of smaller antibody fragments is particularly efficient in *E. coli*, and this has enabled research on novel fragments such as Fv and scFv to progress rapidly. In addition, production of antibody fragments in *E. coli* is relatively inexpensive compared to mammalian cells, leading to more economic manufacture.

5.3.1 *Intracellular expression of antibody fragments in* E. coli

Intracellular expression of antibody fragments can lead to the rapid accumulation of large amounts of insoluble protein and has been used to prepare a range of antibody fragments including Fv, scFv and Fab fragments (Field *et al.*, 1989; Bird *et al.*, 1988; Buchner and Rudolph, 1991). However, refolding of solubilised protein is complex and often has to be developed specifically for individual antibody fragments. Therefore it is often the optimisation of a refolding procedure and the demonstration of a homogeneous, fully functional state for the refolded product which limits the speed and efficiency of the production of antibody fragments by this route. Nevertheless, this method of production can be valuable for antibody fragments for which the secretion route fails, or for the production of fusion proteins which are toxic to the cell in native form, such as immunotoxins.

In common with expression of any foreign gene in *E. coli*, cDNA for the antibody fragment is inserted into the expression cassette of an expression vector. Suitable vectors contain a plasmid origin of replication, an antibiotic selectable marker and an expression cassette comprising a strong promoter and transcriptional terminator separated by a multi-cloning site with an efficient translational initiation region. Expression of foreign proteins exerts a considerable strain on the cells and leads to strong selection pressure in favour of cells which have lost the expression plasmid (segregational instability) or which have deletions in the expression cassette (structural instability). It is therefore desirable to use a tightly controlled promoter to prevent expression during most of the growth phase of the culture. There is little difference in the transcriptional rate achieved from a number of strong promoters available such as ptac, λpL and T7. The method of transcriptional regulation varies, however, with expression from lac based promoters such as tac induced by IPTG (isopropyl-β-D-thiogalactopyranoside) or by a switch of carbon source to lactose, whereas the λ promoters are controlled by a temperature switch. Translation of mRNA is achieved efficiently through use of an appropriately positioned ribosome binding site sequence which is complementary to ribosomal RNA. A spacing of 6–10 base pairs between the ribosome binding site and the initiation codon usually allows efficient translation. Many commercially available expression vectors contain these features, and allow the accumulation of large amounts of expressed protein.

Inclusion bodies are readily isolated by cell lysis and centrifugation, and antibody fragments solubilised using high concentrations of denaturants such as guanidine hydrochloride or urea, usually in the presence of reducing agents to ensure that all incorrectly formed disulphide bonds are reduced. Refolding is then achieved by a variety of methods depending on the particular antibody fragment expressed. Many variables need to be examined to identify the optimal procedure for refolding, including solubilisation agent, pH and redox conditions, protein purity and concentration, temperature and rate of change to native conditions. The main problem is preventing the re-aggregation of partially unfolded protein before the native state can be achieved. For some scFvs, renaturation has been achieved by simple dilution or dialysis into a non-denaturing buffer, although the exact procedures required for individual scFvs vary and yields are often low, in the range of 5–15% of the solubilised material (Bird *et al.*, 1988; Huston *et al.*, 1988). The inclusion of a redox coupling system (comprising reduced and oxidised glutathione in a ratio of 10 : 1) to catalyse disulphide interchange may allow improved yields through permitting reshuffling of disulphides in incorrectly folding structures. After optimisation of refolding conditions using such a procedure, native Fab has been recovered with a yield of 40% of the expressed protein (Buchner and Rudolph, 1991). In cases of refolding Fv or Fab′ fragments with two protein chains it has generally been found advantageous to purify the individual chains before attempting final refolding and association to active material (Field *et al.*, 1989; Buchner and Rudolph, 1991).

An alternative method, developed for scFv, has attempted to form disulphide bonds before refolding into the active conformation. The presence of correct disulphides might be expected to improve the recovery of native material, while incorrect disulphide bonds will lead to non-native material. scFv solubilised in sodium lauroylsarcosine was allowed to form disulphide bonds by air oxidation before gradual removal of denaturant, resulting in a yield of 50% of the solubilised protein (Kurucz *et al.*, 1995).

After refolding, the purified material must be characterised to demonstrate that soluble material is in the native state. Soluble, inactive forms of the protein can be produced through alternative protein folding pathways, and may be difficult to separate. In many cases it is difficult to determine the presence of these. For example, when producing an scFv for the first time, a loss of antigen-binding activity is sometimes observed compared to the parent IgG or monomeric antibody fragments produced from the IgG (e.g. Bird *et al.*, 1988). In cases where the scFv has been produced by refolding it is difficult to ascertain whether this is a result of interference in the binding site conformation as a result of the scFv format or a proportion of incorrectly refolded protein.

The uncertainties in refolding efficiencies which can be achieved make it difficult to compare yields achieved using this system. The yield reported may often depend on the procedure used and the amount of development work carried out to achieve high-efficiency refolding. Yields of active protein reported vary from less than a milligram per litre to typically 1–10 mg per litre. It has been reported that higher yields of dsFv can be obtained compared to scFv (Webber *et al.*, 1995). A yield of 1.6 mg/litre has been reported for anti-Tac scFv whereas the same antibody in dsFv format could be refolded with a yield of 7 mg/litre. Similarly, fusion proteins of this antibody with the bacterial toxin PE38 could be produced at 10–40 mg/litre in the scFv format, whereas 30–70 mg/litre was achieved with dsFv-PE38 (Reiter *et al.*, 1994).

5.3.2 Secretion of antibody fragments from *E. coli*

High-yielding systems have been described for the secretion of Fv, scFv, Fab and other antibody fragments to the *E. coli* periplasm (Skerra and Pluckthun, 1988; Better *et al.*, 1988). Soluble, active protein can often be purified directly from the periplasm and in some cases from the medium. Recovery from the medium is generally a result of accumulation of the antibody fragment in the periplasm causing the cell outer membrane to become leaky, leading to leakage of the antibody fragment into the medium (e.g. King *et al.*, 1993). A signal sequence from an efficiently exported bacterial protein is fused to each antibody gene to direct secretion. There appears to be little difference in the effectiveness of those signal sequences commonly used including pelB, ompA, phoA or stl1.

Antibody fragments are efficiently transported across the inner membrane, the signal sequence is efficiently processed to form the mature protein chains and disulphide bonds appear to form efficiently. There is, however, a tendency for some antibody fragments to accumulate as insoluble protein in the periplasm, probably due to the aggregation of folding intermediates (Whitlow and Filpula, 1991). Aggregation of folding intermediates competes with correct folding, and is often the limiting factor in the yield of soluble antibody fragment (Knappik *et al.*, 1993). The extent of this problem varies markedly between fragments of different antibodies and between different fragments of the same antibody. In general, smaller fragments are less prone to insolubility in the periplasm, with Fv-based fragments produced more efficiently than Fab-based fragments of the same antibody. For example, B72.3 Fv could be expressed at 40 mg/litre in soluble form in shake-flask culture whereas a Fab fragment of the same antibody was expressed at 5 mg/litre (King *et al.*, 1992a). The main determinant of insolubility appears to be the primary sequence of the antibody fragment itself. After comparing the sequences of antibody fragments which fold well or poorly, point mutation studies could identify residues which, when changed, improved folding and consequently the yield of soluble protein recovered (Knappik and Pluckthun, 1995). Similar findings have been made with other residues which affect folding pathways (Kipriyanov *et al.*, 1997b).

Several other protein engineering strategies have been used to improve expression of soluble antibody fragments. Switching Fab constant regions has been shown to result in higher yields in some cases (MacKenzie *et al.*, 1994). scFv variants have been engineered by mutation of residues in the hydrophobic patch created in the scFv where the variable/constant domain interface is normally present. Engineered scFv variants were isolated with improved folding properties resulting in higher yields of soluble protein (Nieba *et al.*, 1997). A particularly useful strategy for achieving high-level secretion of active antibody fragments has been exemplified by Carter *et al.* (1992a). A well expressed, soluble humanised Fab′ fragment has been identified which allowed yields of 1–2 grams per litre to be accumulated in fermentation culture. The subsequent CDR-grafting of other antibodies onto the same Fab′ framework region allowed high-level secretion of other binding specificities (Carter *et al.*, 1992a).

Control of cell growth parameters can also aid expression of soluble secreted protein. Growth at lower temperature during gene expression can lead to improved yields, probably due to decreased periplasmic aggregation (Skerra and Pluckthun, 1991). Also, the addition of certain non-metabolisable sugars may be beneficial in reducing aggregation. The yield of soluble scFv was increased over 100-fold by addition of 0.4 M sucrose to the growth medium in shake-flask culture (Kipriyanov *et al.*, 1997a).

The rate of protein synthesis is critical to achieving high yields of soluble secreted protein. This can be modulated either through control of induction of transcription or by altering translational efficiency. Maximal induction of gene expression may lead to protein synthesis being too rapid to allow the antibody fragment to fold efficiently, leading to increased insolubility through aggregation of rapidly produced folding intermediates. Similarly, translational rate has been demonstrated to be of critical importance for the secretion of heterologous proteins from *E. coli* (Simmons and Yansura, 1996), the optimal rate for secretion of active material being less than the maximum achievable, and variable for different proteins. The optimal yield is therefore often achieved through reducing the rate of protein synthesis by partial induction of expression at a rate which is optimal for secretion and folding. Lower inducer concentrations have been shown to allow higher levels of soluble scFv to be accumulated (Sawyer *et al.*, 1994), and Fab fragments can also be accumulated in higher yields when expression is induced in a controlled fashion over a prolonged period (Yarranton and Mountain, 1992). Accumulation of high levels of foreign protein in the periplasm often leads to the outer membrane becoming leaky, and eventually to cell lysis. Suitable control of induction can also reduce the amount of cell lysis and lead to improved recoveries.

Suitable promoters which allow tight control of expression during cell growth and which can be induced to allow regulation of transcriptional rate are therefore desirable. These include the wild type lac promoter/operator, lacUV5 and the tac promoter, all of which can be induced with the chemical inducer IPTG or by a switch in carbon source to lactose (e.g. Yarranton and Mountain, 1992), the phoA promoter which is induced by phosphate starvation (Carter *et al.*, 1992a), the araB promoter induced by arabinose (Better *et al.*, 1988), and temperature inducible systems such as that described using the trpE promoter controlled by a temperature-induced increase in plasmid copy number titrating out chromosomally expressed trp repressor (Yarranton and Mountain, 1992).

Co-expression of proteins involved in assisting protein folding in *E. coli* has also been investigated in attempts to boost yields of soluble periplasmic protein. The co-expression of the intracellular chaperonins GroES/L had no effect on soluble scFv expression (Duenas *et al.*, 1994). Similarly, over-expression of the periplasmic proteins proline cis–trans isomerase and dsbA, a protein involved in disulphide bond formation in the periplasm, does not result in improved yields (Knappik *et al.*, 1993). The presence of dsbA is required for assembly of Fab fragments in the periplasm, but does not appear to be limiting. The co-expression of the human enzyme protein disulphide isomerase (PDI) has been shown to improve the yield of a relatively poorly produced Fab′ with γ4 constant regions (Humphreys *et al.*, 1996). The same antibody with γ1 constant regions was produced in higher yield, and in this case there was no effect of PDI co-expression. There are several other periplasmic proteins known to be involved in disulphide formation in *E. coli*, currently dsbA, B, C and D have been described, but as yet studies of their role in folding and assembly of expressed antibody fragments have not been described.

Growth of cells to high cell density in fermentation has led to very high yields of antibody fragments in several cases (Figure 5.2). Fermentation culture also allows greatly improved control of the parameters governing cell physiology such that optimal production levels can be achieved. For example, expression of a chimeric Fab has been improved from 1 mg/litre in shake-flask culture to 500 mg/litre in optimised fermenters (Better *et al.*, 1990), and B72.3 Fv expression from 40 mg/litre to 450 mg/litre (King *et al.*, 1993). Bispecific diabodies have been produced at 935 mg/litre in fermentation culture (Zhu *et al.*, 1996). Yields of over a gram per litre have been observed for humanised Fab′ fragments (Carter *et al.*, 1992a), and a bivalent scFv construct has been expressed at over 3 g/litre in functional form (Horn *et al.*, 1996). In such cases expression of antibody fragments in *E. coli* is a highly competitive option compared to mammalian cell expression.

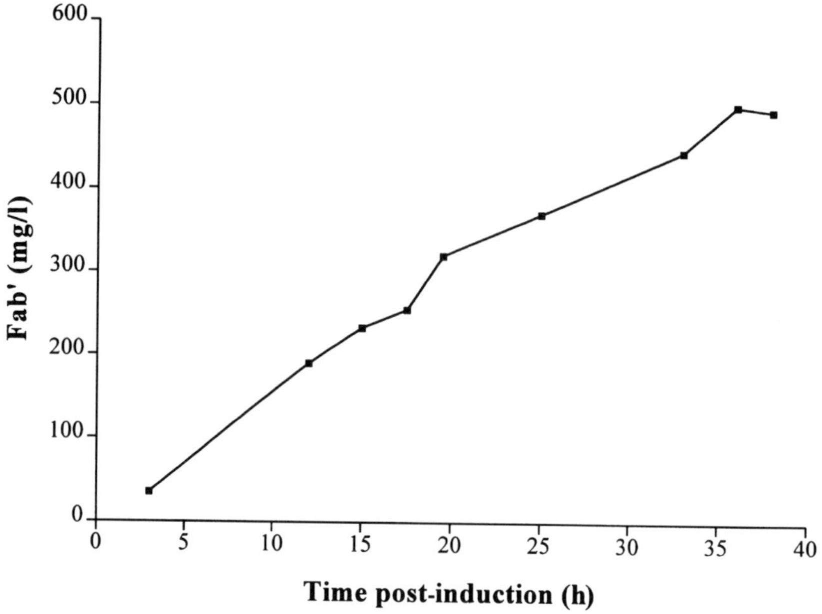

Figure 5.2 Expression of soluble engineered human Fab′ in the periplasm of *E. coli*: data from a 1 litre fermentation by courtesy of Neil Weir and Volker Lang (Celltech Therapeutics)

E. coli grows rapidly; transformed cells can be readily produced and cloned leading to shorter development times than those seen with mammalian cells. Fermentation of microbial cells at very large scale is a well-established industrial procedure which is possible at much larger scale than the scale at which mammalian cells can be grown currently. This could lead to significant economic advantages of using *E. coli* for the expression of antibody fragments at the scale required to satisfy large potential markets.

5.4 Expression in other microbial systems

Alternative microbial hosts have been relatively poorly explored for antibody or antibody fragment expression. The Gram-positive bacterium *Bacillus subtilis* has been shown to be capable of expressing scFv fragments of an antibody to digoxin at 5 mg/litre in shake-flask culture (Wu *et al.*, 1993). In contrast to *E. coli*, *B. subtilis* is capable of secreting large amounts of protein directly into the growth medium, although it can also secrete a variety of extracellular proteases which can lead to breakdown of heterologous secreted proteins. Multiple protease deficient strains have been developed to allow accumulation of intact protein which may be isolated simply from the growth medium. Expressed scFv was isolated and purified in a single step by affinity purification from the growth medium (Wu *et al.*, 1993).

Eukaryotic microorganisms have also been investigated. The yeast *Saccharomyces cerevisiae* can synthesise, process and secrete heavy and light chains. Expression of a mouse IgM resulted in the production of functional antibody, though secretion was relatively inefficient with most of the assembled antibody remaining inside the cell (Wood *et al.*, 1985). Heavy and light chains were expressed from different plasmids under the control of the strong phosphoglycerate kinase promoter, and expressed heavy chain was shown to be glycosylated. Chimeric IgG and Fab have also been expressed in *S. cerevisiae*, resulting in the secretion of antibody and Fab fragment with full antigen binding activity (Horwitz *et al.*, 1988). Secreted levels of antibody were low, with light chain measured at 100 ng/ml and heavy chain at 50–80 ng/ml, 50–70% of which was associated with light chain. The expressed IgG was able to mediate ADCC but could not activate complement, probably as a result of the 'yeast type' carbohydrate added to the constant region.

The methylotrophic yeast *Pichia pastoris* has been developed as a high yielding expression system for several heterologous proteins. *P. pastoris* combines fast growth properties with the general features of eukaryotic protein expression and secretion. The alcohol oxidase gene (AOX1) is used as the basis of expression, and can be used to direct either intracellular expression or secretion. The AOX1 promoter is tightly regulated and drives rapid transcription on addition of methanol, the substrate of alcohol oxidase. Use of the AOX1 promoter therefore allows regulation of expression by supplementation of methanol as the carbon source. Use of this system has allowed secretion of an scFv of a rabbit MAb to more than 100 mg/litre, greater than 100-fold more than could be produced with the same scFv from *E. coli* (Ridder *et al.*, 1995). Two murine scFv fragments have also been expressed, in one case reaching 250 mg/litre, which could be easily recovered from the cell growth medium (Eldin *et al.*, 1997). The expression of other formats of antibody fragments in *P. pastoris* has not yet been reported, but such results with scFv expression merit investigation of this system as a potentially productive alternative.

The filamentous fungus *Trichoderma reesei* has also been investigated for the secretion of antibody fragments (Nyyssonen *et al.*, 1993). This organism has a long history in industrial production of hydrolytic enzymes, which it is capable of secreting at levels of 40 g/litre of culture, and has therefore been investigated as a host for heterologous protein production. The major cellulase, cellobiohydrolase, cbh1 represents 50% of secreted protein and is produced from a single-copy gene, and thus the cbh1 promoter is very strong. This has been used along with the cbh1 signal sequence to direct expression and secretion of a Fab fragment (Nyyssonen *et al.*, 1993). Fab was expressed at only approx. 1 mg/litre, but when a cbh1-Fd fusion protein was co-expressed with light chain, secreted levels reached 150 mg/litre. A *T. reesei* protease was able to cleave some of the cbh1-Fd fusion protein in the culture medium, but resulted in only small amounts of assembled Fab.

5.5 Expression in plants

Transgenic plants have been shown to be capable of producing and assembling both antibody fragments and intact antibodies, including the complex secretory IgA molecule. Plant systems are currently being investigated both as a potential source of large quantities of antibodies for diagnostic or therapeutic application and for potential *in situ* uses within the plant itself (Franken *et al.*, 1997). Such plant-derived antibodies, often termed 'plantibodies', have the potential to be manufactured very cheaply as plant agriculture is the most economical system for production of large quantities of biomass. It has been estimated that expression of antibody in soybean at a level of 1% of total protein (a level already achieved in some cases – see below) could result in production of antibody at a cost of $100 per kilogram (Hiatt, 1990). However, estimation of total cost is difficult as extraction and purification costs following harvest are likely to be significantly higher than other types of production techniques.

The first report of expression of functional antibody from plant cells described the production of intact murine IgG at a level of 1.3% of total protein (Hiatt *et al.*, 1989). The most common method of transforming plant cells is to introduce the genes of interest into the Ti plasmid of *Agrobacterium tumefaciens*. Recombinant bacteria can then be used to infect plant cells, introducing the gene of interest. Tobacco plant (*Nicotiana*) cells are often used due to their relative ease of transformation using this system. Heavy and light chain cDNAs including the signal sequences were introduced into individual leaf segments and used to regenerate mature plants. The plants expressing heavy and light chain were then crossed and progeny producing assembled antibody isolated (Hiatt *et al.*, 1989). The signal sequences were required for production of functional assembled antibody, and resulted in accumulation of antibody in the intercellular spaces, which is a stable environment in plants and may facilitate antibody isolation. In plant cell suspension cultures the antibody was produced at approx. 20 mg/l in the culture medium.

Both scFv and Fab′ fragments have also been produced in plants at levels similar to those achieved with intact IgGs (Owen *et al.*, 1992; DeNeve *et al.*, 1993). It has also proved possible to produce secretory IgA in tobacco plants which requires the co-expression of four protein chains: heavy chain, light chain, joining chain and secretory component (Ma *et al.*, 1995). Each protein chain was introduced into separate plants and a series of crosses carried out to produce plants capable of producing assembled functional sIgA. The antibody used recognises *Streptococcus mutans*, which is largely responsible for dental caries in man. Oral application of the antibody leads to cross-linking of bacteria and protects against dental caries. Secretory IgA is more stable at mucosal surfaces than IgG and may be a suitable molecule for passive immunotherapy under such conditions. The production of such antibodies in plants offers the intriguing prospect of an oral delivery route in edible plant tissue. If expressed in edible plant tissue, a food formulation of antibody may be possible to exert a therapeutic or prophylactic effect in the mouth or gastrointestinal tract, avoiding the need for isolation and purification of antibody (Ma and Hein, 1995).

An alternative plant source of antibody may be seeds. scFv fragments have been shown to be capable of production in seeds at 0.67% of total seed protein (Fiedler and Conrad, 1995). This also appears to be a stable method of storage of scFv, as seeds containing scFv were stable for a year at room temperature without any loss of scFv protein or antigen-binding ability.

The production of antibodies for therapeutic application in plants has not yet been attempted and thus the regulatory implications are at present unknown. Carbohydrate of plant-produced antibodies is different to that attached to antibodies by mammalian cells, as is also the case with several other expression systems as described above. Some sugar residues may be unique to plants and lead to immunogenicity of administered antibody. However, antigen-binding fragments such as Fab and scFv are not normally glycosylated and their use may overcome such concerns. Plant-specific contaminants such as secondary metabolites may be a cause for concern and would need to be demonstrated to be removed from an antibody product. On the other hand, plant-derived antibodies are not likely to be contaminated with animal viruses or other animal-derived agents. It will be interesting to monitor the progress of transgenic plants as an antibody production system over the next few years.

5.6 Production in transgenic animals

The production of MAbs in the milk of transgenic animals is also a potentially low-cost route to the production of antibodies which is currently under development. Genes are introduced into the fertilised egg by microinjection. A number of heterologous proteins have been expressed in the milk of lactating transgenic animals and current research is engaged in the development of such systems to generate proteins for therapeutic use (Colman, 1996). Transgenic sheep, goats and cows are being investigated for their ability to produce large volumes of milk containing the expressed protein. High-level expression of functional antibody in transgenic goats has been reported, with levels of 10 g/litre produced in some animals (Parkinson, 1995); however, full details have not yet been published. One drawback of this approach is the long lag time required to generate a reasonable number of mature animals capable of high-level expression of foreign protein.

5.7 Expression in insect cells

Insect cell expression systems have been developed using baculovirus expression vectors which allow rapid production of recombinant proteins. High-efficiency expression and secretion can be achieved in a cost-effective manner using the Sf9 insect cell line. Expression vectors have been developed using the promoter from the baculovirus polyhedrin gene and insect cells transfected using this system are capable of signal peptide cleavage, glycosylation and efficient secretion from the cell. The baculovirus expression system produces a lytic infection which kills the cells after a short time and therefore production levels are relatively low, typically in the range 5–10 mg/l (Hu *et al.*, 1995). Advantages of insect cell expression compared to mammalian cells include the absence of mammalian viruses and any possible mammalian DNA encoding oncogenes.

Antibodies have been expressed in insect cells by several groups. Heavy and light chains are efficiently processed and assembled into intact antibody when expressed by coinfection with two recombinat viruses or when heavy and light chain are introduced on a dual transfer vector containing the two transcription units in opposite orientations (Hasemann and Capra, 1990; Zu Pulitz *et al.*, 1990). Insect cell-specific carbohydrate is added to the antibody in place of that seen from hybridoma cells, and effector functions of antibodies produced in this system have not always been maintained (Poul *et al.*, 1995). Nevertheless, expression in insect cells can be useful for rapid generation of milligram quantities of antibody, particularly in laboratories that have the system set up for other purposes.

Murine, chimeric and humanised IgG have been expressed in insect cells, as has functional human IgA (Carayannopoulos *et al.*, 1994). scFv fragments have also been expressed at 32 mg/litre in insect cell fermentation culture (Kretzschmar *et al.*, 1996). scFv from insect cells could be easily purified by a two-step method without the need for affinity purification.

5.8 Production of monoclonal antibodies – cell culture

MAbs from hybridoma cells or recombinant antibodies from alternative mammalian cell types can be grown using a number of different methods. The scale and characterisation of the production process is determined by the requirements of the intended application. For example, the production of MAbs for use in humans requires stringent control of the entire production process from preparation of cell banks to characterisation of the purified product, whereas small-scale preparation of antibody for laboratory use presents a different range of problems, such as requiring rapid production often of many different antibodies for evaluation purposes.

Cell culture at small scale (up to a few litres) can produce reasonable amounts of material from cells grown in either attached or suspension culture. More concentrated antibody can be derived by growing hybridoma cells as ascites tumours in mice or rats (Brodeur and Tsang, 1986). Typically antibody is produced rapidly within 1–2 weeks following hybridoma inoculation into mice at concentrations of 1–10 mg/ml. However, the use of animals for antibody production in this way is difficult to justify on ethical grounds when alternative methods are available, and it is prohibited or restricted in several countries. There is also a risk of introducing adventitious agents from the animal into the preparation of antibody. Many *in vitro* cell culture techniques are available including the use of attached cells in roller bottles, shake-flask or spinner cultures of cells in suspension, and a range of bioreactors at either small or large scale, such as hollow fibre perfusion systems and air-lift fermentors (Jackson *et al.*, 1996; Birch *et al.*, 1987). Bioreactors can also be scaled up for industrial manufacture, and both airlift fermentors and perfusion systems are currently used for commercial scale manufacture of MAbs with batch sizes of several thousand litres.

Process development to maximise the productivity of a cell line in fermentation becomes crucial to allow economic manufacture. Such development requires optimisation of culture media, nutrient supplementation throughout the fermentation and key enviromental parameters such as pH, dissolved oxygen concentration and temperature (Brown *et al.*, 1992; Bibila and Robinson, 1995). The degree of process development undertaken in the manufacture of a particular MAb will depend on the balance between the requirement to develop the best process at the lowest cost and the time available.

5.9 Purification of monoclonal antibodies

5.9.1 Purification of IgG

For most applications purified MAbs are required, and for therapeutic puposes they must meet a very high purity specification as described below. Purification of MAbs uses the molecular properties of the immunoglobulin molecule in the same way as any protein purification process relies on the molecular properties of the protein of interest. Affinity chromatography is commonly used because of the high degree of purification which can be achieved in one step. This can be based on the specific binding properties of the immunoglobulin either through Fc-binding ligands such as the bacterial immunoglobulin-binding proteins, Staphylococcal protein A and Streptococcal protein G, or using the antigen-binding site itself through binding to immobilised antigen. Alternatively a number of lower resolution methods can be used either alone or in combination, such as ion-exchange purification based on overall molecular charge, hydrophobic interaction chromatography based on hydrophobicity and gel filtration based on the size of the molecule.

The design of the purification process can be considered in two stages: sample preparation, also known as primary recovery, and purification. After production in cell culture the antibody sample must be prepared by removal of the cells and particulate matter, typically by centrifugation and filtration. At a small scale such steps are relatively trivial and easily accomplished using standard laboratory equipment. At large scale such steps may represent more of a challenge, requiring the use of continuous centrifuges which are not capable of delivering the high *g* forces obtained in laboratory equipment. Precipitation of antibodies, for example, with ammonium sulphate may also be used in primary recovery. This allows concentration of the antibody in the sample as well as a partial

Table 5.1 Binding of immunoglobulin G subclasses to Staphylococcal protein A and Streptococcal protein G

Species	IgG subclass	Protein A	Protein G
Mouse	IgG1	+	+
	IgG2a	++	++
	IgG2b	++	++
	IgG3	++	++
Rat	IgG1	+	+
	IgG2a	—	++
	IgG2b	—	+
	IgG2c	++	++
Human	IgG1	++	++
	IgG2	++	++
	IgG3	—	++
	IgG4	++	++

purification and is often used on a small scale. This method is difficult to use on a large scale because of the requirement for handling and centrifuging the volumes involved. Before purification, the sample may also need to be in solution under certain conditions of pH or ionic strength, depending on the particular method to be used, and therefore the solution may need to be adjusted (conditioned) before purification takes place. Purification is usually achieved through column chromatography using one or more steps. For most laboratory purposes a single step of affinity chromatography is usually sufficient, typically producing MAb at >90% purity. For therapeutic purposes it is necessary to improve the purity to >95% and this is commonly achieved by a combination of affinity chromatography with other techniques such as ion-exchange chromatography or gel filtration.

The best place to start in designing a small-scale purification scheme for IgG is with the bacterial immunoglobulin binding proteins. Both protein A from *Staphylococcus aureus* and protein G from group C and G Streptococci bind to the Fc region of antibodies and are widely available immobilised to suitable solid phases for chromatography applications. Protein G is most useful in recombinant form in which the albumin binding domains, also present on natural protein G, have been removed. Different subclasses of antibody bind to protein A and protein G with different affinities, and thus the subclass is of critical importance in selection of the purification method (Table 5.1). The binding sites for both protein A and protein G for IgG Fc have been defined, and mapped to overlapping sites at the CH2/CH3 domain interface (Deisenhofer, 1981; Sauer-Ericksson *et al.*, 1995).

Clarified culture supernatant is applied directly to a column of protein A or protein G under conditions in which the immunoglobulin will bind. Antibodies which bind strongly to protein A or G often require little pre-conditioning of clarified culture supernatant, whereas those which bind weakly may require special buffer conditions to promote binding. For example, mouse IgG1 requires the inclusion of 4 M sodium chloride in the sample to allow binding to protein A. After washing, antibody is eluted from the column, commonly using a decrease in pH (Figure 5.3). Some antibodies may be unstable at low pH and thus require the use of alternative eluents such as chaotropic ions or agents such

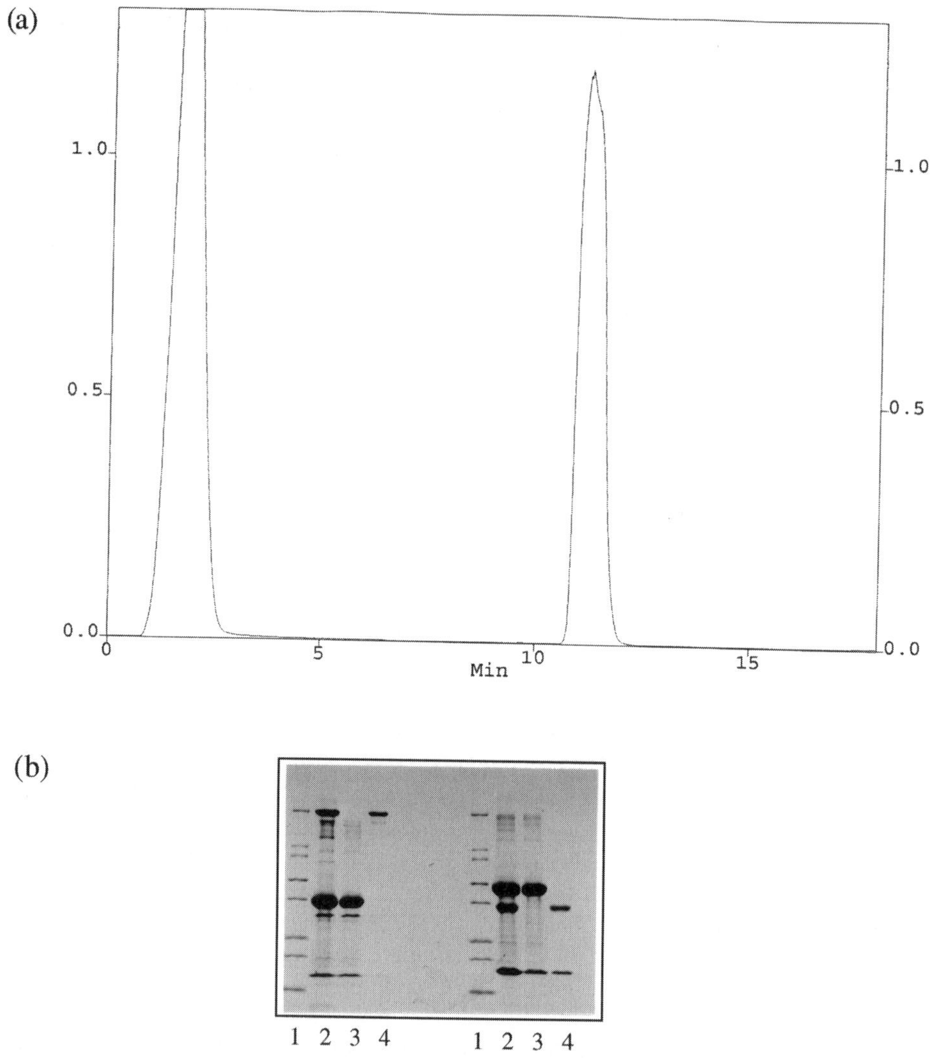

Figure 5.3 Purification of engineered human IgG1 from NS0 cell culture supernatant. Cell culture supernatant was adjusted to pH8 and applied to a column of Poros 50A, immobilised protein A (Perseptive Biosystems) at 600 cm/h which had been pre-equilibrated with 50 mM glycine/glycinate buffer pH8.8. After washing with equilibration buffer, IgG was eluted with 0.1 M glycine–HC1 pH3.5. (a) Absorbance at 280 nm, showing peaks due to non-bound protein (flow through) and eluted IgG. (b) SDS-PAGE under non-reducing and reducing conditions. Lane 1, molecular weight markers; lane 2, culture supernatant sample applied to column; lane 3, flow through peak; lane 4, eluted IgG.

as ethylene glycol (Bywater *et al*., 1983). Most MAbs will be prepared from clonal cell lines and thus there is little contamination with other rodent or human antibodies when prepared in cell culture. Therefore preparation of total immunoglobulin will allow a high degree of purity to be obtained. Some bovine IgG may be present, however, from the bovine serum often used in cell culture techniques. This can also be removed using bacterial immunoglobulin binding proteins, depending on the subclass of the antibody.

Removal is straightforward from subclasses of antibody which bind strongly to protein A, but more difficult from weakly binding antibodies such as mouse IgG1. In this case it may be necessary to include a separate step specifically to remove bovine IgG, such as ion-exchange chromatography, or the use of anti-bovine IgG affinity chromatography. Purification of antibodies from ascitic fluid may be more difficult for two principal reasons. Firstly, ascitic fluid contains a high proportion of cells and particulate material and also lipid which must be removed by high speed centrifugation, or other means, before purification. Secondly, there is considerable contamination with host mouse IgG which may be difficult to remove from the MAb.

Alternatively, affinity chromatography using antigen can be considered. If sufficient antigen is available, antigen can be immobilised onto a solid phase and used to purify the antibody using immunoaffinity chromatography in the reverse of the immunopurification operation described in Section 3.7. Immobilised antigen can result in a high degree of purification and is one of the few techniques capable of resolving mixtures where several antibodies may be present, such as during the preparation of bispecific antibodies. However, antigen is rarely available in sufficient quantity for more than small-scale operations, and elution conditions required are often harsh for high-affinity antibodies which can lead to losses in affinity of the isolated antibody.

Although the use of protein A or protein G is the first choice purification method in most cases, in others alternative methods may be preferable. The main reasons for choosing an alternative method are the instability of the antibody to elution conditions, and cost. In practice it is usually possible to develop elution conditions which are suitable for retention of activity of the antibody, and antibodies for which this cannot be achieved are rare. Affinity materials using protein A or G are relatively expensive, however, with other materials such as ion-exchange resins being an order of magnitude less expensive in many cases. There is often, therefore, commercial pressure to remove affinity steps from a purification process to minimise costs. However, it is important to consider the cost of the affinity matrix compared to the time and cost of several steps (with consequent yield losses) which might be needed to replace it. It may also be possible to reduce costs through use of repeated cycles on a relatively small column.

Alternatives to affinity chromatography rely on conventional purification techniques. Ion-exchange purification of antibody is a useful technique either as an additional step to affinity chromatography or as an alternative. The isoelectric point of MAbs varies widely from antibody to antibody, in my experience from 4.5 to 8.5, and thus a general method for purification cannot be universally applied. Anion exchange is particularly useful as a second step in purification protocols after affinity chromatography and can lead to the generation of highly purified product. In addition DNA binds very strongly to anion exchange materials and the inclusion of anion exchange steps in purification schemes is useful for therapeutic antibodies, as separation of any residual DNA in the preparation will be achieved.

As a first line antibody purification step anion exchange is often problematical, due to binding of serum albumin which is present in most cell culture fluid at high concentration, and which reduces the capacity of the anion exchange material to bind antibody. In addition, phenol red which is added to small-scale cell cultures as a pH indicator also binds very strongly. Cation exchange has been found to be more suitable for antibody purification, although this is highly dependent on the pI of the particular antibody. In some cases extensive conditioning of the material is needed prior to purification to reduce pH and/or conductivity, often resulting in the need for considerable dilution or the use of diafiltration. The purity of the antibody is not as high as that which is obtained from affinity chromatography, but the use of high-resolution anion exchange materials which are now available from several suppliers may still allow acceptable purity to be reached in one or two steps (Carlsson *et al.*, 1985). Mixed-mode ion-exchange is also a particularly useful technique. Mixed-mode ion-exchangers, such as Bakerbond Abx (J.T. Baker), contain both weakly anionic and cationic groups. However, over the pH range used for antibody purification (pH5–7 in most cases) the operation of such materials resembles cation exchange. Antibody purity of 70–95% is readily achievable using an Abx mixed-mode ion-exchanger, and mixtures of antibodies may also be resolvable (Ross *et al.*, 1987). Hydroxylapatite, a form of calcium phosphate, is thought to operate along similar principles and can also result in high-purity antibodies after a single step (Stanker *et al.*, 1985). Hydroxylapatite is more difficult to use, however, as lower flow rates must be used and the resin has lower capacity than many other materials. Ion-exchange methods operate without the need for harsh conditions and are therefore particularly useful for antibodies which are sensitive to low pH or other reagents required for elution from affinity chromatography columns.

Hydrophobic interaction chromatography is a commonly used protein purification procedure which separates proteins on the basis of hydrophobicity. The resolution achievable with this technique is relatively low, and therefore it is useful only in combination with purification steps based on other modalities such as ion-exchange. The combination of ion-exchange with hydrophobic interaction chromatography is useful, however, and can result in the generation of highly purified material, equivalent to affinity purified antibody, without the high cost of expensive affinity resins. One problem with some MAbs is precipitation during sample preparation. Binding to hydrophobic interaction columns requires the use of high concentrations of salts such as ammonium sulphate which can cause antibody precipitation and in some cases loss of antigen-binding activity.

Gel filtration chromatography is also widely used as a final clean-up step following purification by affinity chromatography or other methods. It is particularly useful in removing antibody aggregates which are a common problem in antibody preparations. However, gel filtration is time-consuming and difficult to scale-up, and thus processes which can avoid the use of such steps are preferred at large scale.

Thiophilic adsorption of MAbs has also been developed as a method capable of single-step purification (Belew *et al.*, 1987). In this method high salt concentrations promote binding of a wide range of antibody types to a thiophilic resin produced by coupling mercaptoethanol to divinylsulphone activated agarose. Although similar to hydrophobic interaction chromatography in being salt-promoted, binding to this material is believed to be by a different, though unknown, mechanism. Elution from the adsorbent is achieved by reduction of the salt concentration. The preferential affinity of this material for immunoglobulins has allowed mouse MAbs to be obtained directly from culture medium or ascites fluid at purity levels similar to those achieved with ion-exchange chromatography methods.

One approach to simplify purification processes is to remove the need for primary recovery through the use of fluidised, or expanded, bed adsorption. The use of such techniques allows the application of whole cell fermentation harvest directly onto the expanded bed column without removal of cells and particulates, thus avoiding any yield loss due to centrifugation and filtration of the sample. Expanded bed chromatography operates through the use of a dense solid phase which can be suspended in solution, or expanded, such that particulate matter can flow through the column while the antibody binds to the dense particles and remains within the column. The column material is then packed into a conventional packed bed by reversing the flow, allowing elution of bound material in a sharp peak. Both ion-exchange and protein A materials are available which allow efficient purification of MAbs without clarification (Thommes *et al.*, 1995; Chaplin, 1996). Purification of a murine IgG2a from hybridoma culture was achieved by direct application of the culture to an expanded bed column using immobilised protein A. Material was recovered at a high yield at >95% purity with less than 0.3% aggregates in this single step (Chaplin, 1996). This technology can result in a significant improvement in the efficiency of large-scale operations and is likely to become increasingly important over the next few years.

5.9.2 Purification of IgM

The purification of MAbs of the IgM class is more difficult than IgG, substantially due to their lower stability. However, a number of methods have been devised to obtain pure material, including to the exacting standards required for clinical trials. Ion-exchange chromatography using mixed-mode ion exchangers such as Abx is an efficient method although care must be taken to avoid antibody precipitation at low ionic strength (Chen and Epstein, 1988). The high molecular weight of IgM at 900 kDa also means that gel filtration is an efficient purification method. Few contaminants have similar sizes and thus pure material can be obtained. Selective precipitation can also be used to prepare IgM in its pentameric form, taking advantage of the precipitation of IgM at low ionic strength (Vollmers *et al.*, 1996). In this method culture medium is simply concentrated and dialysed against distilled water, resulting in the selective precipitation of IgM. However, the efficiency of recovery using this precipitation process may be low and a more reliable method may be to use ammonium sulphate precipitation followed by gel filtration.

An affinity chromatography method for purification of IgM is also available based on mannan binding protein. Mannan binding protein (MBP) is a mannose and *N*-acetylglucosamine a specific lectin found in mammalian sera which is structurally similar to the complement component C1q, and binds to IgM and not IgG. Immobilised MBP has been developed in kit form for the purification of IgM (Nevens *et al.*, 1992). Binding to MBP is calcium-dependent and thus elution can be achieved under gentle conditions by use of EDTA. Also, some IgM molecules are capable of purification through the Fab region binding sites on protein A and protein L (see below).

The human monoclonal IgM COU-1 has been purified for clinical use using a four-step chromatographic procedure (Tornoe *et al.*, 1997). In this case antibody was purified from hybridoma cell supernatant using sequential chromatography with hydroxylapatite, hydrophobic interaction, cation exchange and anion exchange. The antibody was substantially pure with regard to protein after step 3 but the fourth step was required to remove contaminating DNA to acceptable levels for human administration. An overall yield of 57% was reported.

5.9.3 Purification of monoclonal antibody fragments

The purification of MAb fragments uses a range of techniques, many of which are similar to those used for IgG purification. The purification required depends on the source of the antibody fragment, with those prepared by digestion of IgG having different requirements to recombinant antibody fragments expressed by mammalian cells or *E. coli*.

After digestion of purified IgG the objective of the purification scheme is to separate the required antibody Fab or $F(ab')_2$ fragment away from any undigested IgG, other antibody fragments and the enzyme used for digestion. Removal of the enzyme is simplest when it is used in immobilised form linked to agarose beads which can be simply removed by centrifugation or filtration at the end of the digest period. Pepsin and papain are readily available in immobilised form, though the use of alternative enzymes such as bromelain may require use of the enzyme in solution. In some cases a simple negative purification can be successfully carried out using protein A or protein G to bind undigested IgG and Fc fragments and collecting Fab or $F(ab')_2$ in the flow-through. This often results in Fab or $F(ab')_2$ which is pure enough for many purposes. However, some Fab and $F(ab')_2$ fragments bind to protein A and/or protein G (see below) such that this strategy can be unsuccessful. Also a mixture of Fab, $F(ab')_2$ and smaller fragments may be present in the flow through from the column, requiring a further purification step to be carried out. Ion-exchange chromatography is usually the most reliable purification method, though some optimisation of the method may be required for each individual antibody. Both anion exchange and cation exchange may be useful depending on the particular antibody and fragment (e.g. Mhatre *et al.*, 1995). Ion-exchange also has the advantage of separating soluble enzymes used in the preparation. Gel filtration is also a useful method at small scale, with good separation of Fab and $F(ab')_2$ from IgG and small fragments achieved with either HPLC or conventional gel filtration columns. Separation of Fc fragments from Fab will not be achieved, however, and if Fc is present, an affinity chromatography or ion-exchange step will be required to remove it.

Purification of recombinant antibody fragments is more of a challenge, with the antibody fragment usually present in a complex mixture of proteins either from mammalian cell supernatant or from an *E. coli* extract. Some Fab and $F(ab')_2$ fragments, including several human and humanised fragments, are capable of binding to low-affinity binding sites on protein A and/or protein G. The successful purification of recombinant humanised Fab′s expressed in both mammalian cells and *E. coli* has been demonstrated using immobilised protein A (Carter et *al.*, 1992b; King *et al.*, 1995). Similarly, protein G can be successfully used for Fab purification from either of these expression systems (Proudfoot *et al.*, 1992; King *et al.*, unpublished data). Protein A binds to Fab′s, particularly of the human gene family VH3, and the sequences involved in binding have been localised to the second CDR region and framework regions 1 and 3 in the heavy chain V domain (Potter *et al.*, 1996). Some Fab fragments from other species including mouse, rabbit and guinea pig have also been shown to bind to protein A (Young *et al.*, 1984). The region of Fab interaction with protein G has been defined by structural studies and shown to lie within the CH1 domain of the heavy chain (Derrick and Wigley, 1992). Approximately 50% of human Fab fragments are believed to bind to protein G (Erntell *et al.*, 1983). At present it is difficult to determine which Fab fragments will bind to these proteins, which would greatly simplify purification, except by trial and error. In the author's laboratory we routinely test all newly expressed Fab fragments for their ability to bind to protein A and protein G before beginning purification studies.

For Fab fragments which do not bind protein A or G, an alternative bacterial Ig binding protein may be used, protein L. Protein L from *Peptostreptococcus magnus* binds specifically to the variable region of light chains and can be used for the purification of many human and mouse IgG, IgM and IgA antibodies and Fab fragments (Nilson *et al.*, 1993). In addition, humanised antibodies could be produced using light chain variable region frameworks designed to bind to protein L. Protein L has also been shown to be useful for the purification of scFv fragments of several, though not all, antibody types. Human light chain subtypes of the κ1, λ2 and λ3 families were found to bind to protein L, whereas κ4 and λ1 scFvs were unable to bind (Akerstrom *et al.*, 1994). As expected,

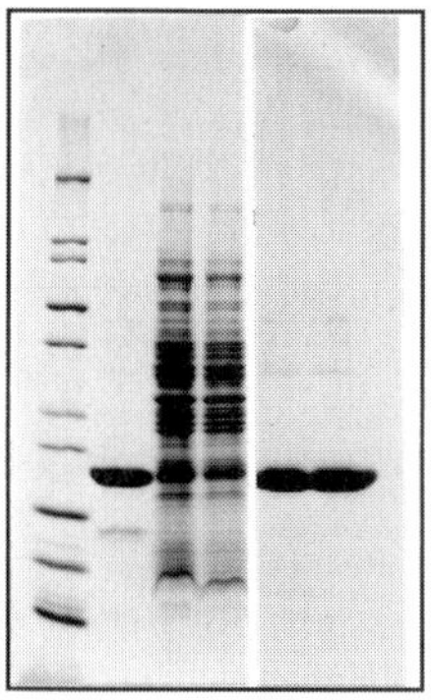

1 2 3 4 5 6

Figure 5.4 Reducing SDS-PAGE analysis of the purification of engineered human Fab′ from *E. coli* cell extract using Streamline A (Pharmacia) in expanded bed mode: lane 1, molecular weight markers; lane 2, Fab′ standard; lanes 3 and 4, cell extract; lanes 5 and 6, eluted Fab′

many scFv fragments containing VH3 heavy chain domains were also able to be purified using protein A. Novel Ig binding proteins have also been produced by gene fusion techniques in attempts to widen the number of Ig molecules which can be purified using a single material. A fusion of protein G and protein L, protein LG, has been prepared which can be used to purify a wide range of MAbs and fragments (Vola *et al.*, 1995). It is likely that the use of such proteins for small-scale purification of antibodies and their fragments will become more widespread as they become increasingly commercially available.

The use of immobilised protein A in expanded bed mode is particularly valuable for the purification of Fab′ fragments from *E. coli.* It is possible to apply crude *E. coli* extract to an expanded bed column and recover pure Fab′ in a single step (Figure 5.4) (Chapman *et al.*, 1996). Extracts from *E. coli* often contain a large amount of particulate material, and although this can be removed by a combination of centrifugation and filtration, this is often at risk of a large yield loss. For Fab′ and other antibody fragments which do not bind to protein A, expanded bed absorption using an ion-exchange material is a valuable first step in the purification process.

Ion-exchange chromatography can also be used to purify recombinant antibody fragments, though this requires a purification protocol involving several steps to be developed. A combination of cation exchange and anion exchange methods may be used to give pure Fab′ fragment expressed from mammalian cells (King *et al.*, 1992a). Alternatively, combinations of ion-exchange and hydrophobic interaction which are suitable for scale-up may be developed which allow high-purity material to be purified from *E. coli*, suitable for clinical use (King *et al.*, unpublished data). Although more steps are required, such protocols are likely to be considerably cheaper than the use of protein A or other Ig binding proteins. This is particularly important when considering the economic advantages of expression in bacterial systems.

Affinity chromatography using the antigen-binding specificity of the antibody fragment to bind to immobilised antigen is an effective purification method and is particularly useful for small antigen-binding fragments such as Fv or scFv which may be difficult to purify by other means (King *et al.*, 1993). As mentioned above, the disadvantages of such purification methods are that antigen may be expensive or simply not available in sufficient quantity and that elution conditions required to dissociate the antibody fragment are

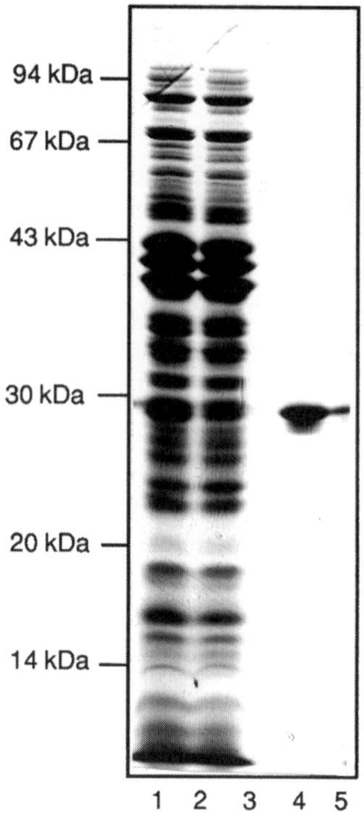

Figure 5.5 SDS-PAGE analysis of the purification of scFv from *E. coli* using immobilised antigen: lane 1, cell extract; lane 2, column flow-through; lane 3, wash fraction; lanes 4 and 5, eluted scFv (positions of molecular weight markers are indicated)

partially denaturing using extremes of pH or chaotropic agents. Monomeric antigen-binding fragments have an advantage in this respect as they can often be eluted under gentler conditions than those required for divalent fragments. When suitable methods are developed, the degree of purification which can be achieved is very high, as demonstrated for the purification of an Fv fragment expressed in *E. coli* (Figure 5.5). Alternatively, immunopurification using an antibody directed against the antibody fragment to be purified can be used. The use of an immobilised anti-human light chain antibody to purify a chimeric Fab′ fragment has been described (King *et al.*, 1992a).

The purification of scFv and other small antigen-binding fragments where antigen is not available can be difficult by conventional means. The most popular approach to achieving their purification has become the use of purification 'tags' which are engineered into the protein specifically for the purpose. A short peptide sequence can be added to the antibody fragment designed to allow simple purification under gentle conditions. Suitable peptides include hexa-histidine which allows purification in a single step of immobilised metal ion affinity chromatography (Skerra *et al.*, 1991); a biotin mimetic peptide, known as strep-tag, which can be used to bind to immobilised streptavidin (Schmidt and Skerra, 1994); and peptides recognised by specific antibodies for which gentle elution systems have been developed, such as the FLAG system described in Section 3.7. Of these the hexa-histidine tag appears to be the most universally applicable, and it has allowed the simple preparation of scFv for clinical application (Casey *et al.*, 1995). Nickel, copper or zinc ions can be used which are bound to the resin through the use of immobilised chelators such as iminodiacetic acid. Clusters of histidine residues allow specific binding to such metal ions through coordination to unused metal coordination sites. Bound protein can then be eluted under gentle conditions with excess imidazole to compete for metal-binding sites.

5.9.4 Purification for therapeutic use

When antibodies and antibody fragments are purified for use in humans, particular care must be taken to ensure safety by achieving high purity and the absence of potentially harmful contaminants. The stringent purity requirements normally require several purification steps to be integrated into a process designed to achieve high purity in a reproducible manner. Guidelines for the preparation of antibodies for use in humans are available from the regulatory authorities. Particular care must be taken not just to remove protein contaminants but also DNA and endotoxin. DNA and endotoxin are both highly charged materials and thus ion-exchange chromatography is useful for their removal. For mammalian cell derived products, specific viral inactivation and removal steps may also be required, even though cell banks need to be characterised with respect to any adventitious agents present. Many viruses are inactivated by low pH, such as used in elution from protein A columns. Specific filtration steps may also be used to improve clearance (Maerz *et al.*, 1996). During process development small-scale replicas of the process can be set up and spiked with large amounts of DNA, viruses or endotoxin to ensure they can be efficiently removed. Processes also need to be carried out under clean conditions such that contamination does not occur, and sterile product can be generated. Good manufacturing practice (GMP) is required as this ensures the preparation of high-quality material reproducibly from batch to batch. Additionally, purified material needs to be well characterised with specific assays required for antibody identity, purity and activity as well as checking for specific contaminants. Specific contaminants may include materials used during the purification. For example, in the case of protein A purified antibodies, specific measurement of protein A levels in the product may be required.

6

Prospects for Engineered Antibodies in Biotechnology

6.1 Gene therapy

6.1.1 Intracellular antibodies

New avenues are already under investigation to extend the range of applications for which the exquisite specificity of MAbs can be exploited. The use of antibodies to interfere with cellular processes inside the cell, i.e. intracellular targeting, is one new area in which the potential of antibodies is being explored for both research and therapeutic purposes. Antibody genes can be introduced into a cell and expressed to produce intracellular antibody which can be used, for example, to modify cellular properties through blocking interactions between macromolecules, to modulate enzyme action through fixing the enzyme in an active or inactive state, or to bind proteins and prevent them from reaching their normal cellular compartments. Attempts to express intact IgG within the cell have been less successful than the expression of a single gene encoded molecule such as a single-chain Fv. scFv can be directed to different cellular compartments using intracellular trafficking signals. For example, the sequence Lys-Asp-Glu-Leu (KDEL) at the C-terminus can retain the scFv in the ER (Beerli *et al.*, 1994), while removal of the signal sequence will direct the protein to the cell cytosol (Biocca *et al.*, 1994). In some cases cytoplasmic Fabs may be more suitable than scFvs due to increased stability in the cytoplasm (Levin *et al.*, 1997). Intracellular antibodies can also be directed to the nuclear compartment through the use of nuclear localisation signals (Mhashilkar *et al.*, 1995).

The specific inhibition of expression of cell surface molecules has important potential applications both in the analysis of the role of cell surface molecules and for therapeutic effects through down-regulation of cell surface receptors. The erb-b2 receptor tyrosine kinase, important in the growth of many breast and ovarian tumour cells, can be inhibited by expression of an ER-located scFv which prevents the transit of erb-b2 through the ER to its cell surface location, resulting in reversal of the transformed phenotype (Beerli *et al.*, 1994). Similarly, cell surface expression of the IL-2 receptor α chain can be inhibited using an ER-located scFv, resulting in cells which are no longer responsive to IL-2 (Richardson *et al.*, 1995), therefore opening up the possibility of controlling the growth of IL-2 receptor-dependent tumour cells. Inhibition of cytosolic oncoproteins such as p21^{ras} has also been demonstrated to be a feasible approach, through directing a suitable antibody to interfere with cytoplasmic protein function (Biocca *et al.*, 1994). However, the development of gene therapy techniques which can be used to achieve *in vivo* transfection of large numbers of tumour cells is beyond current technology, preventing therapeutic use of such approaches at present. Nevertheless, such techniques are powerful tools in the analysis of the function of cytosolic or cell surface proteins, complementing other techniques such as antisense RNA and gene disruption. For example, an scFv against the integrin VLA-4 has been used to prevent cell surface expression and analyse adhesive interactions (Yuan *et al.*, 1996). The KDEL ER retention signal is not always necessary, and secreted scFv molecules which are well expressed dwell sufficiently in the ER to prevent cell surface expression in some cases (Greenman *et al.*, 1996).

Intracellular antibodies are also under investigation as potential therapeutics against infectious agents such as HIV. Expression of an intracellular scFv against the HIV envelope protein gp120 resulted in disruption of virus assembly and a large decrease in the infectivity of virus particles released (Marasco *et al.*, 1993). Other HIV targets have also been studied including the regulatory proteins *rev* and *tat*, reverse transcriptase and the matrix protein p17 (Duan *et al.*, 1994; Mhashilkar *et al.*, 1995; Maciejewski *et al.*, 1995; Levin *et al.*, 1997). Not all antibodies are effective, however, and the choice of antibody or epitope may be critical to blocking function. For example, one scFv against tat was effective in blocking HIV replication and a second one to a different epitope was ineffective (Mhashilkar *et al.*, 1995). Again, this approach is currently limited by the availability of effective gene transfer methods.

6.1.2 Other applications of MAbs in gene therapy

The development of gene therapy in general, as well as the case of intracellular antibodies above, has been limited by the ability to produce vectors which can deliver DNA to cells *in vivo* in a form which can be subsequently expressed to produce a therapeutic effect. MAbs themselves may also play a role in improving selective gene delivery. At present two types of vector are under investigation for the delivery of genes: engineered viruses and non-viral systems such as DNA bound to polycations or liposomes. A major problem with both of these systems is the ability to target the vector to the required tissue *in vivo*. The targeting specificity of antibodies is one approach which is being investigated to overcome this. Antibodies coupled to the vector of choice, which specifically bind to, and are internalised into, target cells can be used to deliver the therapeutic gene. Antibodies have been linked to engineered viruses (Roux *et al.*, 1989), to liposomes encapsulating DNA (Leonetti *et al.*, 1990), to polycationic substances which bind DNA tightly, such as polylysine (Merwin *et al.*, 1995), or directly to DNA (Poncet *et al.*, 1996). All of these approaches have advantages and disadvantages, and at present it is too early to say which may be most successful in achieving targeted gene delivery *in vivo*. Although preliminary results have shown that all of these approaches may be successful at low-efficiency gene transfer, high-efficiency delivery has not been achieved. Engineered antibodies can also be designed with desirable properties, for example a Fab′–protamine fusion protein can bind plasmid DNA encoding a toxin gene which can be used to achieve selective cell killing (Chen *et al.*, 1995).

Gene therapy may open up alternative formats for using the specificity of antibody variable regions. For example, the delivery of genes for antibody variable regions in the form of novel fusion proteins has been investigated. Fusions of scFv to intracellular signalling domains can lead to expression of novel cell surface receptors which can use the specificity of the antibody-binding region to activate intracellular pathways. One example of this is the construction of chimeric T cell receptors for the activation of cytotoxic T lymphocytes (Eshhar *et al.*, 1993). This is a similar therapeutic approach to the use of bispecific antibodies to re-target effector cells to tumours as discussed in Section 4.2.3. In the case of chimeric receptors the T cell is the object of gene transfer, and is transfected with a gene encoding an scFv to a tumour-associated antigen fused directly to transmembrane and signalling domains of the T cell receptor. In this way a T cell is produced which is directly activated by binding of the scFv to its antigen. Such transfected T cells can lyse tumour cells in an antigen-dependent manner. For example, chimeric receptors produced from an scFv against $p185^{HER2}$ and the ζ-signal-transducing subunit of the T cell receptor or the γ-signal-transducing subunit of the Ig Fc receptor CD16 were transfected into T cells and shown to be able to lyse $p185^{HER2}$ expressing cells (Stancovski *et al.*, 1993). Similarly, a chimeric receptor produced from an scFv against a renal carcinoma antigen fused to the γ chain of the FcεRI receptor was transfected into human T cells using a retrovirus, and shown to be capable of tumour cell lysis (Weijtens *et al.*, 1996). T cells transfected with such chimeric receptors have been re-introduced into animal models and have shown some anti-tumour activity, but the extension of this technology to transfection of T cells *in vivo*, resulting in anti-tumour activity, has yet to be achieved.

6.2 Applications of antibodies in plants

In situ applications of antibodies in plants are being increasingly investigated. Resistance to viral attack may be conferred by antibody expression in the plant. A cytoplasmically expressed scFv against artichoke mottled crinkle virus could reduce infection and delay onset of symptoms even though expression levels were low (Tavladoraki *et al.*, 1993). An intact IgG has also been shown to protect against tobacco mosaic virus attack when secreted from plant cells (Voss *et al.*, 1995). However, secreted antibody has not been shown to be effective at protecting plants in all cases. For example, antibody against the harmful salivary secretions of a nematode could not protect the plant from attack (Baum *et al.*, 1996). This may be because the nematode saliva is injected directly into the cell cytoplasm, and a cytoplasmically expressed antibody may be more effective. Research is ongoing on the use of antibodies to protect against attack from a range of plant pathogens, and intracellular antibodies may also be used to affect the action of plant hormones. Expression of an scFv to abscisic acid, targeted to the endoplasmic reticulum of the plant cell, led to a similar wilty phenotype to that seen in abscisic acid deficient mutants (Artsaenko *et al.*, 1995).

6.3 Catalytic antibodies

Antibodies are thought of principally as binding molecules which can target a range of effector functions to their antigen. However, antibodies are also capable of catalysing a number of chemical reactions, in a similar manner to enzymes. Such catalytic antibodies, also known as abzymes, were first prepared by raising MAbs to transition state analogues of hydrolytic reactions (Tramontano *et al.*, 1986; Pollack *et al.*, 1986). For example, during hydrolysis of an ester bond, the reaction mechanism requires attack by a hydroxide anion at the carbonyl carbon atom, resulting in a tetrahedral transition state before cleavage to the carboxylic acid and alcohol products. The production of the transition state is the rate-determining step in the reaction and requires the input of sufficient activation energy. The stabilisation of a transition state lowers the activation energy required and thus catalyses the rate of reaction. Antibodies which bind to, and stabilise, the transition state would therefore be expected to be catalytic. It is usually not possible to raise antibodies directly to transition states of reactions, as by their very nature they are unstable, but antibodies raised against structural analogues of the transition state, which cross-react with the reactants at the transition stage, can be used as catalysts.

Since the early reports the range of chemical reactions which can be catalysed has been greatly extended and the efficiency of many reactions has been improved (Schultz and Lerner, 1995). However, catalytic antibodies remain relatively poor catalysts compared to enzymes, with most rate enhancements in the order of 10^3–10^5 over the uncatalysed reaction, compared to 10^{11} or more for comparable enzymes. Several strategies are being pursued in attempts to improve efficiency, including both random and site-directed mutagenesis, as well as screening large combinatorial libraries of antibodies directly for catalytic activity (Janda *et al.*, 1997). Several chemical transformations which are difficult to achieve by conventional chemistry have already been achieved as well as reactions which are not catalysed by any known enzymes (Schultz and Lerner, 1995), suggesting that catalytic antibodies may find a role in organic chemistry. A particular advantage may be the ability to carry out reactions for selective enantiomers. The ability to produce large amounts of antibody fragments cheaply in bacteria and other systems may be particularly relevant here (see Chapter 5). Applications for catalytic antibodies have also been suggested for converting prodrugs to drugs in cancer therapy (Wentworth *et al.*, 1996; see Section 4.2.8), for other *in vivo* applications such as inactivation of cocaine (Landry *et al.*, 1993), and for *in vitro* applications such as biosensors (Blackburn *et al.*, 1990). However, most of these require a further leap forward in current technology. Whether such applications will ever be realised or whether catalytic antibodies will remain a research curiosity, useful only for studying the mechanisms of catalysis, remains to be seen.

6.4 Towards drug design

As well as being drugs themselves, antibodies can also be used in strategies to design low molecular weight, organic drugs. Antibodies are protein molecules, which restricts their delivery to parenteral routes of administration. For many therapies, low molecular weight drugs which may be orally active are desirable. Antibodies may also assist the development of such drugs in some cases. The desire to produce smaller antigen-binding compounds as a starting point for drug design has prompted several studies of CDR-based peptides. Peptides can be designed based on CDR sequences and structures, or constrained analogues can be produced based on modelling or structural information. Peptides have been designed from an anti-lysozyme antibody based on CDR2 of the heavy chain and CDR3 of the light chain (Welling *et al.*, 1991). Despite relatively low binding affinity, both peptides were immobilised and could be successfully used to 'immunopurify' lysozyme. In a study of an anti-HIV antibody, all six CDR peptides were prepared and compared for their ability to inhibit binding of the parent antibody to the HIV envelope protein (Levi *et al.*, 1993). CDR3 of the heavy chain was most effective, though combinations of the other CDR peptides were synergistic. A cyclised version of the peptide was also produced with improved binding affinity which was able to inhibit HIV replication. CDR-based peptides have also been designed for *in vivo* imaging applications. Peptides based on the heavy chain CDR3 of an anti-platelet antibody were designed and shown to be able to detect vascular thrombi in animal models (Knight *et al.*, 1994), and a peptide also based on CDRH3, in this case of an antibody to a tumour-associated antigen, has been shown to be able to image breast cancer sites in patients (Sivolapenko *et al.*, 1995). These studies demonstrate that some structural properties of the antigen-binding loops are retained in short peptides, which may have practical applications.

One approach to design is to use the antibody to gather structural information about a target molecule, which may be inaccessible to direct structural analysis, and use the information in attempts to design a peptide or peptidomimetic drug. Antibodies can be raised to the key drug action sites of receptors, enzymes or viruses which contain a complementary image of the structure in question. Raising anti-idiotype antibodies to these may result in antibodies with a positive image of the target site. Information can then be gathered by sequencing the antibody and carrying out structural studies either by modelling or, for example, by solving the three-dimensional structure of the Fab fragment by X-ray crystallography. In some cases sequences of the CDR regions of anti-idiotype antibodies have homology to the initial antigen which can locate key regions. For example, a 17 amino acid peptide from the CDRL2 of an anti-reovirus receptor MAb was found to inhibit binding of the parent antibody to receptor, and sequences within the CDR were shown to be similar to those of the viral haemaglutinin (Williams *et al.*, 1991). Organic mimetic compounds could then be designed using a macrocyclic structure as a framework and introducing synthetic modifications. In this way inhibitors with affinities in the low nM range have been reported (Dougall *et al.*, 1994). Similarly, antibodies to GM-CSF have been used to generate a library of anti-idiotype light chains which could be screened for inhibition of binding of the original antibody to GM-CSF (Monfardini *et al.*, 1995). Weak structural homology was found between the CDR1, CDR2 and framework 3 sequences of an antibody light chain from the screen with helices on GM-CSF. A synthetic peptide based on the CDR1 sequence could then act as an antagonist of GM-CSF binding to cells and blocking bioactivity.

In all cases the binding of the CDR peptides is weak compared to the parent antibody as a consequence of the loss of the large contact region between antibody and antigen. One approach to help maintain binding affinity is to begin with a high-affinity single domain antibody which uses three CDRs rather than six to contact antigen. Both heavy and light chain domains can have high antigen-binding affinity alone in some cases, although isolated domains are poorly soluble due to exposed hydrophobic residues which normally pair with the other V domain (Ward *et al.*, 1989; Masat *et al.*, 1994). Camels have been found to produce antibodies which do not have light chains and thus have evolved single V domains which are more soluble. A crystal structure of a camel antibody fragment in complex with its antigen, lysozyme, has revealed that contacts between antibody and antigen in this case were dominated by one CDR, CDR3, which may be a more suitable starting point for small drug design (Desmyter *et al.*, 1996). Phage display libraries set up to display 'camelised' VH domains have been used to identify binding domains (Davies and Riechmann, 1995). For example, a camelised VH against hepatitis C virus protease was isolated to serve as a potential pharmacophore model to design antiviral compounds (Martin *et al.*, 1997).

6.5 Improving affinity

The ability to modulate affinity of antibodies, particularly to increase affinity by protein engineering, has been a desirable objective for some time. Attempts to rationally design antibodies with increased affinity have been made using modelling techniques and structural information, including crystallographic structures of antibody–antigen complexes. The first report of engineering an antibody which resulted in increased affinity for antigen resulted from a mutation which was predicted from modelling studies to decrease affinity (Roberts *et al.*, 1987). There have subsequently been some modest successes in redesigning the binding site of anti-hapten antibodies to increase affinity based on structural data. The affinity of an anti-phenyloxazolone antibody was increased three-fold by a single amino acid substitution predicted from modelling and NMR studies (Riechmann *et al.*, 1992). However, attempts to rationally design antibodies with increased affinity have not generally been successful. This is probably as a result of our limited understanding of antibody–antigen interactions at the structural level, which may differ between different antibody–antigen pairs. Combinations of hydrophobic interactions and electrostatic interactions may well be involved in antigen binding to different extents in each case, and in some antigen-binding events conformational changes to the antibody variable region are known to take place (see Chapter 1). The ability to redesign variable regions for increased affinity therefore remains a goal for improved modelling based on further understanding of antibody–antigen interaction. Eventually the design of an antibody variable region to an antigen without a pre-existing antibody may even be possible (Rees *et al.*, 1994).

At present the best methodology for improving antibody affinity is the use of phage display to select improved variants. This is achieved by screening large numbers of changes introduced by random mutagenesis of all or part of the variable regions, or by chain shuffling techniques (see Chapter 1). The affinity of antibodies isolated by phage display have been improved in this way, including to very high affinity (Schier *et al.*, 1996; Yang *et al.*, 1995), although, as with any mutagenesis strategy, it is important to ensure that the specificity of the antibody is not altered during this process, as changes to the fine specificity of the antibody can result (Ohlin *et al.*, 1996).

6.6 Summary and prospects

As I have attempted to illustrate, the ability to design and produce antibody-based molecules for specific applications is leading to an increase in both the range and the number of applications in which MAbs can be successfully used. In addition, the ability to generate MAbs in new ways through technologies such as phage display opens up the possibility of generating antibody specificities which were not previously obtainable. New production methods through expression in recombinant cells, or transgenic organisms, may allow the generation of low-cost MAb-based reagents, and the combination of genetic engineering and chemical modification approaches allows a wide range of molecules to be made and tested. As more is understood of the nature of antibody structure and function and the basis of specific interactions, design of molecules with optimal properties becomes achievable, allowing the development of a generation of new reagents for diagnostic and therapeutic applications. In this way many of the problems which have hampered the development of MAb-based therapeutics may be overcome, and entirely new applications such as those described in this chapter will be developed.

The range of applications of MAbs now extends from the research laboratory and the diagnostic laboratory to diagnosis in the home, therapy in the clinic, purification in the biotechnology production plant and potentially in drug design, delivery of therapeutic genes and plant protection.

As therapeutic agents, engineered antibodies will no doubt be of value for a limited time. As injectables, antibodies are not ideal therapeutics, and small molecules which can be taken orally will likely be developed to take their place in many cases. However, the unique targeting ability of antibodies may not easily be replaced with current technology and thus a role for antibody-based therapeutics seems likely for the foreseeable future. Indeed, at present the number of antibody-based therapeutic agents under development is increasing as the benefits of antibody engineering in reducing immunogenicity, targeting novel effector functions and designing appropriate pharmacokinetic properties are realised. In addition, the ability of antibodies to bind cell surface receptors and directly activate intracellular signalling mechanisms is of great potential for developing further therapeutic agents.

Another hurdle in the development of antibody-based agents as therapeutics as well as in other applications has been the cost of antibody production. New methods of production at large scale in microbial fermentation or in transgenic organisms are likely to make antibody-based reagents available at lower cost and in larger quantities than possible previously, and this is likely to be a major factor in developing the role of antibodies in new areas. With such developments and the continued increase in our understanding of the nature of the antibody:antigen interaction and in antibody effector mechanisms, the prospects for further application of engineered antibodies in biotechnology appear bright.

7

References

ABUCHOWSKI, A., VAN ES, T., PALCZUK, N.C. and DAVIS, F.F., 1977, Alteration of immunological properties of bovine serum albumin by covalent attachment of polyethylene glycol, *J. Biol. Chem.*, **252**, 3578–3581.

ADAIR, J.R., 1997, Engineered molecules – isolation of antibody variable regions, *The Immunology Methods Manual* (Lefkovits, I., ed.), Academic Press, San Diego.

ADAIR, J.R., ATHWAL, D.S., BODMER, M.W., BRIGHT, S.M., COLLINS, A.M., PULITO, V.L., RAO, P.E., REEDMAN, R., ROTHERMEL, A.L., XU, D., ZIVIN, R.A. and JOLLIFFE, L.K., 1995, Humanization of the murine anti-human CD3 monoclonal antibody OKT3, *Hum. Antibod. Hybridomas*, **5**, 41–47.

ADAIR, J.R., ATHWAL, D.S. and EMTAGE, J.S., 1991, *Humanised antibodies*, International patent application WO91/09967.

ADAIR, J.R., BAKER, T.S., BEBBINGTON, C.R., BEELEY, N.R.A., EMTAGE, J.S., KING, D.J., LAWSON, A.D.G., MILLICAN, T.A., MOUNTAIN, A., OWENS, R.J. and YARRANTON, G.T., 1993, Engineering monoclonal antibody B72.3 for cancer therapy, *Protein Engineering of Antibody Molecules for Prophylactic and Therapeutic Applications* (Clark, M., ed.) pp.145–158, Academic Titles, Nottingham.

ADAIR, J.R. and BRIGHT, S.M., 1995, Progress with humanised antibodies – an update, *Exp. Opin. Invest. Drugs*, **4**, 863–870.

ADAMS, G.P., MCCARTNEY, J.E., TAI, M.S., OPPERMANN, H., HUSTON, J.S., STAFFORD, W.F., BOOKMAN, M.A., FAND, I., HOUSTON, L.L. and WEINER, L.M., 1993, Highly specific in vivo tumor targeting by monovalent and divalent forms of 741F8 anti-c-erbB-2 single-chain Fv, *Cancer Res.*, **53**, 4026–4034.

AFEYAN, N.B., GORDON, N.F. and REGNIER, F.E., 1992, Automated real-time immunoassay of biomolecules, *Nature*, **358**, 603–604.

AIZAWA, M., MORIOKA, A., SUZUKI, S. and NAGAMURA, Y., 1979, Amperometric determination of human choriogonadotrophin by membrane bound antibody, *Anal. Biochem.*, **94**, 22–28.

AKERSTROM, B., NILSON, B.H.K., HOOGENBOOM, H.R. and BJORCK, L., 1994, On the interaction between single chain Fv antibodies and bacterial immunoglobulin-binding proteins, *J. Immunol. Methods*, **177**, 151–163.

AKHAVAN-TAFTI, H., SUGIOKA, K., ARGHAVANI, Z., DESILVA, R., HANDLEY, R.S., SUGIOKA, Y., EICKHOLT, R.A., PERKINS, M.P. and SCHAPP, A.P., 1995, Chemiluminescent detection of horseradish peroxidase by enzymatic generation of acridinium esters, *Clin. Chem.*, **41**, 1368–1369.

ALAUDDIN, M.M., KHAWLI, L.A. and EPSTEIN, A.L., 1992, An improved method of direct labeling monoclonal antibodies with ^{99m}Tc, *Nucl. Med. Biol.*, **19**, 445–454.

ALBRECHT, S., BRANDL, H., STEINKE, M. and FREIDT, T., 1994, Chemiluminescent enzyme immunoassay of prostate-specific antigen based on indoxyl phosphate substrate, *Clin. Chem.*, **40**, 1970–1971.

ALEGRE, M.L., COLLINS, A.M., PULITO, V.L., BROSIUS, R.A., OLSON, W.C., ZIVIN, R.A., KNOWLES, R., THISTLETHWAITE, J.R., JOLLIFFE, L.K. and BLUESTONE, J.A., 1992, Effect of a single amino acid mutation on the activating and immunosuppressive properties of a humanized OKT3 monoclonal antibody, *J. Immunol.*, **148**, 3461–3468.

ALI, S.A., WARREN, S.D., RICHTER, K.Y., BADGER, C.C., EARY, J.F., PRESS, O.W., KROHN, K.A., BERNSTEIN, I.D. and NELP, W.B., 1990, Improving the tumour retention of radioiodinated antibody:aryl carbohydrate adducts, *Cancer Res.*, **50**, 783s–788s.

ALLARD, W.J., MORAN, C.A., NAGEL, E., COLLINS, G. and LARGEN, M.T., 1992, Antigen binding properties of highly purified bispecific antibodies, *Mol. Immunol.*, **29**, 1219–1227.

ALLEN, T.M., AHMAD, I., LOPES DE MENEZES, D.E. and MOASE, E.H., 1995, Immunoliposome-mediated targeting of anti-cancer drugs in vivo, *Biochem. Soc. Trans.*, **23**, 1073–1079.

AMES, R.S., TORNETTA, M.A., MCMILLAN, L.J., KAISER, K.F., HOLMES, S.D., APPELBAUM, E., CUSIMANO, D.M., THEISEN, T.W., GROSS, M.S., JONES, C.S., SILVERMAN, C., PORTER, T.G., COOK, R.M., BENNETT, D. and CHAIKEN, I.M., 1995, Neutralizing murine monoclonal antibodies to human IL-5 isolated from hybridomas and a filamentous phage Fab display library, *J. Immunol.*, **154**, 6355–6364.

ANASETTI, C., HANSEN, J.A., WALDMANN, T.A., APPLEBAUM, F.R., DAVIS, J., DEEG, H.J., DONEY, K., MARTIN, P.J., NASH, R., STORB, R., SULLIVAN, K.M., WITHERSPOON, R.P., BINGER, M.-H., CHIZZONITE, R., HAKIMI, J., MOULD, D., SATOH, H. and LIGHT, S.E., 1994, Treatment of acute graft versus host disease with humanized anti-Tac: an antibody that binds to the interleukin-2 receptor, *Blood*, **84**, 1320–1327.

ANDERSON, C.J., CONNETT, J.M., SCHWARZ, S.W., ROCQUE, P.A., GUO, L.W., PHILPOTT, G.W., ZINN, K.R., MEARES, C.F. and WELCH, M.J., 1992, Copper-64 labeled antibodies for PET imaging, *J. Nucl. Med.*, **33**, 1685–1691.

ANDERSON, C.J., SCHWARZ, S.W., CONNETT, J.M., CUTLER, P.D., GUO, L.W., GERMAIN, C.J., PHILPOTT, G.W., ZINN, K.R., GREINER, D.P., MEARES, C.F. and WELCH, M.J., 1995, Preparation, biodistribution and dosimetry of copper-64-labelled anti-colorectal carcinoma monoclonal antibody fragments 1A3-F(ab′)$_2$, *J. Nucl. Med.*, **36**, 850–858.

ANGAL, S., KING, D.J., BODMER, M.W., TURNER, A., LAWSON, A.D.G., ROBERTS, G., PEDLEY, B. and ADAIR, J.R., 1993, A single amino acid substitution abolishes the heterogeneity of chimeric mouse/human IgG4 antibody, *Mol. Immunol.*, **30**, 105–108.

ANSELL, S.M., TARDI, P.G. and BUCHOWSKY, S.S., 1996, 3-(2-pyridyldithio)propionic acid hydrazide as a cross-linker in the formation of liposome–antibody conjugates, *Bioconj. Chem.*, **7**, 490–496.

ANTON, R. and BEAN, P., 1994, Two methods for measuring carbohydrate-deficient transferrin in inpatient alcoholics and healthy controls compared, *Clin. Chem.*, **40**, 364–368.

ANTONIW, P., FARNSWORTH, A.P.H., TURNER, A., HAINES, A.M.R., MOUNTAIN, A., MACKINTOSH, J., SHOCHAT, D., HUMM, J., WELT, S., OLD, L.J., YARRANTON, G.T. and KING, D.J., 1996, Radioimmunotherapy of colorectal carcinoma xenografts in nude mice with yttrium-90 A33 IgG and tri-Fab (TFM), *Br. J. Cancer*, **74**, 513–524.

APPLEBAUM, F.R., MATTHEWS, D.C., EARY, J.F., BADGER, C.C., KELLOG, M., PRESS, O.W., MARTIN, P.J., FISHER, D.R., NELP, W.B., THOMAS, E.D. and BERNSTEIN, I.D., 1992, The use of radiolabelled anti-CD33 antibody to augment marrow irradiation prior to marrow transplantation for acute myelogenous leukemia, *Transplantation*, **54**, 829–833.

ARMBRUSTER, D.A., HUBSTER, E.C., KAUFMAN, M.S. and RAMON, M.K., 1995, Cloned enzyme donor immunoassay (CEDIA) for drugs of abuse screening, *Clin. Chem.*, **41**, 92–98.

ARMBRUSTER, D.A., SCHWARZHOFF, R.H., HUBSTER, E.C. and LISERIO, M.K., 1993, Enzyme immunoassay, kinetic microparticle immunoassay, radioimmunoassay and fluorescence polarization immunoassay compared for drugs of abuse screening, *Clin. Chem.*, **39**, 2137–2146.

ARNOLD, M.W., SCHNEEBAUM, S., BERENS, A., PETTY, L., MOJZISIK, C., HINCKLE, G. and MARTIN, E.W., 1992, Intraoperative detection of colorectal cancer with radioimmunoguided surgery and CC49, a second generation monoclonal antibody, *Ann. Surgery*, **216**, 11–16.

Artsaenko, O., Peisker, M., zur Nieden, U., Fiedler, U., Weiler, E.W., Muntz, K. and Conrad, U., 1995, Expression of a single-chain Fv antibody against abscisic acid creates a wilty phenotype in transgenic tobacco, *Plant J.*, **8**, 745–750.

Asano, M., Yukita, A., Matsumoto, T., Kondo, S. and Suzuki, H., 1995, Inhibition of tumour growth and metastasis by an immunoneutralizing monoclonal antibody to human vascular endothelial growth factor/vascular permeability factor 121, *Cancer Res.*, **55**, 5296–5301.

Axworthy, D.B., Fritzberg, A.R., Hylarides, M.D., Mallett, R.W., Theodore, L.J., Gustavson, L.M., Su, F.M., Beaumier, P.L. and Reno, J.M., 1994, Preclinical evaluation of an anti-tumour monoclonal antibody/streptavidin conjugate for pretargeted ^{90}Y radioimmunotherapy in a nude mouse xenograft model, *J. Immunother.*, **16**, 158.

Bachmann, M.F., Kalinke, U., Althage, A., Freer, G., Burkhart, C., Roost, H.P., Aguet, M., Hengartner, H. and Zinkernagel, R.M., 1997, The role of antibody concentration and avidity in antiviral protection, *Science*, **276**, 2024–2027.

Badger, C.C., 1990, Bone marrow toxicity for ^{131}I-labeled antibodies, *Antibody Immunoconj. Radiopharm.*, **3**, 281–287.

Bagshawe, K.D., Sharma, S.K., Springer, C.J. and Antoniw, P., 1995, Antibody directed enzyme prodrug therapy: a pilot-scale clinical trial, *Tumor Targeting*, **1**, 17–29.

Bailon, P. and Roy, S.K., 1990, Recovery of recombinant proteins by immunoaffinity chromatography. *Protein Purification* (Ladisch, M.R., Willson, R.C., Painton, C.C. and Builder, S.E., eds), pp. 150–167, ACS, Washington, DC.

Baker, T.S., Begent, R.H.J., Dewji, M.R., Conlan, J. and Secher, D.S., 1991, Characterization of the antibody response in patients undergoing radioimmunotherapy with chimeric B72.3, *Antibody Immunoconj. Radiopharm.*, **4**, 799–809.

Baker, T.S., Bose, C.C., Caskey-Finney, H.M., King, D.J., Lawson, A.D.G., Lyons, A., Mountain, A., Owens, R.J., Rolfe, M.R., Sehdev, M., Yarranton, G.T. and Adair, J.R., 1994, Humanization of an anti-mucin antibody for breast and ovarian cancer therapy, *Antigen and Antibody Molecular Engineering in Breast Cancer Diagnosis and Treatment*, (Ceriani, R.L., ed.), pp. 61–82, Plenum Press, New York.

Barbas, C.F., III, 1995, Synthetic human antibodies, *Nature Med.*, **1**, 837–839.

Barbas, C.F. III and Burton, D.R., 1996, Selection and evolution of high affinity human anti-viral antibodies, *Trends Biotechnol.*, **14**, 230–234.

Barbas, C.F., III, Crowe, J.E., Cababa, D., Jones, T.M., Zebedee, S.L., Murphy, B.R., Chanock, R.M. and Burton, D.R., 1992, Human monoclonal Fab fragments derived from a combinatorial library bind to respiratory syncytial virus F glycoprotein and neutralise infectivity, *Proc. Natl Acad. Sci. (USA)*, **89**, 10164–10168.

Barbas, C.F., III, Kang, A.S., Lerner, R.A. and Benkovic, S.J., 1991, Assembly of combinatorial antibody libraries on phage surfaces, *Proc. Natl Acad. Sci. (USA)*, **88**, 7978–7982.

Barnard, G., Kohen, F., Mikola, H. and Lovgren, T., 1989, Measurement of estrone-3-glucuronide in urine by rapid, homogeneous time-resolved fluoroimmunoassay, *Clin. Chem.*, **35**, 555–559.

Baselga, J., Tripathy, D., Mendelsohn, J., Baughman, S., Benz, C.C., Dantis, L., Sklarin, N.T., Seidman, A.D., Hudis, C.A., Moore, J., Rosen, P.P., Twaddell, T., Henderson, I.C. and Norton, L., 1996, Phase II study of weakly intravenous recombinant humanized anti-p185^{HER2} monoclonal antibody in patients with HER2/*neu*-overexpressing metastatic breast cancer, *J. Clin. Oncol.*, **14**, 737–744.

Batra, J.K., Kasturi, S., Gallo, M.G., Voorman, R.L., Maio, S.M., Chaudhary, V.K. and Pastan, I., 1993, Insertion of constant region domains of IgG1 into CD4-PE40 increases its plasma half-life, *Mol. Immunol.*, **30**, 379–386.

Baum, T.J., Hiatt, A., Parrott, W.A., Pratt, L.H. and Hussey, R.S., 1996, Expression in tobacco of a functional monoclonal antibody specific to stylet secretions of the root-knot nematode, *Mol. Plant Microbe Interact.*, **9**, 382–387.

Beaumier, P.L., Venkatesan, P., Vanderheyden, J.L., Burgua, W.D., Kunz, L.L., Fritzberg, A.R., Abrams, P.G. and Morgan, A.C., 1991, ^{186}Re radioimmunotherapy of small cell lung carcinoma xenografts in nude mice, *Cancer Res.*, **51**, 676–681.

BEBBINGTON, C.R., 1991, Expression of antibody genes in nonlymphoid cells, *Methods*, **2**, 136–145.

BEBBINGTON, C.R., RENNER, G., THOMSON, S., KING, D., ABRAMS, D. and YARRANTON, G.T., 1992, High-level expression of a recombinant antibody from myeloma cells using a glutamine synthetase gene as an amplifiable marker, *Bio/Technol.*, **10**, 169–175.

BECKER, W., BAIR, J., BEHR, T., REPP, R., STRECKENBACH, H., BECK, H., GRAMATZKI, M., WINSHIP, M.J., GOLDENBERG, D.M. and WOLF, F., 1994, Detection of soft tissue infections and osteomyelitis using a technetium-99m-labelled anti-granulocyte monoclonal antibody fragment, *J. Nucl. Med.*, **35**, 1436–1443.

BEERLI, R.R., WELS, W. and HYNES, N.E., 1994, Intracellular expression of single-chain antibodies reverts erb-b2 transformation, *J. Biol. Chem.*, **269**, 23931–23936.

BEESLEY, J.E., ed., 1993, *Immunocytochemistry: a Practical Approach*, IRL Press, Oxford.

BEGENT, R.H.J., KEEP, P.A., GREEN, A.J., SEARLE, F., BAGSHAWE, K.D., JEWKES, R.F., JONES, B.E., BARRATT, G.M. and RYMAN, B.E., 1982, Liposomally entrapped second antibody improves tumour imaging with radiolabeled (first) antitumour antibody, *Lancet*, **ii**, 739–742.

BEGENT, R.H.J., VERHAAR, M.J., CHESTER, K.A., CASEY, J.L., GREEN, A.J., NAPIER, M.P., HOPE-STONE, L.D., CUSHEN, N., KEEP, P.A., JOHNSON, C.J., HAWKINS, R.E., HILSON, A.J.W. and ROBSON, L., 1996, Clinical evidence of efficient tumor targeting based on a single chain Fv antibody from a combinatorial library, *Nature Med.*, **2**, 979–984.

BEHR, T.M., BECKER, W.S., BAIR, H.J., KLEIN, M.W., STUHLER, C.M., CIDLINSKY, K.P., WITTEKIND, C.W., SCHEELE, J.R. and WOLF, F.G., 1995, Comparison of complete versus fragmented technetium-99m-labeled anti-CEA monoclonal antibodies for immunoscintigraphy in colorectal cancer, *J. Nucl. Med.*, **36**, 430–441.

BELEW, M., JUNTTI, N., LARSSON, A. and PORATH, J., 1987, A one-step purification method for monoclonal antibodies based on a salt-promoted adsorption chromatography on a thiophilic adsorbent, *J. Immunol. Methods*, **102**, 173–182.

BENHAR, I., PADLAN, E.A., JUNG, S.H., LEE, B. and PASTAN, I., 1994, Rapid humanisation of the Fv of monoclonal antibody B3 by using framework exchange of the recombinant immunotoxin B3(Fv) PE38, *Proc. Natl Acad. Sci. (USA)*, **91**, 12051–12055.

BENHAR, I. and PASTAN, I., 1995, Characterization of B1(Fv)PE38 and B1(dsFv)PE38: single-chain and disulphide-stabilized Fv immunotoxins with increased activity that cause complete remissions of established human carcinoma xenografts in nude mice, *Clin. Cancer Res.*, **1**, 1023–1029.

BERNSTEIN, I., 1996, Presented at American Society of Haematology Meeting, Orlando, FL, December.

BERRY, M.J. and DAVIES, J., 1992, Use of antibody fragments in immunoaffinity chromatography. Comparison of Fv fragments, VH fragments and paralog peptides, *J. Chromatogr.*, **597**, 239–245.

BERRY, M.J., DAVIES, J., SMITH, C.G. and SMITH, I., 1991, Immobilization of Fv antibody fragments on porous silica and their utility in affinity chromatography, *J. Chromatogr.*, **587**, 161–169.

BERRY, M.J. and PIERCE, J.J., 1993, Stability of immunoadsorbents comprising antibody fragments: comparison of Fv fragments and single-chain Fv fragments, *J. Chromatogr.*, **629**, 161–168.

BETTER, M., BERNHARD, S.L., LEI, S.P., FISHWILD, D.M., LANE, J.A., CARROLL, S.F. and HORWITZ, A.H., 1993, Potent anti-CD5 ricin A chain immunoconjugates from bacterially produced Fab′ and $F(ab')_2$, *Proc. Natl Acad. Sci. (USA)*, **90**, 457–461.

BETTER, M., BERNHARD, S.L., WILLIAMS, R.E., LEIGH, S.D., BAUER, R.J., KUNG, A.H.C., CARROLL, S.F. and FISHWILD, D.M., 1995, T-cell targeted immunofusion proteins from *Escherichia coli*, *J. Biol. Chem.*, **270**, 14951–14957.

BETTER, M., CHANG, C.P., ROBINSON, R.R. and HORWITZ, A.H., 1988, *Escherichia coli* secretion of an active chimeric antibody fragment, *Science*, **240**, 1041–1043.

BETTER, M., WEICKMANN, J. and LIN, Y.L., 1990, Production and scale-up of a chimeric fab from bacteria, *Proceedings of the 1990 Miami Bio/Technology Winter Symposium*, 105.

BEUTLER, B., MISARK, I.W. and CERAMI, A., 1985, Passive immunization against cachectin/tumour necrosis factor protects mice from lethal effects of endotoxin, *Science*, **229**, 869–871.

BHAT, T.N., BENTLEY, G.A., FISCHMANN, T.O., BOULOT, G. and POLJAK, R.J., 1990, Small rearrangements in structures of Fv and Fab fragments of antibody D1.3 on antigen binding, *Nature*, **347**, 483–485.

BIBILA, T.A. and ROBINSON, D.K., 1995, In pursuit of the optimal fed-batch process for monoclonal antibody production, *Biotechnol. Progress*, **11**, 1–13.

BICKEL, U., LEE, V.M., TROJANOWSKI, J.Q. and PARDRIDGE, W.M., 1994, Development and *in vitro* characterization of a cationized monoclonal antibody against beta A4 protein: a potential probe for Alzheimer's disease, *Bioconj. Chem.*, **5**, 119–125.

BIOCCA, S., PIERANDREI-AMALDI, P., CAMPIONI, N. and CATTANEO, A., 1994, Intracellular immunization with cytosolic recombinant antibodies, *Bio/Technol.*, **12**, 396–399.

BIRCH, J.R., LAMBERT, K., THOMPSON, P.W., KENNEY, A.C. and WOOD, L.A., 1987, Antibody production with airlift fermentors, *Large Scale Cell Culture Technology* (Lyderson, B.K., ed.), pp. 1–20, Hanser, New York.

BIRD, R.E., HARDMAN, K.D., JACOBSON, J.W., JOHNSON, S., KAUFMAN, B.M., LEE, S.M., LEE, T., POPE, S.H., RIORDAN, G.S. and WHITLOW, M., 1988, Single chain antigen binding proteins, *Science*, **242**, 423–426.

BISCHOFF-DELALOYE, A., DELALOYE, B., BUCHEGGER, F., VOGEL, C.A., GILLET, M., MACH, J.P., SMITH, A. and SCHUBIGER, P.A., 1997, Comparison of copper-67 and iodine-125-labeled anti-CEA monoclonal antibody biodistribution in patients with colorectal tumours, *J. Nucl. Med.*, **38**, 847–853.

BITOH, S., LANG, G.M. and SEHON, A., 1993, Suppression of human anti-mouse idiotypic antibody responses in hu-PBL-SCID mice, *Hum. Antibod. Hybridomas*, **4**, 144–151.

BLACKBURN, G.F., TALLEY, D.B., BOOTH, P.M., DURFOR, C.N., MARTIN, M.T., NAPPER, A.D. and REES, A.R., 1990, Potentiometric biosensor employing catalytic antibodies as the molecular recognition element, *Anal. Chem.*, **62**, 2211–2216.

BLOOM, J.W., MADANAT, M.S., MARRIOTT, D., WONG, T. and CHAN, S.Y., 1997, Intrachain disulfide bond in the core region of human IgG4, *Protein Sci.*, **6**, 407–415.

BLUMENTHAL, R.D., SHARKEY, R.M., SNYDER, D. and GOLDENBERG, D.M., 1989, Reduction by anti-antibody administration of the radiotoxicity associated with ^{131}I-labeled antibody to carcinoembryonic antigen in cancer radioimmunotherapy, *J. Natl Cancer Inst.*, **81**, 194–199.

BODE, C., RUNGE, M.S., SCHONERMARK, S., EBERLE, T., NEWELL, J.B., KUBLER, W. and HABER, E., 1990, Conjugation to antifibrin Fab′ enhances fibrinolytic potency of single-chain urokinase plasminogen activator, *Circulation*, **81**, 1974–1980.

BOLTON, A.E. and HUNTER, W.M., 1973, The labelling of proteins to high specific radioactivities by conjugation to a ^{125}I-containing acylating agent. Application to the radioimmunoassay, *Biochem. J.*, **133**, 529–539.

BONARDI, M.A., FRENCH, R.R., AMLOT, P., GROMO, G., MODENA, D. and GLENNIE, M.J., 1993, Delivery of saporin to human B-cell lymphoma using bispecific antibody: targeting via CD22 but not CD19, CD37 or immunoglobulin results in efficient killing, *Cancer Res.*, **53**, 3015–3021.

BONE, R.C., BALK, R.A., FEIN, A.M., PERL, T.M., WENZEL, R.P., REINES, H.D., QUENZER, R.W., IBERTI, T.J., MACINTYRE, N., SCHEIN, R.M.H. and the E5 SEPSIS STUDY GROUP, 1995, A second large controlled clinical study of E5, a monoclonal antibody to endotoxin, *Crit. Care Med.*, **23**, 994–1006.

BORGSTROM, P., HILLAN, K.J., SRIRAMARAO, P. and FERRARA, N., 1996, Complete inhibition of angiogenesis and growth of microtumours by anti-vascular endothelial growth factor neutralizing antibody: novel concepts of angiostatic therapy from intravital videomicroscopy, *Cancer Res.*, **56**, 4032–4039.

BORREBAECK, C.A.K, 1989, Strategy for production of human monoclonal antibodies using in vitro activated B cells, *J. Immunol. Methods*, **123**, 157–165.

BORREBAECK, C.A.K., DANIELSSON, L. and MOLLER, S.A., 1988, Human monoclonal antibodies produced by primary in vitro immunization of peripheral blood lymphocytes, *Proc. Natl Acad. Sci. (USA)*, **85**, 3995–3999.

BORREBAECK, C.A.K., MALMBORG, A.C. and OHLIN, M., 1993, Does endogenous glycosylation prevent the use of mouse monoclonal antibodies as cancer therapeutics?, *Immunol. Today*, **14**, 477–479.

BOSCHETTI, N., BRODBECK, U., JENSEN, S.P., KOCH, C. and NORGAARD-PEDERSEN, B., 1996, Monoclonal antibodies against a C-terminal peptide of human brain acetylcolinesterase distinguish between erythrocyte and brain acetylcolinesterases, *Clin. Chem.*, **42**, 19–23.

BOSS, M.A., KENTEN, J.H., WOOD, C.R. and EMTAGE, J.S., 1984, Assembly of functional antibodies from immunoglobulin heavy and light chains synthesised in *E. coli*, *Nucl. Acids Res.*, **12**, 3791–3806.

BOSSLET, K., CZECH, J. and HOFFMANN, D., 1994, Tumor-selective prodrug activation by fusion protein-mediated catalysis, *Cancer Res.*, **54**, 2151–2159.

BOSSLET, K., CZECH, J., LORENZ, P., SEDLACEK, H.H., SCHUERMANN, M. and SEEMAN, G., 1992, Molecular and functional characterization of a fusion protein suited for tumour specific prodrug activation, *Br. J. Cancer*, **65**, 234–238.

BOTHWELL, A.L.M., PASKIND, M., RETH, M., IMANISHI-KARI, T., RAJEWSKY, K. and BALTIMORE, D., 1981, Heavy chain variable region contribution to the NP^b family of antibodies: somatic mutation evident in a γ2a variable region, *Cell*, **24**, 625–637.

BOXER, G.M., BEGENT, R.H.J., KELLY, A.M.B., SOUTHALL, P.J., BLAIR, S.B., THEODOROU, N.A., DAWSON, P.M. and LEDERMANN, J.A., 1992, Factors influencing variability of localisation of antibodies to carcinoembryonic antigen (CEA) in patients with colorectal cancer: implications for radioimmunotherapy, *Br. J. Cancer*, **65**, 825–831.

BRAMBELL, F.W.R., HEMMINGS, W.A. and MORRIS, I.G., 1964, A theoretical model of γ-globulin catabolism, *Nature*, **203**, 1352–1355.

BRECHBIEL, M.W., GANSOW, O.A., ATCHER, R.A., SCHLOM, J., ESTEBAN, J., SIMPSON, D. and COLCHER, D., 1986, Synthesis of 1-(p-isothiocyanatobenzyl) derivatives of DTPA and EDTA. Antibody labeling and tumor imaging studies, *Inorg. Chem.*, **25**, 2772–2781.

BREITZ, H.B., WEIDEN, P.L., VANDERHEYEN, J.L., APPLEBAUM, J.W., BJORN, M.J., FER, M.F., WOLF, S.B., RATLIFF, B.A., SEILER, C.A., FOISIE, D.C., FISHER, D.R., SCHROFF, R.W., FRITZBERG, A.R. and ABRAMS, P.G., 1992, Clinical experience with rhenium-186-labeled monoclonal antibodies for radioimmunotherapy: results of phase I trials, *J. Nucl. Med.*, **33**, 1099–1112.

BRENNAN, M., DAVISON, P.F. and PAULUS, H., 1985, Preparation of bispecific antibodies by chemical recombination of monoclonal immunoglobulin G1 fragments, *Science*, **229**, 81–83.

BRETT, S., BAXTER, G., COOPER, H., JOHNSTON, J.M., TITE, J. and RAPSON, N., 1996, Repopulation of blood lymphocyte sub-populations in rheumatoid arthritis patients treated with the depleting humanised monoclonal antibody, CAMPATH–1H, *Immunol.*, **88**, 13–19.

BRIDEY, F., PHILIPOTTEAU, C., DREYFUS, M. and SIMONNEAU, G., 1989, Plasma D-dimer and pulmonary embolism, *Lancet*, **I**, 791–792.

BRINKMANN, U., PAI, L.H., FITZGERALD, D.J., WILLINGHAM, M. and PASTAN, I., 1991, B3(Fv)-PE38KDEL, a single chain immunotoxin that causes regression of a human carcinoma in mice, *Proc. Natl Acad. Sci. (USA)*, **88**, 8616–8620.

BRODEUR, B.R. and TSANG, P.S., 1986, High yield monoclonal antibody production in ascites, *J. Immunol. Methods*, **86**, 239–241.

BRONSTEIN, I., VOYTA, J.C., THORPE, G.H.G., KRICKA, L.J. and ARMSTRONG, G., 1989, Chemiluminescent assay of alkaline phosphatase applied in an ultrasensitive immunoassay of thyrotropin, *Clin. Chem.*, **35**, 1441–1446.

BROOKS, P.C., MONTGOMERY, A.M.P., ROSENFELD, M., REISFELD, R.A., HU, T., KLIER, G. and CHERESH, D.A., 1994, Integrin $\alpha_v\beta_3$ antagonists promote tumor regression by inducing apoptosis of angiogenic blood vessels, *Cell*, **79**, 1157–1164.

BROWN, B.A., COMEAU, R.D., JONES, P.L., LIBERATORE, F.A., NEACY, W.P., SANDS, H. and GALLAGHER, B.M., 1987, Pharmacokinetics of the monoclonal antibody B72.3 and its fragments labelled with either ^{125}I or ^{111}In, *Cancer Res.*, **47**, 1149–1154.

BROWN, M.E., RENNER, G., FIELD, R.P. and HASSELL, T., 1992, Process development for the production of recombinant antibodies using the glutamine synthetase (GS) system, *Cytotechnol.*, **9**, 231–236.

BROWN, P.M., TAGARI, P., ROWAN, K.R., YU, V.L., O'NEILL, G.P., MIDDAUGH, C.R., SANYAL, G., FORD-HUTCHINSON, A.W. and NICHOLSON, D.W., 1995, Epitope-labelled soluble human interleukin-5 receptors. Affinity cross-link labelling, IL-5 binding and biological activity, *J. Biol. Chem.*, **270**, 29236–29243.

BROWN, P.S., PARENTEAU, G.L., DIRBAS, F.M., GARSIA, R.J., GOLDMAN, C.K., BUKOWSKI, M.A., JUNGHANS, R.P., QUEEN, C., HAKIMI, J., BENJAMIN, W.R., CLARK, R.E. and WALDMANN, T.A., 1991, Prolongation of primate renal allograft survival by anti-Tac, an anti-human IL-2 receptor monoclonal antibody, *Proc. Natl. Acad. Sci. (USA)*, **88**, 2663–2667.

BRUGGEMANN, M., WILLIAMS, G.T., BINDON, C.I., CLARK M.R., WALKER, M.R., JEFFERIS, R., WALDMANN, H. and NEUBERGER, M.S., 1987, Comparison of the effector functions of human immunoglobulins using a matched set of chimeric antibodies, *J. Exp. Med.*, **166**, 1351–1361.

BUCHEGGER, F., MACH, J.P., PELEGRIN, A., GILLET, M., VOGEL, C.A., BUCLIN, T., RYSER, J.E., DELALOYE, B. and DELALOYE, A.B., 1995, Radiolabeled chimeric anti-CEA monoclonal antibody compared with the original mouse monoclonal antibody for surgically treated colorectal carcinoma, *J. Nucl. Med.*, **36**, 420–429.

BUCHEGGER, F., PELEGRIN, A., DELALOYE, B., BISCHOFF-DELALOYE, A. and MACH, J.P., 1990, Iodine-131 labelled Mab $F(ab')_2$ fragments are more effective and less toxic than intact anti-CEA antibodies in radioimmunotherapy of large human colon carcinoma grafted in nude mice, *J. Nucl. Med.*, **31**, 1035–1044.

BUCHEGGER, F., PELEGRIN, A., HARDMAN, N., HEUSSER, C., LUKAS, J., DOLCI, W. and MACH, J.P., 1992, Different behaviour of mouse–human chimeric antibody $F(ab')_2$ fragments of IgG1, IgG2 and IgG4 subclass *in vivo*, *Int. J. Cancer*, **50**, 416–422.

BUCHNER, J. and RUDOLPH, R., 1991, Renaturation, purification and characterization of recombinant Fab fragments produced in *Escherichia coli*, *Bio/Technol.*, **9**, 157–162.

BUCKLEY, D.T.J., ROBINS, A.R. and DURRANT, L.G., 1995, Clinical evidence that the human monoclonal anti-idiotypic antibody 105AD7 delays tumor growth by stimulating anti-tumour T-cell responses, *Hum. Antibod. Hybridomas*, **6**, 68–72.

BUIST, M.R., KENEMANS, P., DEN HOLLANDER, W., VERMORKEN, J.B., MOLTHOFF, C.J.M., BURGER, C.W., HELMERHORST, T.J.M., BAAK, J.P.A. and ROOS, J.C., 1993, Kinetics and tissue distribution of the radiolabelled chimeric monoclonal antibody MOv18 IgG and $F(ab')_2$ fragments in ovarian carcinoma patients, *Cancer Res.*, **53**, 5413–5418.

BUIST, M.R., KENEMANS, P., VAN KAMP, G.J. and HAISMA, H.J., 1995, Minor human antibody response to a mouse and chimeric monoclonal antibody after a single i.v. infusion in ovarian carcinoma patients: a comparison of five assays, *Cancer Immunol. Immunother.*, **40**, 24–30.

BURROWS, F.J., DERBYSHIRE, E.J., TAZZARI, P.L., AMLOT, P., GAZDAR, A.F., KING, S.W., LETARTE, M., VITETTA, E.S. and THORPE, P.E., 1995, Up-regulation of endoglin on vascular endothelial cells in human solid tumours: implications for diagnosis and therapy, *Clin. Cancer Res.*, **1**, 1623–1634.

BURROWS, F.J. and THORPE, P.E., 1993, Eradication of large solid tumours in mice with an immunotoxin directed against tumour vasculature, *Proc. Natl Acad. Sci. (USA)*, **90**, 8996–9000.

BURTON, D.R. and BARBAS, C.F., III, 1994, Human antibodies from combinatorial libraries, *Adv. Immunol.*, **57**, 191–280.

BYWATER, R., ERIKSSON, G. and OTTOSON, T., 1983, Desorption of immunoglobulins from protein A-Sepharose CL-4B under mild conditions, *J. Immunol. Methods*, **64**, 1–6.

CABILLY, S. and RIGGS, A.D., 1985, Immunoglobulin transcripts and molecular history of a hybridoma that produces antibody to carcinoembryonic antigen, *Gene*, **40**, 157–161.

CABILLY, S., RIGGS, A.D., PANDE, H., SHIVELY, J.E., HOLMES, W.E., REY, M., PERRY, L.J., WETZEL, R. and HEYNECKER, H.L., 1984, Generation of antibody activity from immunoglobulin polypeptide chains produced in *Escherichia coli*, *Proc. Natl Acad. Sci. (USA)*, **81**, 5369–5373.

CALIFF, R.M. for the EPIC investigators, 1994, Use of a monoclonal antibody directed against the platelet glycoprotein IIb/IIIa receptor in high risk angioplasty, *N. Engl. J. Med.*, **330**, 956–1007.

CALVIN, J., BURLING, K., BLOW, C., BARNES, I. and PRICE, C.P., 1986, Evaluation of fluorescence excitation transfer immunoassay for measurement of specific proteins, *J. Immunol. Methods*, **86**, 249–256.

CAMERA, L., KINUYA, S., GARMESTANI, K., WU, C., BRECHBIEL, M.W., PAI, L.H., MCMURRY, T.J., GANSOW, O.A., PASTAN, I., PAIK, C.H. and CARRASQUILLO, J.A., 1994, Evaluation of the serum stability and *in vivo* biodistribution of CHX-DTPA and other ligands for yttrium labeling of monoclonal antibodies, *J. Nucl. Med.*, **35**, 882–889.

CANFIELD, S.M. and MORRISON, S.L., 1991, The binding affinity of human IgG for its high affinity Fc receptor is determined by multiple amino acids in the CH2 domain and is modulated by the hinge region, *J. Exp. Med.*, **173**, 1483–1491.

CAPON, D.J., CHAMOW, S.M., MORDENTI, J., MARSTERS, S.A., GREGORY, T., MITSUYA, H., BYRN, R.A., LUCAS, C., WURM, F.M., GROOPMAN, J.E., BRODER, S. and SMITH, D.H., 1989, Designing CD4 immunoadhesin for AIDS therapy, *Nature*, **337**, 525–531.

CARAYANNOPOULOS, L., MAX, E.E. and CAPRA, J.D., 1994, Recombinant human IgA expressed in insect cells, *Proc. Natl Acad. Sci. (USA)*, **91**, 8348–8352.

CARLSSON, M., HEDIN, A., INGANAS, M., HARFAST, B. and BLOMBERG, F., 1985, Purification of in vitro produced mouse monoclonal antibodies. A two-step procedure using cation exchange chromatography and gel filtration, *J. Immunol. Methods*, **79**, 89–98.

CARLSSON, R., MARTENSSON, C., KALLIOMAKI, S., OHLIN, M. and BORREBAECK, C.A.K., 1992, Human peripheral blood lymphocytes transplanted into scid mice constitute an *in vivo* culture system exhibiting several parameters found in a normal humoral immune response and are a source of immunocytes for the production of human monoclonal antibodies, *J. Immunol.*, **148**, 1065–1071.

CARON, P.C., CO, M.S., BULL, M.K., AVDALOVIC, N.K., QUEEN, C. and SCHEINBERG, D.A., 1992, Biological and immunological features of humanized M195 (anti-CD33) monoclonal antibodies, *Cancer Res.*, **52**, 6761–6767.

CARON, P.C., JURCIC, J.G., SCOTT, A.M., FINN, R.D., DIVGI, C.R., GRAHAM, M.C., JUREIDINI, I.M., SGOUROS, G., TYSON, D., OLD, L.J., LARSON, S.M. and SCHEINBERG, D.A., 1994, A phase 1B trial of humanized monoclonal antibody M195 (anti-CD33) in myeloid leukemia: specific targeting without immunogenicity, *Blood*, **83**, 1760–1768.

CARRIER, A., DUCANCEL, F., SETTIAWAN, N.B., CATTOLICO, L., MAILLERE, B., LEONETTI, M., DREVET, P., MENEZ, A. and BOULAIN, J.C., 1995, Recombinant antibody–alkaline phosphatase conjugates for diagnosis of human IgGs: application to anti-HBsAg detection, *J. Immunol. Methods*, **181**, 177–186.

CARRIO, I., BERNA, L., BALLESTER, M., ESTORCH, M., OBRADOR, D., CLADELLAS, M., ABADAL, L. and GINJAUNE, M., 1988, Indium-111 antimyosin scintigraphy to assess myocardial damage in patients with suspected myocarditis and cardiac rejection, *J. Nucl. Med.*, **29**, 1893–1900.

CARRIO, I., ESTORCH, M., BERNA, L., LOPEZ-POUSA, J., TABERNO, J. and TORRES, G., 1995, Indium-111-antimyosin and iodine-123-MIBG studies in early assessment of doxorubicin cardiotoxicity, *J. Nucl. Med.*, **36**, 2044–2049.

CARROLL, W.L., MENDEL, E. and LEVY, S., 1988, Hybridoma fusion cell lines contain an aberrant kappa transcript, *Mol. Immunol.*, **25**, 991.

CARTER, P., KELLEY, R.F., RODRIGUES, M.L., SNEDCOR, B., COVARRUBIAS, M., VELLIGAN, M.D., WONG, W.L.T., ROWLAND, A.M., KOTTS, C.E., CARVER, M.E., YANG, M., BOURELL, J.H., SHEPARD, H.M. and HENNER, D., 1992a, High level *Escherichia coli* expression and production of a bivalent humanized antibody fragment, *Bio/Technol.*, **10**, 163–167.

CARTER, P., PRESTA, L., GORMAN, C.M., RIDGWAY, J.B.B., HENNER, D., WONG, W.L.T., ROWLAND, A.M., KOTTS, C., CARVER, M.E. and SHEPHARD, H.M., 1992b, Humanization of an anti-p185HER2 antibody for human cancer therapy, *Proc. Natl Acad. Sci. (USA)*, **89**, 4285–4289.

CARTNER, A.M., CONRY, R.M., SAFAVY, A., KHAZAELI, M.B., SUMEREL, L.A. and LOBUGLIO, A.F., 1993, An animal model to predict the immunogenicity of murine V regions in humans, *Hum. Antibod. Hybridomas*, **4**, 174–180.

CASADEI, J., POWELL, M.J. and KENTEN, J.H., 1990, Expression and secretion of aequorin as a chimeric antibody by means of a mammalian expression vector, *Proc. Natl Acad. Sci. (USA)*, **87**, 2047–2051.

CASEY, J.L., KEEP, P.A., CHESTER, K.A., ROBSON, L., HAWKINS, R.E. and BEGENT, R.H.J., 1995, Purification of bacterially expressed single chain Fv antibodies for clinical applications using metal chelate chromatography, *J. Immunol. Methods*, **179**, 105–116.

CASEY, J.L., KING, D.J., CHAPLIN, L.C., HAINES, A.M.R., PEDLEY, R.B., MOUNTAIN, A., YARRANTON, G.T. and BEGENT, R.H.J., 1996, Preparation, characterisation and tumour targeting of cross-linked divalent and trivalent anti-tumour Fab′ fragments, *Br. J. Cancer*, **74**, 1397–1405.

CATTANEO, A. and NEUBERGER, M.S., 1987, Polymeric immunoglobulin M is secreted by transfectants of non-lymphoid cells in the absence of immunoglobulin J chain, *EMBO J.*, **6**, 2753–2758.

CHAN, D.W. (ed.), 1996, *Immunoassay Automation: an Updated Guide to Systems*, Academic Press, San Diego.

CHAPLIN, L.C., 1996, Streamline rProtein A for expanded bed adsorption of mabs, *Pharmacia Downstream*, **22**, 22–23.

CHAPMAN, A.P., CHAPLIN, L.C., SPITALI, M., WEIR, N., SEHDEV, M., LAWSON, A.D.G. and KING, D.J., 1996, Purification of recombinant Fab′ fragments using Streamline A in expanded or batch mode. Presented at *1st International Conference on Expanded Bed Adsorption*, Cambridge.

CHARI, R.V.J., JACKEL, K.A., BOURRET, L.A., DERR, S.M., TADAYONI, B.M., MATTOCKS, K.M., SHAH, S.A., LIU, C., BLATTLER, W.A. and GOLDMACHER, V.S., 1995, Enhancement of the selectivity and antitumour efficacy of a CC-1065 analogue through immunoconjugate formation, *Cancer Res.*, **55**, 4079–4084.

CHARI, R.V.J., MARTELL, B.A., GROSS, J.L., COOK, S.B., SHAH, S.A., BLATTLER, W.A., MCKENZIE, S.J. and GOLDMACHER, V.S., 1992, Immunoconjugates containing novel maytansinoids: promising anticancer drugs, *Cancer Res.*, **52**, 127–131.

CHARNOW, S.M. and ASHKENAZI, A., 1996, Immunoadhesins: principles and applications, *Trends Biotechnol.*, **14**, 52–60.

CHEN, F. and EPSTEIN, A.L., 1988, Preparative separation of IgM monoclonal antibody from mouse ascites and purification of F(ab′)2 fragments – the use of Abx and Superose columns, *Antibod. Immunoconj. Radiopharm.*, **1**, 333–341.

CHEN, S.Y., ZANI, C., KHOURI, Y. and MARASCO, W.A., 1995, Design of a genetic immunotoxin to eliminate toxin immunogenicity, *Gene Therapy*, **2**, 116–123.

CHIANG, C.M. and ROEDER, R.G., 1993, Expression and purification of general transcription factors by FLAG epitope tagging and peptide elution, *Peptide Res.*, **6**, 62–64.

CHRISTOFIDES, N.D. and SHEEHAN, C.P., 1995, Enhanced chemiluminescence labeled-antibody immunoassay (Amerlite-MAB^{TM}) for free thyroxine: design, development and technical validation, *Clin. Chem.*, **41**, 17–23.

CLACKSON, T., HOOGENBOOM, H.R., GRIFFITHS, A.D. and WINTER, G., 1991, Making antibody fragments in phage display libraries, *Nature*, **352**, 624–628.

CLARK, P.M.S. and PRICE, C.P., 1986, Enzyme amplified immunoassays: a new ultrasensitive assay of thyrotropin evaluated, *Clin. Chem.*, **32**, 88–92.

CO, M.S., AVDALOVIC, N.M., CARON, P.C., AVDALOVIC, M.V., SCHEINBERG, D.A. and QUEEN, C., 1992, Chimeric and humanized antibodies with specificity for the CD33 antigen, *J. Immunol.*, **148**, 1149–1154.

CO, M.S., SCHEINBERG, D.A., AVDALOVIC, N.M., MCGRAW, K., VASQUEZ, M., CARON, P.C. and QUEEN, C., 1993, Genetically engineered deglycosylation of the variable domain increases the affinity of an anti-CD33 monoclonal antibody, *Mol. Immunol.*, **30**, 1361–1367.

COCKETT, M.I., BEBBINGTON, C.R. and YARRANTON, G.T., 1991, The use of engineered E1a genes to transactivate the hCMV-MIE promoter in permanent CHO cell lines, *Nucl. Acids Res.*, **19**, 319–325.

COLBERT, D.L., GALLACHER, G. and MAINWARING-BURTON, R.W., 1985, Single reagent polarization fluoroimmunoassay for amphetamine in urine, *Clin. Chem.*, **31**, 1193–1195.

COLE, W.C., DENARDO, S.J., MEARES, C.F., MCCALL, M.J., DENARDO, G.L., EPSTEIN, A.L., O'BRIEN, H.A. and MOI, M.K., 1987, Comparative serum stability of radiochelates for antibody radiopharmaceuticals, *J. Nucl. Med.*, **28**, 83–90.

COLMAN, A., 1996, Production of proteins in the milk of transgenic livestock: problems, solutions and successes, *Am. J. Clin. Nutrition*, **63**, 639S–645S.

COLOMA, M.J. and MORRISON, S.L., 1997, Design and production of novel tetravalent bispecific antibodies, *Nature Biotechnol.*, **15**, 159–163.

CONNETT, J.M., ANDERSON, C.J., GUO, L.W., SCHWARZ, S.W., ZINN, K.R., ROGERS, B.E., SIEGAL, B.A., PHILPOTT, G.W. and WELCH, M.J., 1996, Radioimmunotherapy with a ^{64}Cu-labeled monoclonal antibody: a comparison with ^{67}Cu, *Proc. Natl Acad. Sci. (USA)*, **93**, 6814–6818.

COOK, A.G. and WOOD, P.J., 1994, Chemical synthesis of bispecific monoclonal antibodies: potential advantages in immunoassay systems, *J. Immunol. Methods*, **171**, 227–237.

COOK, C.J., 1997, Real-time measurements of corticosteroids in conscious animals using an antibody-based electrode, *Nature Biotechnol.*, **15**, 467–471.

COOK, D.B. and SELF, C.H., 1993, Determination of one thousandth of an attomole (1 zeptomole) of alkaline phosphatase: application in an immunoassay of proinsulin, *Clin. Chem.*, **39**, 965–971.

COOPER, L.J.N., SHIKHMAN, A.R., GLASS, D.D., KANGISSER, D., CUNNINGHAM, M.W. and GREENSPAN, N.S., 1993, Role of heavy chain constant domains in antibody–antigen interaction, *J. Immunol.*, **150**, 2231–2242.

COVELL, D.G., BARBET, J., HOLTON, O.D., BLACK, C.D.V., PARKER, R.J. and WEINSTEIN, J.N., 1986, Pharmacokinetics of monoclonal immunoglobulin G1 $F(ab')_2$ and F(ab′) in mice, *Cancer Res.*, **46**, 3969–3978.

COX, J.P., JANKOWSKI, K.J., KATAKY, R., PARKER, D., BEELEY, N.R.A., BOYCE, B.A., EATON, M.A.W., MILLAR, K., MILLICAN, A.T., HARRISON, A. and WALKER, C., 1989, Synthesis of a kinetically stable yttrium-90 labelled macrocycle–antibody conjugate, *J. Chem. Soc. Chem. Commun.*, **1989**, 797–798.

CRAIG, A.S., HELPS, I.A., JANKOWSKI, K.J., PARKER, D., BEELEY, N.R.A., BOYCE, B.A., EATON, M.A.W., MILLICAN, T.A., MILLAR, K., PHIPPS, A., RHIND, S.K., HARRISON, A. and WALKER, C., 1989, Towards tumour imaging with indium-111 labelled macrocycle antibody conjugates, *J. Chem. Soc. Chem. Commun.*, **12**, 794–796.

CRIPPA, F., BOLIS, G., SEREGNI, E., GAVONI, N., SCARFONE, G., FERRARIS, C., BURAGGI, G.L. and BOMBARDIERI, E., 1995, Single-dose intraperitoneal radioimmunotherapy with the murine monoclonal antibody I-131 Mov18, *Eur. J. Cancer*, **31A**, 686–690.

CROWE, J.S., HALL, V.S., SMITH, M.A., COOPER, H.J. and TITE, J.P., 1992, Humanized monoclonal antibody CAMPATH-1: myeloma cell expression of genomic constructs, nucleotide sequence of cDNA constructs and comparison of effector mechanisms of myeloma and Chinese hamster ovary cell-derived material, *Clin. Exp. Immunol.*, **87**, 105–110.

CUMBER, A.J., WARD, E.S., WINTER, G., PARNELL, G.D. and WAWRZYNCZAK, E.J., 1992, Comparative stabilities in vitro and in vivo of a recombinant mouse antibody FvCys and a bisFvCys conjugate, *J. Immunol.*, **149**, 120–126.

CZUCZMAN, M.S., GRILLO-LOPEZ, A.J., JONAS, C., GORDON, L., SALEH, M., WHITE, C.A., VARNS, C. and ROGERS, J., 1995, IDEC-C2B8 and CHOP chemotherapy of low-grade lymphoma, *Blood*, **86**, Suppl. 1, 55A.

DANIELS, P.B., FLETCHER, J.E., O'NEILL, P.M., STAFFORD, C.G., BACARESE-HAMILTON, T. and ROBINSON, G.A., 1995, A comparison of three fluorophores for use in an optical biosensor for the measurement of prostate-specific antigen in whole blood, *Sensors Actuators*, **B26**, 447–451.

DANIELSSON, L., MOLLER, S.A. and BORREBAECK, C.A.K., 1987, Effect of cytokines on specific in vitro immunization of human peripheral B lymphocytes against T-cell dependent antigens, *Immunology*, **61**, 51–55.

DARVEAU, A., CHEVRIER, M.C., NERON, S., DELAGE, R. and LEMIEUX, R., 1993, Efficient preparation of human monoclonal antibody-secreting heterohybridomas using peripheral B lymphocytes cultured in the CD40 system, *J. Immunol. Methods*, **159**, 139–143.

DAS, C., KULKARNI, P.V., CONSTANTINESCU, A., ANTICH, P., BLATTNER, F.R. and TUCKER, P.W., 1992, Recombinant antibody–metallothionein: design and evaluation for radioimmunoimaging, *Proc. Natl Acad. Sci. (USA)*, **89**, 9749–9753.

DATTA, P., HINZ, V. and KLEE, G., 1996, Superior specificity of monoclonal based assay for digoxin, *Clin. Biochem.*, **29**, 541–547.

DAUGHERTY, B.L., DEMARTINO, J.A., LAW, M.F., KAWKA, D.W., SINGER, I.I. and MARK, G.E., 1991, Polymerase chain reaction facilitates the cloning, CDR-grafting and rapid expression of a murine monoclonal antibody directed against the CD18 component of leukocyte integrins, *Nucl. Acids Res.*, **19**, 2471–2476.

DAVIDSON, I.W., 1992, Rapid immunoassays, *Anal. Proc.*, **29**, 459–460.

DAVIES, J. and RIECHMANN, L., 1995, Antibody VH domains as small recognition units, *Bio/Technol.*, **13**, 475–479.

DAVIS, A.C., ROUX, K.H. and SHULMAN, M.J., 1988, On the structure of polymeric IgM, *Eur. J. Immunol.*, **18**, 1001–1008.

DEBINSKI, W. and PASTAN, I., 1995, Recombinant Fab′ C242-Pseudomonas exotoxin, but not the whole antibody based immunotoxin causes regression of a human colorectal tumor xenograft, *Clin. Cancer Res.*, **1**, 1015–1022.

DEFAUCAL, P., PELTIER, P., PLANCHON, B., DUPAS, B., TONZE, M.D., BARON, D., SCIABLE, T., BERGER, H.J. and CHATAL, J.F., 1991, Evaluation of indium-111 labelled antifibrin monoclonal antibody for the diagnosis of venous thrombotic disease, *J. Nucl. Med.*, **32**, 785–791.

DEISENHOFER, J., 1981, Crystallographic refinement and atomic models of a human Fc fragment and its complex with fragment B of protein A from *Staphylococcus aureus* at 2.9 and 2.8 Å resolution, *Biochem.*, **20**, 2361–2370.

DE LALLA, C., TAMBORINI, E., LONGHI, R., TRESOLDI, E., MANONI, M., SICCARDI, A.G., AROSIO, P. and SIDOLI, A., 1996, Human recombinant antibody fragment specific for a rye grass pollen allergen, *Mol. Immunol.*, **33**, 1049–1058.

DELGADO, C., PEDLEY, R.B., HERRAEZ, A., BODEN, R., BODEN, J.A., KEEP, P.A., CHESTER, K.A., FISHER, D., BEGENT, R.H.J. and FRANCIS, G.E., 1996, Enhanced tumour specificity of an anti-carcinoembrionic antigen Fab′ fragment by poly(ethylene glycol) (PEG) modification, *Br. J. Cancer*, **73**, 175–182.

DEMANET, C., BRISSINCK, J., DE JONGE, J. and THIELEMANS, K., 1996, Bispecific antibody-mediated immunotherapy of the BCL1 lymphoma: increased efficacy with multiple injections and CD28 induced costimulation, *Blood*, **10**, 4390–4398.

DEMIDEN, A., HANNA, N., HARIHARAN, H. and BONAVIDA, B., 1995, Chimeric anti-CD20 antibody (IDEC-C2B8) is apoptopic and sensitizes drug resistant human B cell lymphomas and AIDS related lymphomas to the cytotoxic effect of CDDP, VP-16 and toxins, *FASEB J.*, **9**, A206.

DENARDO, S.J., O'GRADY, L.F., RICHMAN, C.M. and DENARDO, G.L., 1994, Overview of radio-immunotherapy in advanced breast cancer using 131-I chimeric L6, *Adv. Exp. Med. Biol.*, **353**, 203–211.

DENARDO, D.A., DENARDO, G.L., YUAN, A., SHEN, S., DENARDO, S.J., MACEY, D.J., LAMBORN, K.R., MAHE, M., GROCH, M.W. and ERWIN, W.D., 1996a, Prediction of radiation doses from therapy using tracer studies with iodine-131-labeled antibodies, *J. Nucl. Med.*, **37**, 1970–1975.

DENARDO, G.L., MIRICK, G.R., KROGER, L.A., O'DONNELL, R.T., MEARES, C.F. and DENARDO, S.J., 1996b, Antibody responses to macrocycles in lymphoma, *J. Nucl. Med.*, **37**, 451–456.

DENEVE, M., DELOOSE, M., JACOBS, A., VAN HOUDT, H., KALUZA, B., WEIDLE, U., VAN MONTAGU, M. and DEPICKER, A., 1993, Assembly of an antibody and its derived antibody fragment in *Nicotiana* and *Arabidopsis*, *Transgenic Res.*, **2**, 227–237.

DERRICK, J.P. and WIGLEY, D.B., 1992, Crystal structure of a streptococcal protein G domain bound to a Fab fragment, *Nature*, **359**, 752–754.

DESHPANDE, S.V., DENARDO, S.J., KUKIS, D.L., MOI, M.K., MCCALL, M.J., DENARDO, G.L. and MEARES, C.F., 1990, Yttrium-90 labelled monoclonal antibody for therapy: labelling by a new macrocyclic bifunctional chelating agent, *J. Nucl. Med.*, **31**, 473–479.

DESMYTER, A., TRANSUE, T.R., GHAHROUDI, M.A., THI, M.H., POORTMANS, F., HAMERS, R., MUYLDERMANS, S. and WYNS, L., 1996, Crystal structure of a camel single-domain VH antibody fragment in complex with lysozyme, *Nature Struct. Biol.*, **3**, 808–811.

DESPLANCQ, D., KING, D.J., LAWSON, A.D.G. and MOUNTAIN, A., 1994, Multimerization behaviour of single chain Fv variants for the tumour binding antibody B72.3, *Protein Engin.*, **7**, 1027–1033.

DIAMANDIS, E.P., 1988, Immunoassays with time-resolved fluorescence spectroscopy: principles and applications, *Clin. Biochem.*, **21**, 139–150.

DIAMANDIS, E., 1991, Multiple labeling and time-resolvable fluorophores, *Clin. Chem.*, **37**, 1486–1491.

DITZEL, H.J., BINLEY, J.M., MOORE, J.P., SODROSKI, J., SULLIVAN, N., SAWYER, L.S.W., HENDRY, R.M., YANG, W.P., BARBAS, C.F., III, and BURTON, D.R., 1995, Neutralizing recombinant human antibodies to a conformational V2 and CD4-binding site-sensitive epitope of HIV-1 gp120 isolated by using an epitope masking procedure, *J. Immunol.*, **154**, 893–906.

DOERR, R.J., ABDEL-NABI, H., KRAG, D. and MITCHELL, E., 1991, Radiolabelled antibody imaging in the management of colorectal carcinoma, *Ann. Surgery*, **214**, 118–124.

DOHLSTEIN, M., ABRAHMSEN, L., BJORK, P., LANDO, P.A., HEDLUND, G., FORSBERG, G., BRODIN, T., GASCOIGNE, N.R.J., FORBERG, C., LIND, P. and KALLAND, T., 1994, Monoclonal antibody–superantigen fusion proteins: tumor-specific agents for T-cell based tumor therapy, *Proc. Natl Acad. Sci. (USA)*, **91**, 8945–8949.

DOHLSTEIN, M., HANSSON, J., OHLSSON, L., LITTON, M. and KALLAND, T., 1995, Antibody-targeted superantigens are potent inducers of tumor-infiltrating lymphocytes *in vivo*, *Proc. Natl Acad. Sci. (USA)*, **92**, 9791–9795.

DONOHOE, P.J., MACARDLE, P.J. and ZOLA, H., 1995, Making and using conventional mouse monoclonal antibodies, *Monoclonal Antibodies, the Second generation* (Zola, H., ed.), pp. 15–42, Bios Scientific Publishers, Oxford.

DORAI, H., MCCARTNEY, J.E., HUDZIAK, R.M., TAI, M.S., LAMINET, A.A., HOUSTON, L.L., HUSTON, J.S. and OPPERMANN, H., 1994, Mammalian cell expression of single-chain Fv (sFv) antibody proteins and their C-terminal fusions with interleukin-2 and other effector domains, *Bio/Technol.*, **12**, 890–897.

DORAI, H. and MOORE, G.P., 1987, The effect of dihydrofolate reductase-mediated gene amplification on the expression of transfected immunoglobulin genes, *J. Immunol.*, **139**, 4232–4241.

DOUGALL, W.C., PETERSON, N.C. and GREENE, M.I., 1994, Antibody-structure-based design of pharmacological agents, *Trends Biotechnol.*, **12**, 372–379.

DOUGLAS, A.S. and MONTEITH, C.A., 1994, Improvements to immunoassays by use of covalent binding assay plates, *Clin. Chem.*, **40**, 1833–1837.

DRAGACCI, S., GLEIZES, E., FREMY, J.M. and CANDLISH, A.A., 1995, Use of immunoaffinity chromatography as a purification step for the determination of aflatoxin M1 in cheeses, *Food Additives Contaminants*, **12**, 59–65.

DUAN, L., BAGASRA, O., LAUGHLIN, M.A., OAKES, J.W. and POMERANTZ, R.J., 1994, Potent inhibition of human immunodeficiency virus type 1 replication by an intracellular anti-rev single-chain antibody, *Proc. Natl Acad. Sci. (USA)*, **91**, 5075–5079.

DUCANCEL, F., GILLET, D., CARRIER, A., LAJEUNESSE, E., MENEZ, A. and BOULAIN, J.C., 1993, Recombinant colorimetric antibodies: construction and characterization of a bifunctional $F(ab)_2$/alkaline phosphatase conjugate produced in *Escherichia coli*, *Bio/Technol.*, **11**, 601–605.

DUCHOSAL, M.A., EMING, S.A., FISCHER, P., LETURCQ, D., BARBAS, C.F., III, MCCONAHEY, P.J., CAOTHEIN, R.H., THORNTON, G.B., DIXON, F.J. and BURTON, D.R., 1992, Immunization of hu-PBL-SCID mice and the rescue of human monoclonal Fab fragments through combinatorial libraries, *Nature*, **355**, 258–262.

DUENAS, M., VAZQUEZ, J., AYALA, M., SODRLIND, E., OHLIN, M., PEREZ, L., BORREBAECK, C.A.K. and GAVILONDO, J.V., 1994, Intra- and extracellular expression of an scFv antibody fragment in *E. coli*: effect of bacterial strains and pathway engineering using GroES/L chaperonins, *BioTechniques*, **16**, 476–483.

DUNCAN, A.R. and WINTER, G., 1988, The binding site for C1q on IgG, *Nature*, **332**, 738–740.

DUNCAN, A.R., WOOF, J.M., PARTRIDGE, L.J., BURTON, D.R. and WINTER, G., 1988, Localization of the binding site for the human high affinity Fc receptor on IgG, *Nature*, **332**, 563–564.

DYER, M.J.S., HALE, G., HAYHOE, F.G.J. and WALDMANN, H., 1989, Effects of CAMPATH-1 antibodies *in vivo* in patients with lymphoid malignancies: influence of antibody isotype, *Blood*, **73**, 1431–1439.

EARY, J., SCHROFF, R.W., ABRAMS, P.G., FRITZBERG, A.R., MORGAN, A.C., KASINA, S., RENO, J.M., SRINIVASAN, A., WOODHOUSE, C.S., WILBUR, D.S., NATALE, R.B., COLLINS, C., STEHLIN, J.S., MITCHELL, M. and NELP, W.B., 1989, Successful imaging of malignant melanoma with technetium-99m labeled monoclonal antibodies, *J. Nucl. Med.*, **30**, 25–32.

EGAN, R.W., ATHWAL, D., CHOU, C.C., EMTAGE, S., JEHN, C.H., KUNG, T.T., MAUSER, P.J., MURGOLO, N.J. and BODMER, M.W., 1995, Inhibition of pulmonary eosinophilia and hyperreactivity by antibodies to interleukin-5, *Int. Arch. Allergy Immunol.*, **107**, 321–322.

EISENHUT, M., LEHMANN, W.D., BECKER, W., BEHR, T., ELSER, H., STRITTMATTER, W., STEINSTRASSER, A., BAUM, R.P., VALERIUS, T., REPP, R. and DEO, Y., 1996, Bifunctional NHS-BAT ester for antibody conjugation and stable technetium labeling, *J. Nucl. Med.*, **37**, 362–370.

ELDIN, P., PAUZA, M.E., HIEDA, Y., LIN, G., MUTAUGH, M.P., PENTEL, P.R. and PENNELL, C.A., 1997, High-level secretion of two antibody single chain Fv fragments by *Pichia pastoris*, *J. Immunol. Methods*, **201**, 67–75.

ELLIOTT, M.J., MAINI, R.N., FELDMANN, M., KALDEN, J.R., ANTONI, C., SMOLEN, J.S., LEEB, B., BREEDFELD, F.C., MACFARLANE, J.D., BIJL, H. and WOODY, J.N., 1994, Randomised double-blind comparison of chimeric monoclonal antibody to tumor necrosis factor α (cA2) versus placebo in rheumatoid arthritis, *Lancet*, **344**, 1105–1110.

ELLIS, J.H., BARBER, K.A., TUTT, A., HALE, C., LEWIS, A.P., GLENNIE, M.J., STEVENSON, G.T. and CROWE, J.S., 1995, Engineered anti-CD38 monoclonal antibodies for immunotherapy of multiple myeloma, *J. Immunol.*, **155**, 925–937.

ERNTELL, M., MYHRE, E.B. and KRONVALL, G., 1983, Alternative non-immune IgG $F(ab')_2$ binding to group C and G streptococci, *Scand. J. Immunol.*, **17**, 201–209.

ESHHAR, Z., WAKS, T., GROSS, G. and SCHINDLER, D.G., 1993, Specific activation and targeting of cytotoxic lymphocytes through chimeric single chains consisting of antibody-binding domains and the γ or ζ subunits of the immunoglobulin and T-cell receptors, *Proc. Natl Acad. Sci. (USA)*, **90**, 720–724.

ESKELAND, T. and CHRISTENSEN, T.B., 1975, IgM molecules with and without J chain in serum and after purification, studied by ultracentrifugation, electrophoresis, and electron microscopy, *Scand. J. Immunol.*, **4**, 217–228.

EVANS, M.J., ROLLINS, S.A., WOLFF, D.W., ROTHER, R.P., NORIN, A.J., THERRIEN, D.M., GRIJALVA, G.A., MUELLER, J.P., NYE, S.H., SQUINTO, S.P. and WILKINS, J.A., 1995, In vitro and in vivo inhibition of complement activity by a single chain Fv fragment recognizing human C5, *Mol. Immunol.*, **32**, 1183–1195.

EXLEY, D. and EKEKE, G.I., 1981, Fluoroimmunoassay for 5-dihydrotestosterone, *J. Steroid Biochem.*, **14**, 1297–1302.

FAGERBERG, J., RAGNHAMMAR, P., LILJEFORS, M., HJELM, A.L., MELLSTEDT, H. and FRODIN, J.E., 1996, Humoral anti-idiotypic and anti-anti-idiotypic immune response in cancer patients treated with monoclonal antibody 17-1A, *Cancer Immunol. Immunother.*, **42**, 81–87.

FAGNANI, R., HAGAN, M.S. and BARTHOLOMEW, R., 1990, Reduction of immunogenicity by covalent modification of murine and rabbit immunoglobulins with oxidized dextrans of low molecular weight, *Cancer Res.*, **50**, 3638–3645.

FALLGREEN-GEBAUER, E., GEBAUER, W., BASTIAN, A., KRATZIN, H.D., EIFFERT, H., ZIMMERMAN, B., KARAS, M. and HILSCHMAN, N., 1993, The covalent linkage of secretory component to IgA. Structure of sIgA, *Biol. Chem. Hoppe-Seyler*, **374**, 1023–1028.

FANGER, M.W. and GUYRE, P.M., 1991, Bispecific antibodies for targeted cellular cytotoxicity, *Trends Biotechnol.*, **9**, 375–380.

FEIT, C., BARTAL, A.H., TAUBER, G., DUMBORT, G. and HIRSHAUT, Y., 1983, An enzyme-linked immunosorbent assay (ELISA) for the detection of monoclonal antibodies recognizing surface antigens expressed on viable cells, *J. Immunol. Methods*, **58**, 301–308.

FELDMANN, M., ELLIOT, M.J., WOODY, J.N. and MAINI, R.N., 1997, Anti-tumor necrosis factor – a therapy of rheumatoid arthritis, *Adv. Immunol.*, **64**, 283–350.

FELL, H.P., GAYLE, M.A., GROSMAIRE, L. and LEDBETTER, J.A., 1991, Genetic construction and characterization of a fusion protein consisting of a chimeric Fab′ with specificity for carcinomas and human IL-2, *J. Immunol.*, **146**, 2446–2452.

FERGUSON, R.A., YU, H., KALYVAS, M., ZAMMIT, S. and DIAMANDIS, E.P., 1996, Ultrasensitive detection of prostate-specific antigen by a time-resolved immunofluorometric assay and the immulite immunochemiluminescent third-generation assay: potential application in prostate and breast cancers, *Clin. Chem.*, **42**, 675–684.

FIEDLER, U. and CONRAD, U., 1995, High-level production and long-term storage of engineered antibodies in transgenic tobacco seeds, *Bio/Technol.*, 13, 1090–1093.

FIELD, H., YARRANTON, G.T. and REES, A.R., 1989, Expression of mouse immunoglobulin light and heavy chain variable regions in *Escherichia coli* and reconstitution of antigen binding activity, *Protein Engin.*, **3**, 641–647.

FIELDS, B.A., GOLDBAUM, F.A., YSERN, X., POLJAK, R. and MARIUZZA, R.A., 1995, Molecular basis of antigen mimicry by an anti-idiotope, *Nature*, **374**, 739–742.

FISCH, I., KUNZI, G., ROSE, K. and OFFORD, R.E., 1992, Site-specific modification of a fragment of a chimeric monoclonal antibody using reverse proteolysis, *Bioconj. Chem.*, **3**, 147–153.

FISHWILD, D.M., O'DONNELL, S.L., BENGOECHEA, T., HUDSON, D.V., HARDING, F., BERNHARD, S.L., JONES, D., KAY, R.M., HIGGINS, K.M., SCHRAMM, S.R. and LONBERG, N., 1996, High-avidity human IgGk monoclonal antibodies from a novel strain of minilocus transgenic mice, *Nature Biotechnol.*, **14**, 845–851.

FOOTE, J. and WINTER, G., 1992, Antibody framework residues affecting the conformation of the hypervaraiable loops, *J. Mol. Biol.*, **224**, 487–499.

FOSANG, A.J., LAST, K. and MACIEWICZ, R.A., 1996, Aggrecan is degraded by matrix metalloproteinases in human arthritis, *J. Clin. Invest*, **98**, 2292–2299.

FOWELL, S.A. and CHASE, H.A., 1986, A comparison of some activated matrices for preparation of immunosorbents, *J. Biotechnol.*, **4**, 355–368.

FRANKEN, E., TEUSCHEL, U. and HAIN, R., 1997, Recombinant proteins from transgenic plants, *Curr. Opin. Biotechnol.*, **8**, 411–416.

FRENCH, R.R., HAMBLIN, T.J., BELL, A.J., TUTT, A.L. and GLENNIE, M., 1995, Treatment of B-cell lymphoma with combination of bispecific antibodies and saporin, *Lancet*, **346**, 223–224.

FRIEDMAN, P.N., CHACE, D.F., TRAIL, P.A. and SIEGALL, C.B., 1993, Antitumor activity of the single-chain immunotoxin BR96 sFv-PE40 against established breast and lung tumor xenografts, *J. Immunol.*, **150**, 3054–3061.

FRITSCH, G., STIMPFL, M., KURZ, M., LEITNER, A., PRINTZ, D., BUCHINGER, P., HOECKER, P. and GADNER, H., 1995, Characterization of hematopoietic stem cells, *Ann. NY Acad. Sci.*, **770**, 42–52.

FRITZBERG, A.R., ABRAMS, P.G., BEUMIER, P.L., KASINA, S., MORGAN, A.C., RAO, T.N., RENO, J.M., SANDERSON, J.A., SRINIVASAN, A., WILBUR, D.S. and VANDERHEYDEN, J.L., 1988, Specific and stable labeling of antibodies with technetium-99m with a diamide dithiolate chelating agent, *Proc. Natl Acad. Sci. (USA)*, **85**, 4025–4029.

GADINA, M., NEWTON, D.L., RYBAK, S.M., WU, Y.N. and YOULE, R.J., 1994, Humanized immunotoxins, *Ther. Immunol.*, **1**, 59–64.

GANDECHA, A., OWEN, M.R.L., COCKBURN, W. and WHITELAM, G.C., 1994, Antigen detection using recombinant bifunctional single-chain Fv fusion proteins synthesised in *Escherichia coli*, *Protein Exp. Purific.*, **5**, 385–390.

GARKAVIJ, M., TENNVALL, J., STRAND, S.E., SJOGREN, H.O., JIANCHENG, C., NILSSON, R. and ISAKSSON, M., 1997, Extracorporeal whole-blood immunoadsorption enhances radioimmunotargeting of iodine-125-labeled BR96–biotin monoclonal antibody, *J. Nucl. Med.*, **38**, 895–901.

GARRARD, L.J., YANG, M., O'CONNELL, M.P., KELLEY, R.F. and HENNER, D.J., 1991, Fab assembly and enrichment in a monovalent phage display system, *Bio/Technol.*, **9**, 1373–1377.

GEORGE, A.J.T., JAMAR, F., TAI, M.S., HEELAN, B.T., ADAMS, G.P., MCCARTNEY, J.E., HOUSTON, L.L., WEINER, L.M., OPPERMANN, H., PETERS, A.M. and HUSTON, J.S., 1995, Radiometal labeling of recombinant proteins by a genetically engineered minimal chelation site, *Proc. Natl Acad. Sci. (USA)*, **92**, 8358–8362.

GHETIE, M.A., PODAR, E.M., ILGEN, A., GORDON, B.E., UHR, J.W. and VITETTA, E.S., 1997a, Homodimerization of tumor-reactive monoclonal antibodies markedly increases their ability to induce growth arrest or apoptosis of tumor cells, *Proc. Natl Acad. Sci. (USA)*, **94**, 7509–7514.

GHETIE, V., HUBBARD, J.G., KIM, J.K., TSEN, M.F., LEE, Y. and WARD, E.S., 1996, Abnormally short serum half lives of IgG in β2-microglobulin deficient mice, *Eur. J. Immunol.*, **26**, 690–696.

GHETIE, V., POPOV, S., BORVAK, J., RADU, C., MATESOI, D., MEDESAN, C., OBER, R.J. and WARD, E.S., 1997b, Increasing the serum persistence of an IgG fragment by random mutagenesis, *Nature Biotechnol.*, **15**, 637–640.

GHETIE, V. and VITETTA, E., 1994, Immunotoxins in the therapy of cancer: from bench to clinic, *Pharmac. Ther.*, **63**, 209–234.

GILLIES, S.D., DORAI, H., WESLOWSKI, J., MAJEAU, G., YOUNG, D., BOYD, J., GARDNER, J. and JAMES, K., 1989a, Expression of human anti-tetanus toxoid antibody in transfected murine myeloma cells, *Bio/Technol.*, **7**, 799–804.

GILLIES, S.D., LO, K.M. and WESLOWSKI, J., 1989b, High level expression of chimeric antibodies using adapted cDNA variable region cassettes, *J. Immunol. Methods*, **125**, 191–202.

GILLIES, S.D., REILLY, E.B., LO, K.M. and REISFELD, R.A., 1992, Antibody-targeted interleukin 2 stimulates T-cell killing of autologous tumor cells, *Proc. Natl Acad. Sci. (USA)*, **89**, 1428–1432.

GILLIES, S.D., YOUNG, D., LO, K.M. and ROBERTS, S., 1993, Biological activity and *in vivo* clearance of antitumour antibody/cytokine fusion proteins, *Bioconj. Chem.*, **4**, 230–235.

GINSBERG, J.S., WELLS, P.S., BRILL-EDWARDS, P., DONOVAN, D., PANJU, A., VAN BEEK, J.R. and PATEL, A., 1995, Application of a novel and rapid whole blood assay for D-Dimer in patients with clinically suspected pulmonary embolism, *Thrombosis Haemostasis*, **73**, 35–38.

GIOVAGNOLI, M.R. and VECCHIONE, A., 1996, Immunocytochemistry of cytological specimens as a diagnostic and prognostic tool, *Anticancer Res.*, **16**, 2225–2232.

GLENNIE, M.J., MCBRIDE, H.M., WORTH, A.T. and STEVENSON, G.T., 1987, Preparation and performance of bispecific $F(ab'\gamma)_2$ antibody containing thioether-linked Fab'γ fragments, *J. Immunol.*, **139**, 2367–2375.

GLOCKSHUBER, R., MALIA, M., PFITZINGER, I. and PLUKTHUN, A., 1990, A comparison of strategies to stabilise immunoglobulin Fv fragments, *Biochem.*, **29**, 1362–1367.

GOLDBERG, M.E. and DJAVADI-OHANIANCE, L., 1993, Methods for measurement of antibody/antigen affinity based on ELISA and RIA, *Curr. Opin. Immunol.*, **5**, 278–281.

GOLDENBERG, D.M., GOLDENBERG, H., SHARKEY, R.M., HIGGINBOTHAM-FORD, E., LEE, R.E., SWAYNE, L.C., BURGER, K.A., TSAIN, D., HOROWITZ, J.A., HALL, T.C., PINSKY, C.M. and HANSEN, H.J., 1990, Clinical studies of radioimmunodetection with carcinoembryonic antigen monoclonal antibody fragments labelled with ^{123}I or ^{99m}Tc, *Cancer Res.*, **50**, 909–921.

GOODMAN, G.E., HELLSTROM, I., YELTON, D.E., MURRAY, J.L., O'HARA, S., MEAKER, E., ZEIGLER, L., PALAZOLLO, P., NICAISE, C., USAKEWICZ, J. and HELLSTROM, K.E., 1993, Phase I trial of chimeric (human–mouse) monoclonal antibody L6 in patients with non-small-cell lung, colon, and breast cancer, *Cancer Immunol. Immunother.*, **36**, 267–273.

GOODWIN, D.A., MEARES, C.F., MCCALL, M.J., MCTIGUE, M. and CHAOVAPONG, W., 1988, Pretargeted immunoscintigraphy of murine tumours with indium-111 labeled bifunctional haptens, *J. Nucl. Med.*, **29**, 226–234.

GOODWIN, D.A., MEARES, C.F., WATANABE, N., MCTIGUE, M., CHAOVAPONG, W., RANSONE, C.M., RENN, O., GREINER, D.P., KUKIS, D.L. and KRONENBERGER, S.I., 1994, Pharmacokinetics or pretargeted monoclonal antibody 2D12.5 and ^{88}Y-janus-2-(*p*-nitrobenzyl)-1,4,7,10-tetraazacyclododecanetetraacetic acid (DOTA) in BALB/c mice with KHJ mouse adenocarcinoma: a model for radioimmunotherapy, *Cancer Res.*, **54**, 5937–5946.

GORMAN, C., 1990, Mammalian cell expression, *Curr. Opin. Biotechnol.*, **1**, 36–47.

GORMAN, S.C., CLARK, M.E., ROUTLEDGE, E.G., COBBOLD, S.P. and WALDMANN, H., 1991, Reshaping a therapeutic CD4 antibody, *Proc. Natl Acad. Sci. (USA)*, **88**, 4181–4185.

GOSHORN, S.C., SVENSSON, H.P., KERR, D.E., SOMERVILLE, J.E., SENTER, P. and FELL, H.P., 1993, Genetic construction, expression and characterization of a single-chain anti-carcinoma antibody fused to β-lactamase, *Cancer Res.*, **53**, 2123–2127.

GRAHAM, B.M., PORTER, A.J. and HARRIS, W.J., 1995, Cloning, expression and characterization of a single-chain antibody fragment to the herbicide paraquat, *J. Chem. Technol. Biotechnol.*, **63**, 279–289.

GREEN, L.L., HARDY, M.C., MAYNARD-CURRIE, C.E., TSUDA, H., LOUIE, D.M., MENDEZ, M.J., ABDERRAHIM, H., NOGUCHI, M., SMITH, D.H., ZENG, Y., DAVID, N.E., SASAI, H., GARZA, D., BRENNER, D.G., HALES, J.F., MCGUINNESS, R.P., CAPON, D.J., KLAPHOLZ, S. and JAKOBOVITS, A., 1994, Antigen-specific human monoclonal antibodies from mice engineered with human Ig heavy and light chain YACs, *Nature Genetics*, **7**, 13–21.

GREENMAN, J., JONES, E., WRIGHT, M.D. and BARCLAY, A.N., 1996, The use of intracellular single-chain antibody fragments to inhibit specifically the expression of cell surface molecules, *J. Immunol. Methods*, **194**, 169–180.

GREENWOOD, J., CLARK, M. and WALDMANN, H., 1993, Structural motifs involved in human IgG antibody effector functions, *Eur. J. Immunol.*, **23**, 1098–1104.

GRIFFITHS, A.D., MALMQVIST, M., MARKS, J.D., BYE, J.M., EMBLETON, M.J., MCCAFFERTY, J., BAIER, M., HOLLIGER, K.P., GORICK, B.D., HUGHES-JONES, N.C., HOOGENBOOM, H.R. and WINTER, G., 1993, Human anti-self antibodies with high specificity from phage display libraries, *EMBO J.*, **12**, 725–734.

GRIFFITHS, A.D., WILLIAMS, S.C., HARTLEY, O., TOMLINSON, I.M., WATERHOUSE, P., CROSBY, W.L., KONTERMANN, R.E., JONES, P.T., LOW, N.M., ALLISON, T.J., PROSPERO, T.D., HOOGENBOOM, H.R., NISSIM, A., COX, J.P.L., HARRISON, J.L., ZACCOLO, M., GHERARDI, E. and WINTER, G., 1994, Isolation of high affinity human antibodies directly from large synthetic repertoires, *EMBO J.*, **13**, 3245–3260.

GRUBER, M., SCHODIN, B.A., WILSON, E.R. and KRANZ, D.M., 1994, Efficient tumour cell lysis mediated by a bispecific single chain antibody expressed in *Escherichia coli*, *J. Immunol.*, **152**, 5368–5374.

GUAN, X.M., KOBILKA, T.S. and KOBILKA, B.K., 1992, Enhancement of membrane insertion and function in a type IIIb membrane protein following introduction of a cleavable signal peptide, *J. Biol. Chem.*, **267**, 21995–21998.

GUDDAT, L.W., SHAN, L., ANCHIN, J.M. and EDMUNDSON, A.B., 1994, Local and transmitted conformational changes on complexation of an anti-sweetener Fab, *J. Mol. Biol.*, **236**, 247–274.

GURTLER, L., 1996, Difficulties and strategies of HIV diagnosis, *Lancet*, **348**, 176–179.

GUSTAFSSON, B. and HINKULA, J., 1994, Antibody production of a human EBV-transformed B cell line and its heterohybridoma and trioma cell line descendants in different culture systems, *Human Antibod. Hybridomas*, **5**, 98–104.

HAKIMI, J., CHIZZONITE, R., LUKE, D.R., FAMILLETTI, P.C., BAILON, P., KONDAS, J.A., PILSON, R.S., LIN, P., WEBER, D.V., SPENCE, C., MONDINI, L.J., TSIEN, W.-H., LEVIN, J.E., GALLATI, V.H., KORN, L., WALDMANN, T.A., QUEEN, C. and BENJAMIN, W.R., 1991, Reduced immunogenicity and improved pharmacokinetics of humanized anti-Tac in cynomolgus monkeys, *J. Immunol.*, **147**, 1352–1359.

HALE, G., DYER, M., CLARK, M.R., PHILLIPS, J.M., MARCUS, R., RIECHMANN, L., WINTER, G. and WALDMANN, H., 1988, Remission induction in non-Hodgkin lymphoma with reshaped human monoclonal antibody CAMPATH-1H, *Lancet*, **ii**, 1394–1399.

HANK, J.A., SURFUS, J.E., GAN, J., JAEGER, P., GILLIES, S.D., REISFELD, R.A. and SONDEL, P.M., 1996, Activation of human effector cells by a tumor reactive recombinant anti-ganglioside GD2 interleukin-2 fusion protein (ch14.18-IL2), *Clin. Cancer Res.*, **2**, 1951–1959.

HANSSON, J., OHLSSON, L., PERSSON, R., ANDERSSON, G., ILBACK, N.G., LITTON, M.J., KALLAND, T. and DOHLSTEIN, M., 1997, Genetically engineered superantigens as tolerable antitumor agents, *Proc. Natl Acad. Sci. (USA)*, **94**, 2489–2494.

HARLOW, E. and LANE, D., 1988, *Antibodies: a Laboratory Manual*, Cold Spring Harbor Laboratory, Cold Spring Harbor, NY.

HARRIS, L.J., LARSON, S.B., HASEL, K.W. and MCPHERSON, A., 1997, Refined structure of an intact IgG2a monoclonal antibody, *Biochem.*, **36**, 1581–1597.

HARRISON, A., WALKER, C.A., PARKER, D., JANKOWSKI, K.J., COX, J.P.L., CRAIG, A.S., SANSOM, J.M., BEELEY, N.R.A., BOYCE, R.A., CHAPLIN, L.C., EATON, M.A.W., FARNSWORTH, A.P.H., MILLAR, K., MILLICAN, A.T., RANDALL, A.M., RHIND, S.K., SECHER, D.S. and TURNER, A., 1991, The *in vivo* release of ^{90}Y from cyclic and acyclic ligand–antibody conjugates, *Nucl. Med. Biol.*, **18**, 469–476.

HARTMANN, F., HORAK, E.M., GARMESTANI, K., WU, C., BRECHBIEL, M.W., KOZAK, R.W., TSO, J., KOSTEINY, S.A., GANSOW, O.A., NELSON, D.L. and WALDMANN, T.A., 1994, Radioimmunotherapy of nude mice bearing a human interleukin 2 receptor α-expressing lymphoma utilizing the α-emitting radionuclide-conjugated monoclonal antibody ^{212}Bi–anti-Tac, *Cancer Res.*, **54**, 4362–4370.

HASEMAN, C.A. and CAPRA, J.D., 1990, High-level production of a functional immunoglobulin heterodimer in a baculovirus expression system, *Proc. Natl Acad. Sci. (USA)*, **87**, 3942–3946.

HAUNSCHILD, J., FARO, H.P., PACK, P. and PLUCKTHUN, A., 1995, Pharmacokinetic properties of bivalent miniantibodies and comparison to other immunoglobulin forms, *Antibod. Immunoconj. Radiopharm.*, **8**, 111–128.

HEMMINKI, A., HOFFREN, A.M., TAKKINEN, K., VEHNIAINEN, M., MAKINEN, M.L., PETTERSSON, K., TELEMAN, O., SODERLUND, H. and TEERI, T.T., 1995, Introduction of lysine residues on the light chain constant domain improves the labelling properties of a recombinant Fab′ fragment, *Protein Engin.*, **8**, 185–191.

HENDERSHOT, L.M., TING, J. and LEE, A.S., 1988, Identity of the immunoglobulin heavy chain binding protein with the 78,000 dalton glucose-regulated protein and the role of posttranslational modifications in its binding function, *Mol. Cell Biol.*, **8**, 4250–4256.

HENDERSON, D.R., FRIEDMAN, S.B., HARRIS, J.D., MANNING, W.B. and ZOCCOLO, M.A., 1986, CEDIA™, a new homogeneous immunoassay system, *Clin. Chem.*, **32**, 1637–1641.

HENDRICKSON, E.R., HATFIELD-TRUBY, T.M., JOERGER, R.D., MAJARIAN, W.R. and EBERSOLE, R.C., 1995, High sensitivity multianalyte immunoassay using covalent DNA labelled antibodies and polymerase chain reaction, *Nucl. Acids Res.*, **23**, 522–529.

HERLYN, D., SOMASUNDARAM, R., LI, W. and MARAYUMA, H., 1996, Anti-idiotype cancer vaccines: past and future, *Cancer Immunol. Immunother.*, **43**, 65–76.

HIATT, A., 1990, Antibodies produced in plants, *Nature*, **344**, 469–470.

HIATT, A., CAFFERKEY, R. and BOWDISH, K., 1989, Production of antibodies in transgenic plants, *Nature*, **342**, 76–78.

HINKLE, G.H., MOJZISIK, C.M., LOESCH, J.A., HILL, T.L., THURSTON, M.O., SAMPSEL, J., OLSEN, J. and MARTIN, E.W., 1991, The evolution of the radioimmunoguided surgery system: an innovative technique for the intraoperative detection of tumour, *Antibody Immunoconj. Radiopharm.*, **4**, 339–358.

HINMAN, L.M., HAMANN, P.R., WALLACE, R., MENENDEZ, A.T., DURR, F.E. and UPESLACIS, J., 1993, Preparation and characterization of monoclonal antibody conjugates of the calicheamicins: a novel and potent family of antitumor antibiotics, *Cancer Res.*, **53**, 3336–3342.

HIRD, V., MARAVEYAS, A., SNOOK, D., DHOKIA, B., SOUTTER, W.P., MEARES, C., STEWART, J.S., MASON, P., LAMBERT, H.E. and EPENETOS, A.A., 1993, Adjuvant therapy of ovarian cancer with radioactive monoclonal antibody, *Br. J. Cancer*, **68**, 403–406.

HIRD, V., VERHOEYEN, M., BADLEY, R.A., PRICE, D., SNOOK, D., KOSMAS, C., GOODEN, C., BAMIAS, A., MEARES, C., LAVENDAR, J.P. and EPENETOS, A.A., 1991, Tumour localisation with a radioactively labelled reshaped human monoclonal antibody, *Br. J. Cancer*, **64**, 911–914.

HNATOWICH, D.J., CHILDS, R.L., LANTEIGNE, D. and NAJAFI, A., 1983, The preparation of DTPA coupled antibodies radiolabelled with metallic radionuclides; an improved method, *J. Immunol. Methods*, **65**, 911–914.

HNATOWICH, D.J., MARDIROSSIAN, G., RUSCKOWSKI, M., FOGARASI, M., VIRZI, F. and WINNARD, P., 1993, Directly and indirectly technetium-99m-labeled antibodies – a comparison of *in vitro* and *in vivo* properties, *J. Nucl. Med.*, **34**, 109–119.

HNATOWICH, D.J., VIRZI, F. and RUCHOWSKI, M., 1987, Investigation of avidin and biotin for imaging applications, *J. Nucl. Med.*, **28**, 1294–1302.

HOLLIGER, P., BRISSINCK, J., WILLIAMS, R.L., THIELEMANS, K. and WINTER, G., 1996, Specific killing of lymphoma cells by cytotoxic T-cells mediated by a bispecific diabody, *Protein Engin.*, **9**, 299–305.

HOLLIGER, P., PROSPERO, T. and WINTER, G., 1993, 'Diabodies': small bivalent and bispecific antibody fragments, *Proc. Natl Acad. Sci. (USA)*, **90**, 6444–6448.

HOLLIGER, P., WING, M., POUND, J.D., BOHLEN, H. and WINTER, G., 1997, Retargeting serum immunoglobulin with bispecific diabodies, *Nature Biotechnol.*, **15**, 632–636.

HOLLIS, G.F. and MARK, G.E., 1995, *Homologous recombination antibody expression system for murine cells*, International patent application, WO95/17516.

HOLVOET, P., LAROCHE, Y., LIJNEN, H.R., VAN HOEF, B., BROUWERS, E., DE COCK, F., LAUWEREYS, M., GANSEMANS, Y. and COLLEN, D., 1991, Biochemical characterization of single-chain chimeric plasminogen activators consisting of single-chain Fv fragment of a fibrin-specific antibody and single-chain urokinase, *Eur. J. Biochem.*, **210**, 945–952.

HOLWILL, I., GILL, A., HARRISON, J., HOARE, M. and LOWE, P.A., 1996, Rapid analysis of biosensor data using the initial rate determination and its application to bioprocess monitoring, *Process Control Quality*, **8**, 133–145.

HOOGENBOOM, H.R., GRIFFITHS, A.D., JOHNSON, K.S., CHISWELL, D.J., HUDSON, P. and WINTER, G., 1991, Multi-subunit proteins on the surface of filamentous phage: methodologies for displaying antibody (Fab) heavy and light chains, *Nucl. Acids Res.*, **19**, 4133–4137.

HOPP, T.P., GALLIS, B. and PRICKETT, K.S., 1996, Metal-binding properties of a calcium dependent monoclonal antibody, *Mol. Immunol.*, **33**, 601–608.

HOPP, T.P., PRICKETT, K.S., PRICE, V., LIBBY, R.T., MARCH, C.J., CERRETTI, P., URDAL, D.L. and CONLON, P.J., 1988, A short polypeptide marker sequence useful for recombinant protein identification and purification, *Bio/Technol.*, **6**, 1204–1210.

HORI, A., SASADA, R., MATSUTANI, E., NAITO, K., SAKURA, Y., FUJITA, T. and KOZAI, Y., 1991, Suppression of solid tumour growth by immunoneutralising monoclonal antibody against human basic fibroblast growth factor, *Cancer Res.*, **51**, 6180–6184.

HORN, U., STRITTMATTER, W., KREBBER, A., KNUPFER, U., KUJAU, M., WENDEROTH, R., MULLER, K., MATZKU, S., PLUCKTHUN, A. and RIESENBERG, D., 1996, High volumetric yields of functional dimeric miniantibodies in *Escherichia coli* using an optimised expression vector and high cell density fermentation under non-limited growth conditions, *Appl. Microbiol. Biotechnol.*, **46**, 524–532.

HORWITZ, A.H., CHANG, C.P., BETTER, M., HELLSTROM, K.E. and ROBINSON, R.R., 1988, Secretion of functional antibody and Fab fragment from yeast cells, *Proc. Natl Acad. Sci. (USA)*, **85**, 8678–8682.

HOULDEN, B.A., WIDACKI, S.M. and BLUESTONE, J.A., 1991, Signal transduction through class I MHC by a monoclonal antibody that detects multiple murine and human class I molecules, *J. Immunol.*, **146**, 425–430.

HU, P., GLASKY, M.S., YUN, A., ALAUDDIN, M.M., HORNICK, J.L., KHAWLI, L.A. and EPSTEIN, A.L., 1995, A human–mouse chimeric lym-1 monoclonal antibody with specificity for human lymphomas expressed in a baculovirus system, *Hum. Antibod. Hybridomas*, **6**, 57–67.

HU, S., SHIVELY, L., RAUBITSCHEK, A., SHERMAN, M., WILLIAMS, L.E., WONG, J.Y.C., SHIVELY, J.E. and WU, A.M., 1996, Minibody: a novel engineered anti-carcinoembryonic antigen antibody fragment (single-chain Fv-CH3) which exhibits rapid, high-level targeting of xenografts, *Cancer Res.*, **56**, 3055–3061.

HUANG, X., MOLEMA, G., KING, S., WATKINS, L., EDGINGTON, T.S. and THORPE, P.E., 1997, Tumor infarction in mice by antibody-directed targeting of tissue factor to tumor vasculature, *Science*, **275**, 547–550.

HUMPHREYS, D.P., WEIR, N., LAWSON, A., MOUNTAIN, A. and LUND, P.A., 1996, Co-expression of human protein disulphide isomerase (PDI) can increase the yield of an antibody Fab′ fragment expressed in *Escherichia coli*, *FEBS Lett.*, **380**, 194–197.

HUSAIN, M. and BIENARZ, C., 1994, Fc site-specific labeling of immunoglobulins with calf intestinal alkaline phosphatase, *Bioconjugate Chem.*, **5**, 482–490.

HUSE, W.D., SASTRY, L., IVERSON, S.A., KANG, A.S., ALTING-MEES, M., BURTON, D.R., BENKOVIC, S.J. and LERNER, R., 1989, Generation of a large combinatorial library of the immunoglobulin repertoire in the phage lambda, *Science*, **246**, 1275–1281.

HUSTON, J.S., LEVINSON, D., MUDGETT-HUNTER, M., TAI, M.S., NOVOTNY, J., MARGOLIES, M.N., RIDGE, R.J., BRUCCOLERI, R.E., HABER, E., CREA, R. and OPPERMANN, H., 1988, Protein engineering of antibody binding sites: recovery of specific activity in an anti-digoxin single-chain Fv analogue produced in *Escherichia coli*, *Proc. Natl Acad. Sci. (USA)*, **85**, 5879–5883.

HUTZELL, P., KASHMIRI, S., COLCHER, D., PRIMUS, F.J., HORAN HAND, P., ROSELLI, M., FINCH, M., YARRANTON, G., BODMER, M., WHITTLE, N., KING, D., LOULLIS, C.C., MCCOY, D.W., CALLAHAN, R. and SCHLOM, J., 1991, Generation and characterization of a recombinant/chimeric B72.3 (human $\gamma 1$), *Cancer Res.*, **51**, 181–189.

ILIADES, P., KORTT, A.A. and HUDSON, P.J., 1997, Triabodies: single chain Fv fragments without a linker form trivalent trimers, *FEBS Lett.*, **409**, 437–441.

INBAR, D., HOCHMAN, J. and GIVOL, D., 1972, Localization of antibody-combining sites within the variable portions of heavy and light chains, *Proc. Natl. Acad. Sci. (USA)*, **69**, 2569–2662.

IND, T.E., GRANOWSKA, M., BRITTON, K.E., MORRIS, G., LOWE, D.G., HUDSON, C.N. and SHEPHARD, J.H., 1994, Preoperative radioimmunodetection of ovarian carcinoma using a hand-held gamma detection probe, *Br. J. Cancer*, **70**, 1263–1266.

IP, C.C., MILLER, W.J., SILBERKLANG, M., MARK, G.E., ELLIS, R.W., HUANG, L., GLUSHKA, J., VAN HALBEEK, H., ZHU, J. and ALHADEFF, J.A., 1994, Structural characterisation of the N-glycans of a humanised anti-CD18 murine immunoglobulin G, *Arch. Biochem. Biophys.*, **308**, 387–399.

ISAACS, J.D., WATTS, R.A., HAZLEMAN, B.L., HALE, G., KEOGAN, M.T., COBBOLD, S.P. and WALDMANN, H., 1992, Humanised monoclonal antibody therapy for rheumatoid arthritis, *Lancet*, **340**, 748–752.

ISSEKUTZ, A.C., AYER, L., MIYASAKA, M. and ISSEKUTZ, T.B., 1996, Treatment of established adjuvant arthritis in rats with monoclonal antibody to CD18 and very late activation antigen-4 integrins suppresses neutrophil and T-lymphocyte migration to the joints and improves clinical disease, *Immunol.*, **88**, 569–576.

JACKSON, J.R., SATHE, G., ROSENBERG, M. and SWEET, R., 1995, In vitro antibody maturation, improvement of a high affinity neutralizing antibody against IL-1β, *J. Immunol.*, **154**, 3310–3319.

JACKSON, L.R., TRUDEL, L.J., FOX, J.G. and LIPMAN, N.S., 1996, Evaluation of hollow fibre bioreactors as an alternative to murine ascites production for small scale monoclonal antibody production, *J. Immunol. Methods*, **189**, 217–231.

JACOBO-MOLINA, A., DING, J., NANNI, R.G., CLARK, A.D., LU, X., TANTILLO, C., WILLIAMS, R.L., KAMER, G., FERRIS, A.L., CLARK, P., HIZI, A., HUGHES, S.H. and ARNOLD, E., 1993, Crystal structure of human immunodeficiency virus type 1 reverse transcriptase complexed with double-stranded DNA at 3.0 Å resolution shows bent DNA, *Proc. Natl Acad. Sci. (USA)*, **90**, 6320–6324.

JAHN, G. and PLACHTER, B., 1993, Diagnostics of persistent viruses: human cytomegalovirus as an example, *Intervirology*, **35**, 60–72.

JANDA, K.D., LO, L.-C., LO, C.-H.L., SIM, M.-M., WANG, R., WONG, C.H. and LERNER, R.A., 1997, Chemical selection for catalysts in combinatorial antibody libraries, *Science*, **275**, 945–948.

JESPERS, L.S., ROBERTS, A., MAHLER, S.M., WINTER, G. and HOOGENBOOM, H.R., 1994, Guiding the selection of human antibodies from phage display repertoires to a single epitope of an antigen, *Bio/Technol.*, **12**, 899–903.

JOHNSON, D.A., BARTON, R.L., FIX, D.V., SCOTT, W.L. and GUTOWSKI, M.C., 1991a, Induction of immunogenicity of monoclonal antibodies by conjugation with drugs, *Cancer Res.*, **51**, 5774–5776.

JOHNSON, T.K., MADDOCK, S., KSALIWAL, R., BLOEDOW, D., HARTMANN, C., FEYERABEND, A., DIENHART, D.G., THICKMAN, D., GLENN, S., GONZALEZ, R., LEAR, J. and BUNN, P., 1991b, Radioimmunoadsorption of KC-4G3 antibody in peripheral blood: implications for radioimmunotherapy, *Antibod. Immunoconj. Radiopharm.*, **4**, 885–893.

JONES, P.T., DEAR, P.H., FOOTE, J., NEUBERGER, M.S. and WINTER, G., 1986, Replacing the complementarity-determining regions in a human antibody with those from a mouse, *Nature*, **321**, 522–525.

JOST, C.R., TITUS, J.A., KURUCZ, I. and SEGAL, D.M., 1996, A single-chain bispecific Fv2 molecule produced in mammalian cells redirects lysis by activated CTL, *Mol. Immunol.*, **33**, 211–219.

JUNGHANS, R.P., DOBBS, D., BRECHBIEL, M.W., MIRZADEH, S., RAUBITSCHEK, A.A., GANSOW, O.A. and WALDMANN, T.A., 1993, Pharmacokinetics and bioactivity of 1,4,7,10-tetra-azacylododecane *N,N′,N″,N‴*-tetraacetic acid (DOTA)–bismuth-conjugated anti-tac antibody for α-emitter (^{212}Bi) therapy, *Cancer Res.*, **53**, 5683–5689.

JUNGHANS, R.P., WALDMANN, T.A., LANDOLFI, N.F., AVDALOVIC, N.M., SCHNEIDER, W.P. and QUEEN, C., 1990, Anti-Tac-H, a humanized antibody to the interleukin 2 receptor with new features for immunotherapy in malignant and immune disorders, *Cancer Res.*, **50**, 1495–1502.

JURCIC, J.G., CARON, P.C., NIKULA, T.K., PAPADOPOULOS, E.B., FINN, R.D., GANSOW, O.A., MILLER, W.H., GEERLINGS, M.W., WARRELL, R.P., LARSON, S.M. and SCHEINBERG, D.A., 1995, Radiolabelled anti-CD33 monoclonal antibody M195 for myeloid leukemias, *Cancer Res.*, **55**, 5908s–5910s.

JUWEID, M., SHARKEY, R.M., ALAVI, A., SWAYNE, L.C., HERSKOVIC, T., HANLEY, D., RUBIN, A.D., PEREIRA, M. and GOLDENBERG, D.M., 1997, Regression of advanced refractory ovarian cancer treated with iodine-131-labeled anti-CEA monoclonal antibody, *J. Nucl. Med.*, **38**, 257–260.

JUWEID, M., SHARKEY, R.M., BEHR, T., SWAYNE, L.C., DUNN, R., SIEGAL, J. and GOLDENBERG, D.M., 1996, Radioimmunotherapy of patients with small-volume tumours using iodine-131-labeled anti-CEA monoclonal antibody NP-4 F(ab′)$_2$, *J. Nucl. Med.*, **37**, 1504–1510.

KABAT, E.A., WU, T.T., PERRY, H.M., GOTTESMAN, K.S. and FOELLER, C., 1991, *Sequences of Proteins of Immunological Interest*, 5th edn, US Department of Health and Human Services, Public Health Service, National Institutes of Health (NIH Publication 91-3242), Washington, DC.

KAKU, S., KAWASAKI, T., SAKAI, Y., TANIUCHI, Y., YANO, S., SUZUKI, K., TERAZAKI, C., KAWAMURA, K., MASUHO, Y., SATOH, N., TAKENAKA, T., YANAGI, K. and OHSHIMA, N., 1995, Antithrombotic effect of a humanized anti-GPIIb/IIIa monoclonal antibody, YM207, in a photochemically induced thrombosis model in monkeys, *Eur. J. Pharmacol.*, **279**, 115–121.

KALAB, T. and SKLADEL, P., 1995, A disposable amperometric immunosensor for 2,4-dichlorophenoxyacetic acid, *Anal. Chim. Acta*, **304**, 361–368.

KALOFONOS, H.P., KOSMAS, C., HIRD, V., SNOOK, D.E. and EPENETOS, A.A., 1994, Targeting of tumours with murine and reshaped human monoclonal antibodies against placental alkaline phosphatase: immunlocalisation, pharmacokinetics and immune response, *Eur. J. Cancer*, **30A**, 1842–1850.

KAMINSKI, M.S., ZASADNY, K.R., FRANCIS, I.R., FENNER, M.C., ROSS, C.W., MILIK, A.W., ESTES, J., TUCK, M., REGAN, D., FISHERE, S., GLENN, S.D. and WAHL, R.L., 1996, Iodine-131-anti-B1 radioimmunotherapy for B-cell lymphoma, *J. Clin. Oncol.*, **14**, 1974–1981.

KARPOVSKI, B., TITUS, J.A., STEPHANY, D.A. and SEGAL, D.M., 1984, Production of target-specific effector cells using hetero-cross-linked aggregates containing anti-target cell and anti-Fcγ receptor antibodies, *J. Exp. Med.*, **160**, 1686–1701.

Kasina, S., Rao, T.L., Srinavasan, A., Sanderson, J.A., Fitzner, J.N., Reno, J.M., Beaumier, P.L. and Fritzberg, A.R., 1991, Development and biologic evaluation of a kit for preformed chelate technetium-99m radiolabeling of an antibody fab fragment using a diamide dimercaptide chelating agent, *J. Nucl. Med.*, **32**, 1445–1451.

Kaufmann, S.H., Ewing, C.M. and Shaper, J.H., 1987, The erasable western blot, *Anal. Biochem.*, **161**, 89–95.

Kavanaugh, A.F., Davis, L.S., Nichols, L.A., Norris, S.H., Rothlein, R., Scharschmidt, L.A. and Lipsky, P.E., 1994, Treatment of refractory rheumatoid athritis with a monoclonal antibody to intercellular adhesion molecule 1, *Arthritis Rheum.*, **37**, 992–999.

Keelan, E.T.M., Harrison, A.A., Chapman, P.T., Binns, R.M., Peters, A.M. and Haskard, D.O., 1994, Imaging vascular endothelial activation: an approach using radiolabelled monoclonal antibodies against the endothelial cell adhesion molecule E-selectin, *J. Nucl. Med.*, **35**, 276–281.

Kelley, R.F., O'Connell, M.P., Carter, P., Presta, L., Eigenbrot, C., Covarrubias, M., Snedcor, B., Bourell, J.H. and Vetterlein, D., 1992, Antigen binding thermodynamics and anti-proliferative effects of chimeric and humanised anti-p185HER2 antibody Fab fragments, *Biochem.*, **31**, 5434–5441.

Kent, S.J., Karlik, S.J., Cannon, C., Hines, D.K., Yednock, T.A., Fritz, L.C. and Horner, H.C., 1995, a monoclonal antibody to α4 integrin suppresses and reverses experimental allergic encephalomyelitis, *J. Neuroimmunol.*, **58**, 1–10.

Khaw, B.A., Klibanov, A., O'Donnell, S.M., Saito, T., Nossiff, N., Slinkin, M.A., Newell, J.B., Strauss, H.W. and Torchilin, V.P., 1991, Gamma imaging with negatively charge-modified monoclonal antibody: modification with synthetic polymers, *J. Nucl. Med.*, **32**, 1742–1751.

Khaw, B.A., Yasuda, T., Gold, H.K., Leinbach, R.C., Johns, J.A., Kanke, M., Barlai-Kovach, M., Strauss, H.W. and Haber, E., 1987, Acute myocardial infarct imaging with 111-indium labeled monoclonal antimyosin Fab, *J. Nucl. Med.*, **28**, 1671–1678.

Khazaeli, M.B., Conry, R.M. and LoBuglio, A.F., 1994, Human immune response to monoclonal antibodies, *J. Immunotherapy*, **15**, 42–52.

Khazaeli, M.B, Saleh, M.N., Liu, T.P., Meredith, R.F., Wheeler, R.H., Baker, T.S., King, D., Secher, D., Allen, L., Rogers, K., Colcher, D., Schlom, J., Shochat, D. and LoBuglio, A.F., 1991, Pharmacokinetics and immune response of ^{131}I-chimeric mouse/human B72.3 (human γ4) monoclonal antibody in humans, *Cancer Res.*, **51**, 5461–5466.

Kim, H.C., McMillan, C.W., White, G.C., Bergman, G.E., Horton, M.W. and Saidi, P., 1992, Purified factor IX using monoclonal immunoaffinity technique, *Blood*, **79**, 568–575.

Kim, J.K., Li, B., Winer, J., Armanini, M., Gillet, N., Phillips, H.S. and Ferrara, N., 1993, Inhibition of vascular endothelial growth factor-induced angiogenesis suppresses tumour growth in vivo, *Nature*, **362**, 841–844.

Kim, J.K., Tsen, M.F., Ghetie, V. and Ward, E.S., 1994, Identifying residues that influence plasma clearance of murine IgG1 fragments by site-directed mutagenesis, *Eur. J. Immunol.*, **24**, 542–548.

King, D.J., Adair, J.R., Angal, S., Low, D.C., Proudfoot, K.A., Lloyd, J.C., Bodmer, M.W. and Yarranton, G.T., 1992a, Expression, purification and characterization of a mouse–human chimeric antibody and chimeric Fab′ fragment, *Biochem. J.*, **281**, 317–323.

King, D.J., Antoniw, P., Owens, R.J., Adair, J.R., Haines, A.M.R., Farnsworth, A.P.H., Finney, H., Lawson, A.D.G., Lyons, A., Baker, T.S., Baldock, D., Mackintosh, J., Gofton, C., Yarranton, G.T., McWilliams, W., Shochat, D., Leichner, P., Welt, S., Old, L.J. and Mountain, A., 1995, Preparation and preclinical evaluation of humanised A33 immunoconjugates for radioimmunotherapy, *Br. J. Cancer*, **72**, 1364–1372.

King, D.J., Byron, O.D., Mountain, A., Weir, N., Harvey, A., Lawson, A.D.G., Proudfoot, K.A., Baldock, D., Harding, S.E., Yarranton, G.T. and Owens, R.J., 1993, Expression, purification and characterization of B72.3 Fv fragments, *Biochem. J.*, **290**, 723–729.

King, D.J., Mountain, A., Adair, J.R., Owens, R.J., Harvey, A., Weir, N., Proudfoot, K.A., Phipps, A., Lawson, A., Rhind, S.K., Pedley, B., Boden, J., Boden, R., Begent, R.H.J.

and YARRANTON, G.T., 1992b, Tumor localization of engineered antibody fragments, *Antibod. Immunoconj. Radiopharm.*, **5**, 159–170.

KING, D.J., TURNER, A., FARNSWORTH, A.P.H., ADAIR, J.R., OWENS, R.J., PEDLEY, R.B., BALDOCK, D., PROUDFOOT, K.A., LAWSON, A.D.G., BEELEY, N.R.A., MILLAR, K., MILLICAN, T.A., BOYCE, B., ANTONIW, P., MOUNTAIN, A., BEGENT, R.H.J., SHOCHAT, D. and YARRANTON, G.T., 1994, Improved tumour targeting with chemically cross-linked recombinant antibody fragments, *Cancer Res.*, **54**, 6176–6185.

KINNE, R.W., BECKER, W., KOSCHECK, T., KUHLMANN, J., SHARKEY, R.M., BEHR, T., PALOMBO-KINNE, E., GOLDENBERG, D.M., WOLF, F. and EMMRICH, F., 1995, Rat adjuvant arthritis: imaging with technetium-99m-anti-CD4 Fab′ fragments, *J. Nucl. Med.*, **36**, 2268–2275.

KIPRIYANOV, S.M., LITTLE, M., KROPSHOFER, H., BREITLING, F., GOTTER, S. and DUBEL, S., 1996, Affinity enhancement of a recombinant antibody: formation of complexes with multiple valency by a single-chain Fv fragment–core streptavidin fusion, *Protein Engin.*, **9**, 203–211.

KIPRIYANOV, S.M., MOLDENHAUER, G. and LITTLE, M., 1997a, High level production of soluble single chain antibodies in small scale *Escherichia coli* cultures, *J. Immunol. Methods*, **200**, 69–77.

KIPRIYANOV, S.M., MOLDENHAUER, G., MARTIN, A.C.R., KUPRIYANOVA, O.A. and LITTLE, M., 1997b, Two amino acid mutations in an anti-human CD3 single chain Fv antibody fragment that affect the yield on bacterial secretion but not the affinity, *Protein Engin.*, **10**, 445–453.

KIRCHER, V. and PARLAR, H., 1996, Determination of delta 9-tetrahydrocannabinol from human saliva by tandem immunoaffinity chromatography–high-performance liquid chromatography, *J. Chromatogr.*, **677**, 245–255.

KITAMURA, K., TAKAHASHI, T., YAMAGUCHI, T., NOGUCHI, A., NOGUCHI, A., TAKASHINA, K., TSURUMI, T., TOYOKUNI, T. and HAKOMORI, S., 1991, Chemical engineering of the monoclonal antibody A7 by polyethylene glycol for targeting cancer chemotherapy, *Cancer Res.*, **51**, 4310–4315.

KLEYMANN, G., OSTERMEIER, C., HEITMANN, K., HAASE, W. and MICHEL, H., 1995, Use of antibody fragments (Fv) in immunohistochemistry, *J. Histochem. Cytochem.*, **43**, 607–614.

KNAPPIK, A., KREBBER, C. and PLUCKTHUN, A., 1993, The effect of folding catalysts on the *in vivo* folding process of different antibody fragments expressed in *Escherichia coli*, *Bio/Technol.*, **11**, 77–83.

KNAPPIK, A. and PLUCKTHUN, A., 1994, An improved affinity tag based on the FLAG epitope for detection and purification of recombinant antibody fragments, *Biotechniques*, **17**, 754–761.

KNAPPIK, A. and PLUCKTHUN, A., 1995, Engineered turns of a recombinant antibody improve its *in vivo* folding, *Protein Engin.*, **8**, 81–89.

KNIGHT, D.M., WAGNER, C., JORDAN, R., MCALEER, M.F., DERITA, R., FASS, D.N., COLLER, B.S., WEISMAN, H.F. and GHRAYEB, J., 1995, The immunogenicity of the 7E3 murine monoclonal Fab antibody fragment variable region is dramatically reduced in humans by substitution of human for murine constant regions, *Mol. Immunol.*, **32**, 1271–1281.

KNIGHT, L.C., RADCLIFFE, R., MAURER, A.H., RODWELL, J.D. and ALVAREZ, V.L., 1994, Thrombus imaging with technetium-99m synthetic peptides based upon the binding domain of a monoclonal antibody to activated platelets, *J. Nucl. Med.*, **35**, 282–288.

KNOX, S.J., GORIS, M.L., TRISLER, K., NEGRIN, R., DAVIS, T., LILES, T.M., GRILLO-LOPEZ, A., CHINN, P., VARNS, C., NING, S.C., FOWLER, S., DEB, N., BECKER, M., MARQUEZ, C. and LEVY, R., 1996, Yttrium-90 labeled anti-CD20 monoclonal antibody therapy of recurrent B-cell lymphoma, *Clin. Cancer Res.*, **2**, 457–470.

KNOX, S.J., LEVY, R., HODGKINSON, S., BELL, R., BROWN, S., WOOD, G.S., HOPPE, R., ABEL, E.A., STEINMAN, L., BERGER, R.G., GAISER, C., YOUNG, G., BINDL, J., HANHAM, A. and REICHERT, T., 1991, Observations on the effect of chimeric anti-CD4 monoclonal antibody in patients with mycosis fungoides, *Blood*, **77**, 20–30.

KODA, K. and GLASSY, M.C., 1990, In vitro immunization for the production of human monoclonal antibody, *Human Antibod. Hybridomas*, **1**, 15–22.

KOHLER, G. and MILSTEIN, C., 1975, Continuous culture of fused cells secreting antibody of predefined specificity, *Nature*, **265**, 495–497.

KONIG, B. and GRATZEL, M., 1993, Detection of viruses and bacteria with piezoelectric immunosensors, *Anal. Lett.*, **26**, 1567–1585.

KONTERMANN, R.E., WING, M.G. and WINTER, G., 1997, Complement recruitment using bispecific diabodies, *Nature Biotechnol.*, **15**, 629–631.

KOSMAS, K., SNOOK, D., GOODEN, C.S., COURTENAY-LUCK, N.J., MCCALL, M.J., MEARES, C. and EPENETOS, A.A., 1992, Development of humoral immune responses against a macrocyclic chelating agent (DOTA) in cancer patients receiving radioimmunoconjugates for imaging and therapy, *Cancer Res.*, **52**, 904–911.

KOSTELNY, S.A., COLE, M.S. and TSO, J.Y., 1992, Formation of a bispecific antibody by the use of leucine zippers, *J. Immunol.*, **148**, 1547–1553.

KOVARI, L.C., MOMANY, C. and ROSSMAN, M.G., 1995, The use of antibody fragments for crystallization and structure determinations, *Structure*, **3**, 1291–1293.

KOZBOR, D., LAGARDE, A. and RODER, J.C., 1982, Human hybridomas constructed with antigen specific Epstein-Barr virus-transformed cell lines, *Proc. Natl Acad. Sci. (USA)*, **79**, 6651–6655.

KRANENBORG, M.H.G.C., OYEN, W.J.G., CORSTENS, F.H.M., OOSTERWIJK, E., VAN DER MEER, J.W.M. and BOERMAN, O.C., 1997, Rapid imaging of experimental infection with technetium-99m-DTPA after anti-DTPA monoclonal antibody priming, *J. Nucl. Med.*, **38**, 901–906.

KRETZSCHMAR, T., AOUSTIN, L., ZINGEL, O., MARANGI, M., VONACH, B., TOWBIN, H. and GEISER, M., 1996, High-level expression in insect cells and purification of secreted monomeric single-chain Fv antibodies, *J. Immunol. Methods*, **195**, 93–101.

KRICKA, L.J., 1993, Ultrasensitive immunoassay techniques, *Clin. Biochem.*, **26**, 325–331.

KROUWELS, F.H., NOCKER, R.E.T., SNOEK, M., LUTTER, R., VAN DER ZEE, J.S., WELLER, F.R., JANSEN, H.M. and OUT, T.A., 1997, Immunocytochemical and flow cytofluorimetric detection of intracellular IL-4, IL-5 and IFN-γ: applications using blood and airway derived cells, *J. Immunol. Methods*, **203**, 89–101.

KUROKAWA, T., IWASA, S. and KAKINUMA, A., 1989, Enhanced fibrinolysis by a bispecific monoclonal antibody reactive to fibrin and tissue plasminogen activator, *Bio/Technol.*, **7**, 1163–1167.

KURRLE, R., SHEARMAN, C.W., MOORE, G.P. and SEILER, F., 1990, *Improved monoclonal antibodies against the human alpha/beta T-cell receptor, their production and use*, European patent application EP0403156.

KURUCZ, I., TITUS, J.A., JOST, C.R. and SEGAL, D.M., 1995, Correct disulphide pairing and efficient refolding of detergent solubilised single-chain Fv proteins from bacterial inclusion bodies, *Mol. Immunol.*, **32**, 1443–1452.

LAMBERT, J.M., GOLDMACHER, V.S., COLLINSON, A.R., NADLER, L.M. and BLATTLER, W.A., 1991, An immunotoxin prepared with blocked ricin: a natural plant toxin adapted for therapeutic use, *Cancer Res.*, **51**, 6236–6242.

LANDRY, D.W., ZHAO, K., YANG, C.X.Q., GLICKMAN, M. and GEORGIADIS, T.M., 1993, Antibody catalysed degradation of cocaine, *Science*, **259**, 1899–1901.

LANE, D.M., EAGLE, K.F., BEGENT, R.H.J., HOPE-STONE, L.D., GREEN, A.J., CASEY, J.L., KEEP, P.A., KELLY, A.M.B., LEDERMAN, J.A., GLASER, M.G. and HILSON, A.J.W., 1994, Radioimmunotherapy of metastatic colorectal tumours with iodine-131-labelled antibody to carcinoembryonic antigen: phase I/II study with comparative biodistribution of intact and $F(ab')_2$ antibodies, *Br. J. Cancer*, **70**, 521–525.

LANGMUIR, V.K., 1992, Radioimmunotherapy: clinical results and dosimetric considerations, *Nucl. Med. Biol.*, **19**, 213–225.

LAROCHE, Y., DEMAEYER, M., STASSEN, J.M., GANSEMANS, Y., DEMARSIN, E., MATTHYSSENS, G., COLLEN, D. and HOLVOET, P., 1991, Characterization of a recombinant single-chain molecule comprising the variable domains of a monoclonal antibody specific for human fibrin fragment D-dimer, *J. Biol. Chem.*, **266**, 16343–16349.

LARSON, S.M., 1995, Improving the balance between treatment and diagnosis: a role for radioimmunodetection, *Cancer Res.*, **55**, 5756s–5758s.

LAUKKANEN, M.J., ORELLANA, A. and KEINANEN, K., 1995, Use of genetically engineered lipid-tagged antibody to generate functional europium chelate loaded liposomes. Application in fluoroimmunoassay, *J. Immunol. Methods*, **185**, 95–102.

LAUKKANEN, M.J., TEERI, T.T. and KEINANEN, K., 1993, Lipid-tagged antibodies: bacterial expression and characterization of a lipoprotein–single chain antibody fusion protein, *Protein Engin.*, **6**, 449–454.

LAW, S.J., MILLER, T., PIRAN, U., KLUKAS, C., CHANG, S. and UNGER, J., 1989, Novel poly-substituted aryl acridinium esters and their use in immunoassay, *J. Biolumin. Chemilumin.*, **4**, 88–98.

LEDERMANN, J.A., BEGENT, R.H.J., BAGSHAWE, K.D., RIGGS, S.J., SEARLE, F., GLASER, M.G., GREEN, A.J. and DALE, R.G., 1988, Repeated antitumour antibody therapy in man with suppression of the host response by cyclosporin A, *Br. J. Cancer*, **58**, 654–657.

LEDOUSSAL, J.M., CHETANNEAU, A., GRUAZ-GUYON, A., MARTIN, M., GAUTHEROT, E., LEHUR, P.A., CHATAL, J.F., DELAAGE, M. and BARBET, J., 1993, Bispecific monoclonal antibody mediated targeting of an indium-111-labeled DTPA dimer to primary colorectal tumors, *J. Nucl. Med.*, **34**, 1662–1671.

LEDOUSSAL, J.M., MARTIN, M., GAUTHEROT, E., DELAAGE, M. and BARBET, J., 1989, *In vitro* and *in vivo* targeting of radiolabeled monovalent and divalent haptens with dual specificity monoclonal antibody conjugates, *J. Nucl. Med.*, **30**, 1358–1366.

LEGER, O.J.P., YEDNOCK, T.A., TANNER, L., HORNER, H.C., HINES, D.K., KEEN, S., SALDANHA, J., JONES, S.T., FROTZ, L.C. and BENDIG, M.M., 1997, Humanization of a mouse antibody against human alpha-4 integrin: a potential therapeutic for the treatment of multiple sclerosis, *Human Antibodies*, **8**, 3–16.

LEONETTI, J.P., MACHY, P., DEGOLS, G., LEBLEU, B. and LESERMAN, L., 1990, Antibody-targeted liposomes containing oligodeoxyribonucleotides complementary to viral RNA selectively inhibit viral replication, *Proc. Natl Acad. Sci. (USA)*, **87**, 2448–2451.

LEUNG, S., LOSMAN, M.J., GOVIDAN, S.V., GRIFFITHS, G.L., GOLDENBERG, D.M. and HANSEN, H.J., 1995, Engineering a unique glycosylation site for site-specific conjugation of haptens to antibody fragments, *J. Immunol.*, **154**, 5919–5926.

LEVI, M., SALLBERG, M., RUDEN, U., HERLYN, D., MARUYAMA, H., WIGZELL, H., MARKS, J. and WAHREN, B., 1993, A complementarity-determining region synthetic peptide acts as a miniantibody and neutralises human immunodeficiency virus type 1 in vitro, *Proc. Natl Acad. Sci. (USA)*, **90**, 4374–4378.

LEVIN, R., MHASHILKAR, A.M., DORFMAN, T., BUKOVSKY, A., ZANI, C., BAGLEY, J., HINKULA, J., NIEDRIG, M., ALBERT, J., WAHREN, B., GOTTLINGER, H.G. and MARASCO, W.A., 1997, Inhibition of early and late events of the HIV-1 replication cycle by cytoplasmic Fab intrabodies against the matrix protein p17, *Mol. Med.*, **3**, 96–110.

LEVY, R., WEISMAN, M., WIESENHUTTER, C., YOCUM, D., SCHNITZER, T., GOLDMAN, A., SCHIFF, M., BREEDVELD, F., SOLINGER, A., MACDONALD, B. and LIPANI, J., 1996, Results of a placebo-controlled multicenter trial using a primatised non-depleting anti-CD4 monoclonal antibody in the treatment of rheumatoid arthritis, *Arthritis Rheum.*, **39**, S122, abstract 574.

LIEBER, M., 1996, Immunoglobulin diversity: rearranging by cutting and repairing, *Curr. Biol.*, **6**, 134–136.

LIN, L. and PUTNAM, F.W., 1978, Cold pepsin digestion, a novel method to produce the Fv fragment from human immunoglobulin M, *Proc. Natl Acad. Sci. (USA)*, **75**, 2649–2653.

LINDHOFER, H., MOCIKAT, R., STEIPE, B. and THIERFELDER, S., 1995, Preferential species-restricted heavy/light chain pairing in rat/mouse quadromas, *J. Immunol.*, **155**, 219–225.

LIU, A.H., CREADON, G. and WYSLOCKI, L.J., 1992, Sequencing heavy and light chain variable genes of single B-hybridoma cells by total enzymatic application, *Proc. Natl Acad. Sci. (USA)*, **89**, 7610–7614.

LIU, A.Y., MACK, P.W., CHAMPION, C.I. and ROBINSON, R.R., 1987, Expression of a mouse:human immunoglobulin heavy-chain cDNA in lymphoid cells, *Gene*, **54**, 33–40.

LIU, C., TADAYONI, B.M., BOURRET, L.A., MATTOCKS, K.M., DERR, S.M., WIDDISON, W.C., KEDERSHA, N.L., ARINIELLO, P.D., GOLDMACHER, V.S., LAMBERT, J.M., BLATTLER, W.A.

and CHARI, R.V.J., 1996, Eradication of large colon tumour xenografts by targeted delivery of maytansinoids, *Proc. Natl Acad. Sci. (USA)*, **93**, 8618–8623.

LIU, L.F., 1989, DNA topoisomerase poisons as antitumour drugs, *Ann. Rev. Biochem.*, **58**, 351–375.

LOBUGLIO, A.F., WHEELER, R.H., TRANG, J., HAYNES, A., ROGERS, K., HARVEY, E.B., SUN, L., GHRAYEB, J. and KHAZAELI, M.B., 1989, Mouse/human chimeric monoclonal antibody in man: kinetics and immune response, *Proc. Natl Acad. Sci. (USA)*, **86**, 4220–4224.

LOCKWOOD, C.M., THIRU, S., ISAACS, J.D., HALE, G. and WALDMANN, H., 1993, Long term remission of intractable systemic vasculitis with monoclonal antibody therapy, *Lancet*, **341**, 1620–1622.

LONBERG, N., TAYLOR, L.D., HARDING, F.A., TROUNSTINE, M., HIGGINS, K.M., SCHRAMM, S.R., KUO, C.C., MASHAYEKH, R., WYMORE, K., MCCABE, J.G., MUNOZ-O'REGAN, D., O'DONNELL, S.L., LAPACHET, E.S.G., BENGOCHEA, T., FISHWILD, D.M., CARMACK, C.E., KAY, R.M. and HUSZAR, D., 1994, Antigen-specific human antibodies from mice comprising four distinct genetic modifications, *Nature*, **368**, 856–859.

LOVGREN, T., MERIO, L., MITRUMEN, K., MAKINEN, M.L., MAKELA, M., BLOMBERG, K., PALENIUS, T. and PETTERSSON, K., 1996, One-step all-in-one dry reagent immunoassays with fluorescent europium chelate label and time resolved fluorometry, *Clin. Chem.*, **42**, 410–415.

LOVQUIST, A., SUNDIN, A., AHLSTROM, H., CARLSSON, J. and LUNDQVIST, H., 1997, Pharmacokinetics and experimental PET imaging of a bromine-76-labeled monoclonal anti-CEA antibody, *J. Nucl. Med.*, **38**, 395–401.

LUND, J., TAKAHASHI, N., NAKAGAWA, H., GOODALL, M., BENTLEY, T., HINDLEY, S.A., TYLER, R. and JEFFERIS, R., 1993, Control of IgG/Fc glycosylation: a comparison of oligosaccharides from chimeric human/mouse and mouse subclass immunoglobulin Gs, *Mol. Immunol.*, **30**, 741–748.

LUND, J., TAKAHASHI, N., POUND, J.D., GOODALL, M. and JEFFERIS, R., 1996, Multiple interactions of IgG with its core oligosaccharide can modulate recognition by complement and human FcγRI and influence the synthesis of its oligosaccharide chains, *J. Immunol.*, **157**, 4963–4969.

LUND, J., TANAKA, T., TAKAHASHI, N., SARMAY, G., ARATA, Y. and JEFFERIS, R., 1990, A protein structural change in aglycosylated IgG3 correlates with loss of huFcγRI and huFcγRIII binding and/or activation, *Mol. Immunol.*, **27**, 1145–1153.

LUND, J., WINTER, G., JONES, P.T., POUND, J.D., TANAKA, T., WALKER, M.R., ARTYMIUK, P.J., ARATA, Y., BURTON, D.R., JEFFERIS, R. and WOOF, J.M., 1991, Human FcγRI and FcγRII interact with distinct but overlapping sites on human IgG, *J. Immunol.*, **147**, 2657–2662.

LYONS, A., KING, D.J., OWENS, R.J., YARRANTON, G.T., MILLICAN, A., WHITTLE, N.R. and ADAIR, J.R., 1990, Site-specific attachment to recombinant antibodies via introduced surface cysteine residues, *Protein Engin.*, **3**, 703–708.

MA, J.K. and HEIN, M.B., 1995, Immunotherapeutic potential of antibodies produced in plants, *Trends Biotechnol.*, **13**, 522–527.

MA, J.K., HIATT, A., HEIN, M., VINE, N.D., WANG, F., STABILA, P., VAN DOLLEWEERD, C., MOSTOV, K. and LEHNER, T., 1995, Generation and assembly of secretory antibodies in plants, *Science*, **268**, 716–719.

MCCAFFERTY, J., GRIFFITHS, A.D., WINTER, G. and CHISWELL, D.J., 1990, Phage antibodies: filamentous phage displaying antibody variable domains, *Nature*, **348**, 552–554.

MCCLOSKEY, N., TURNER, M.W., STEFFNER, P., OWENS, R. and GOLDBLATT, D., 1996, Human constant regions influence the antibody binding characteristics of mouse–human chimeric IgG subclasses, *Immunology*, **88**, 169–173.

MACGLASHAN, D.W., BOCHNER, B.S., ADELMAN, D.C., JARDIEU, P.M., TOGIAS, A., MCKENZIE-WHITE, J., STERBINSKY, S.A., HAMILTON, R.G. and LICHTENSTEIN, L.M., 1997, Down-regulation of FceRI expression on human basophils during *in vivo* treatment of atopic patients with anti-IgE antibody, *J. Immunol*, **158**, 1438–1445.

MCGUINNESS, B.T., WALTER, G., FITZGERALD, K., SCHULER, P., MAHONEY, W., DUNCAN, A.R. and HOOGENBOOM, H.R., 1996, Phage diabody repertoires for selection of large numbers of bispecific antibody fragments, *Nature Biotechnol.*, **14**, 1149–1154.

Maciejewski, J.P., Wiechold, F.F., Young, N.S., Cara, A., Zella, D., Reitz, M.S. and Gallo, R.C., 1995, Intracelluar expression of antibody fragments against HIV reverse transcriptase prevents HIV infection *in vitro*, *Nature Med.*, **1**, 667–673.

Mack, M., Reithmuller, G. and Kufer, P., 1995, A small bispecific antibody construct expressed as a functional single-chain molecule with high tumour cell cytotoxicity, *Proc. Natl Acad. Sci. (USA)*, **92**, 7021–7025.

MacKenzie, C.R., Clark, I.D., Evans, S.V., Hill, I.E., MacManus, J.P., Dubuc, G., Bundle, D.R., Narang, S.A., Young, N.M. and Szabo, A.G., 1995, Bifunctional fusion proteins consisting of a single-chain antibody and an engineered lanthanide binding protein, *Immunotechnology*, **1**, 139–150.

MacKenzie, C.R., Sharma, V., Brummell, D., Bilous, D., Dubuc, G., Sadowska, J., Young, N.M., Bundle, D.R. and Narang, S.A., 1994, Effect of Cλ–Cκ domain switching on Fab activity and yield in *Escherichia coli*: synthesis and expression of genes encoding two anti-carbohydrate Fabs, *Bio/Technol.*, **12**, 390–395.

MacLean, J.A., Su, Z., Colvin, R.B. and Wong, J.T., 1995, Anti-CD3:anti-IL-2 receptor–bispecific mab-mediated immunomodulation, *J. Immunol.*, **155**, 3674–3682.

Maeda, H., Matsushita, S., Eda, Y., Kimachi, K., Tokiyoshi, S. and Bendig, M., 1991, Construction of reshaped human antibodies with HIV neutralising activity, *Human Antibodies Hybridomas*, **2**, 124–134.

Maerz, H., Hahn, S.O., Maassen, A., Meisel, H., Roggenbuck, D., Sato, T., Tanzmann, H., Emmrich, F. and Marx, U., 1996, Improved removal of viruslike particles from purified monoclonal antibody IgM preparation via virus filtration, *Nature Biotech.*, **14**, 651–652.

Magnani, P., Paganelli, G., Songini, C., Samuel, A., Sudati, F., Siccardi, A.G. and Fazio, F., 1996, Pretargeted immunoscintigraphy in patients with medullary thyroid carcinoma, *Br. J. Cancer*, **74**, 825–831.

Maiorella, B.L., Winkelhake, J., Young, J., Moyer, B., Bauer, R., Hora, M.J., Thomson, J., Patel, T. and Parekh, R., 1993, Effect of culture conditions on IgM antibody structure, pharmacokinetics and activity, *Bio/Technol.*, **11**, 387–392.

Mallender, W.D. and Voss, E.W., 1994, Construction, expression and activity of a bivalent bispecific single-chain antibody, *J. Biol. Chem.*, **269**, 199–206.

Malmborg, A.C. and Borrebaeck, C.A.K., 1995, BIAcore as a tool in antibody engineering, *J. Immunol. Methods*, **183**, 7–13.

Malmqvist, M., 1993, Biospecific interaction analysis using biosensor technology, *Nature*, **361**, 186–187.

Maloney, D.G., Grillo-Lopez, A.J., White, C.A., Bodkin, D., Schilder, R.J., Neidhart, J.A., Janakiraman, N., Foon, K.A., Liles, T.M., Dallaire, B.K., Wey, K., Royston, I., Davis, T. and Levy, R., 1997, IDEC-C2B8 (Rituximab) anti-CD20 monoclonal antibody therapy in patients with relapsed low-grade non-Hodgkin's lymphoma, *Blood*, **90**, 2188–2195.

Maloney, D.G., Liles, T.M., Czerwinski, D.K., Waldichuk, C., Grillo-Lopez, A. and Levy, R., 1994, Phase I clinical trial using escalating single-dose infusion of chimeric anti-CD20 monoclonal antibody (IDEC-C2B8) in patients with recurrent B-cell lymphoma, *Blood*, **84**, 2457–2466.

Marasco, W.A., Haseltine, W.A. and Chen, S.Y., 1993, Design, intracellular expression and activity of a human anti-human immunodeficiency virus type 1 gp120 single-chain antibody, *Proc. Natl Acad. Sci. (USA)*, **90**, 7889–7893.

Mares, A., De Boever, J., Osher, J., Quiroga, S., Barnard, G. and Kohen, F., 1995, A direct non-competitive idiometric enzyme immunoassay for serum oestradiol, *J. Immunol. Methods*, **181**, 83–90.

Mariani, G. and Strober, W., 1990, *Immunoglobulin Metabolism, In Fc Receptors and the Action of Antibodies* (Metzger, H., ed.), pp. 94–177, American Society of Microbiology, Washington, DC.

Mariani, M., Camagna, M., Tarditi, L. and Seccamani, E., 1991, A new enzymatic method to obtain high-yield $F(ab)_2$ suitable for clinical use from mouse IgG1, *Mol. Immunol.*, **28**, 69–77.

MARKS, J.D., GRIFFITHS, A.D., MALMQVIST, M., CLACKSON, T.P., BYE, J.M. and WINTER, G., 1992, By-passing immunization: building high affinity human antibodies by chain shuffling, *Bio/Technol.*, **10**, 779–783.

MARKS, J.D., TRISTREM, M., KARPAS, A. and WINTER, G., 1991, Oligonucleotide primers for polymerase chain reaction amplification of human immunoglobulin variable genes and design of family specific oligonucleotide probes, *Eur. J. Immunol.*, **21**, 985–991.

MARQUET, P.Y., DAVIER, A., SAPIN, R., BRIDGI, B., MURATET, J.P., HARTMANN, D.J., PAOLUCCI, F. and PAU, B., 1996, Highly sensitive immunoradiometric assay for serum thyroglobulin with minimal interference from autoantibodies, *Clin. Chem.*, **42**, 258–262.

MARSHALL, D., PEDLEY, R.B., BODEN, J.A., BODEN, R. and BEGENT, R.H.J., 1994, Clearance of circulating radio-antibodies using streptavidin or second antibodies in a xenograft model, *Br. J. Cancer*, **69**, 502–507.

MARTIN, F., VOLPARI, C., STEINKUHLER, C., DIMASI, N., BRUNETTI, M., BIASIOL, G., ALTAMURA, S., CORTESE, R., DEFRANCESCO, R. and SOLLAZZO, M., 1997, Affinity selection of a camelized VH domain antibody inhibitor of hepatitis C virus NS3 protease, *Protein Engin.*, **10**, 607–614.

MARTIN, P.J., NELSON, B.J., APPELBAUM, F.R., ANASETTI, C., DEEG, H.J., HANSEN, J.A., MCDONALD, G.B., NASH, R.A., SULLIVAN, K.M., WITHERSPOON, R.P., SCANNON, P.J., FRIEDMANN, N. and STORB, R., 1996, Evaluation of a CD5 specific immunotoxin for treatment of acute graft versus host disease after allogenic marrow transplantation, *Blood*, **88**, 824–830.

MASAT, L., WABL, M. and JOHNSON, J.P., 1994, A simpler sort of antibody, *Proc. Natl Acad. Sci. (USA)*, **91**, 893–896.

MATTES, M.J., 1987, Biodistribution of antibodies after intraperitoneal or intravenous injection and effect of carbohydrate modifications, *J. Natl Cancer Inst.*, **79**, 855–863.

MEIJS, W.E., HAISMA, H.J., KLOK, R.P., VAN GOG, F.B., KIEVIT, E., PINEDO, H.M. and HERSCHEID, J.D.M., 1997, Zirconium-labeled monoclonal antibodies and their distribution in tumor-bearing nude mice, *J. Nucl. Med.*, **38**, 112–118.

MELTON, R.G. and SHERWOOD, R.F., 1996, Antibody–enzyme conjugates for cancer therapy, *J. Natl Cancer Inst.*, **88**, 153–165.

MENDEZ, M.J., GREEN, L.L., CORVALAN, J.R., JIA, X.C., MAYNARD-CURRIE, C.E., YANG, X.D., GALLO, M.L., LOUIE, D.M., LEE, D.V., ERICKSON, K.L., LUNA, J., ROY, C.M., ABDERRAHIM, H., KIRSCHENBAUM, F., NOGUCHI, M., SMITH, D.H., FUKUSHIMA, A., HALES, J.F., FINER, M.H., DAVIS, C.G., ZSEBO, K.M. and JAKOBOVITS, A., 1997, Functional transplant of megabase human immunoglobulin loci recapitulates human antibody responses in mice, *Nature Genet.*, **15**, 146–156.

MEREDITH, R.F., KHAZAELI, M.B., LIU, T.P., PLOTT, G., WHEELER, R.H., RUSSELL, C., COLCHER, D., SCHLOM, J., SHOCHAT, D. and LOBUGLIO, A.F., 1992a, Dose fractionation of radiolabelled antibodies in patients with metastatic colon cancer, *J. Nucl. Med.*, **33**, 1648–1653.

MEREDITH, R.F., KHAZAELI, M.B, PLOTT, W.E., SALEH, M.N., LIU, T., ALLEN, L.F., RUSSELL, C.D., ORR, R.A., COLCHER, D., SCHLOM, J., SHOCHAT, D., WHEELER, R.H. and LOBUGLIO, A.F., 1992b, Phase I trial of iodine-131–chimeric B72.3 (human IgG4) in metastatic colorectal cancer, *J. Nucl. Med.*, **33**, 23–32.

MEREDITH, R.F., KHAZAELI, M.B., PLOTT, W.E., SPENCER, S.A., WHEELER, R.H., BRADY, L.W., WOO, D.V. and LOBUGLIO, A.F., 1995, Initial clinical evaluation of iodine-125-labeled chimeric 17-1A for metastatic colon cancer, *J. Nucl. Med.*, **36**, 2229–2233.

MEREDITH, R.F., LOBUGLIO, A.F., PLOTT, W.E., ORR, R.A., BREZOVICH, I.A., RUSSELL, C.D., HARVEY, E.B., YESTER, M.V., WAGNER, A.J., SPENCER, S.A., WHEELER, R.H., SALEH, M.N., ROGERS, K.J., POLANSKY, A., SALTER, M.M. and KHAZAELI, M.B., 1991, Pharmacokinetics, immune response, and biodistribution of iodine-131-labeled chimeric mouse/human IgG1,k 17-1A monoclonal antibody, *J. Nucl. Med.*, **32**, 1162–1168.

MEREDITH, R.F., PARTRIDGE, E.E., ALVAREZ, R.D., KHAZAELI, M.B., PLOTT, G., RUSSELL, C.D., WHEELER, R.H., LIU, T., GRIZZLE, W.E., SCHLOM, J. and LOBUGLIO, A.F., 1996, Intraperitoneal radioimmunotherapy of ovarian cancer with lutetium-177-CC49, *J. Nucl. Med.*, **37**, 1491–1496.

MERWIN, J.R., CARMICHAEL, E.P., NOELL, G.S., DEROME, M.E., THOMAS, W.L., ROBERT, N., SPITALNY, G. and CHIOU, H.C., 1995, CD5-mediated specific delivery of DNA to T lymphocytes: compartmentalization augmented by adenovirus, *J. Immunol. Methods*, **186**, 257–266.

MEUSEL, M., RENNEBERG, R., SPENER, F. and SCHMITZ, G., 1995, Development of a heterogeneous amperometric immunosensor for the determination of apolipoprotein E in serum, *Biosens. Bioelectronics*, **10**, 577–586.

MEYER, K.B. and NEUBERGER, M.S., 1989, The immunoglobulin kappa locus contains a second stronger B-cell specific enhancer which is located downstream of the constant region, *EMBO J.*, **8**, 1959–1964.

MHASHILKAR, A.M., BAGLEY, J., CHEN, S.Y., SZILVAY, A.M., HELLAND, D.G. and MARASCO, W.A., 1995, Inhibition of HIV-1 tat-mediated LTR transactivation and HIV-1 infection by anti-tat single-chain intrabodies, *EMBO J.*, **14**, 1542–1551.

MHATRE, R., NASHABEH, W., SCHMALZING, D., YAO, X., FUCHS, M., WHITNEY, D. and REGNIER, F., 1995, Purification of antibody Fab fragments by cation-exchange chromatography and pH gradient elution, *J. Chromatogr.*, **707**, 225–231.

MILENIC, D.E., ESTEBAN, J.M. and COLCHER, D., 1989, Comparisons of methods for the generation of immunoreactive fragments of a monoclonal antibody (B72.3) reactive with human carcinomas, *J. Immunol. Methods*, **120**, 71–83.

MILENIC, D.E., YOKOTA, T., FILPULA, D.R., FINKELMAN, M.A.J., DODD, S.W., WOOD, J.F.,WHITLOW, M., SNOY, P. and SCHLOM, J., 1991, Construction, binding properties, metabolism and tumor targeting of a single-chain Fv derived from the pancarcinoma monoclonal antibody CC49, *Cancer Res.*, **51**, 6363–6371.

MILES, L.E.M. and HALES, C.N., 1968, Labelled antibodies and immunological assay systems, *Nature*, **219**, 186–189.

MILSTEIN, C. and CUELLO, A.C., 1983, Hybrid hybridomas and their use in immunohistochemistry, *Nature*, **305**, 537–540.

MIURA, N., HIGOBASHI, H., SAKAI, G., TAKEYASHU, A., UDA, T. and YAMAZOE, N., 1993, Piezoelectric crystal immunosensor for sensitive detection of metamphetamine (stimulant drug) in human urine, *Sensors Actuators B Chem.*, **13**, 188–191.

MIYASHITA, H., KARAKI, Y., KIKUCHI, M. and FUJII, I., 1993, Prodrug activation via catalytic antibodies, *Proc. Natl Acad. Sci. (USA)*, **90**, 5337–5340.

MOI, M.K., MEARES, C.F., MCCALL, M.J., COLE, W.C. and DENARDO, S.J., 1985, Copper chelates as probes of biological systems: stable copper complexes with a macrocyclic chelating agent, *Anal. Biochem.*, **148**, 249–253.

MOLLOY, P., BRYDON, L., PORTER, A.J. and HARRIS, W.J., 1995, Separation and concentration of bacteria with immobilized antibody fragments, *J. Appl. Bact.*, **78**, 359–365.

MONFARDINI, C., KIEBER-EMMONS, T., VONFELDT, J.M., O'MALLEY, B., ROSENBAUM, H., GODILLOT, A.P., KAUSHANSKY, K., BROWN, C.B., VOET, D., MCCALLUS, D.E., WEINER, D.B. and WILLIAMS, W.V., 1995, Recombinant antibodies in bioactive peptide design, *J. Biol. Chem.*, **270**, 6628–6638.

MORELOCK, M.M., ROTHLEIN, R., BRIGHT, S.M., ROBINSON, M.K., GRAHAM, E.T., SABO, J.P., OWENS, R., KING, D.J., NORRIS, S.H., SCHER, D.S., WRIGHT, J.L. and ADAIR, J.R., 1994, Isotype choice for chimeric antibodies affects binding properties, *J. Biol. Chem.*, **269**, 13048–13055.

MORGAN, A., JONES, N.D., NESBITT, A.M., CHAPLIN, L., BODMER, M. and EMTAGE, J.S., 1995, The N-terminal end of the C_H2 domain of chimeric human IgG1 anti-HLA-DR is necessary for C1q, FcγRI and FcγRIII binding, *Immunology*, **86**, 319–324.

MORGUET, A.J., SANDROCK, D., STILLE-SIEGENER, M. and FIGULLA, H.R., 1995, Indium-111–antimyosin Fab imaging to demonstrate myocardial involvement in systemic lupus erythematosus, *J. Nucl. Med.*, **36**, 1432–1435.

MORPHY, J.R., PARKER, D., ALEXANDER, R., BAINS, A., CARNE, A.F., EATON, M.A.W., HARRISON, A., MILLICAN, A., PHIPPS, A., RHIND, S.K., TITMAS, R. and WEATHERBY, D., 1988, Antibody labelling with functionalised cyclam macrocycles, *J. Chem. Soc. Chem. Commun.*, 156–158.

MORPHY, J.R., PARKER, D., KATAKY, R., HARRISON, A., EATON, M.A.W., MILLICAN, A., PHIPPS, A. and WALKER, C., 1989, Towards tumour targeting with copper-radiolabelled macrocycle–antibody conjugates, *J. Chem Soc. Chem. Commun.*, 792–794.

MORRISON, S.L., JOHNSON, M.J., HERZENBERG, L.A. and OI, V.T., 1984, Chimeric human antibody molecules: mouse antigen-binding domains with human constant region domains, *Proc. Natl Acad. Sci. (USA)*, **81**, 6851–6855.

MORTON, H.C., ATKIN, J.D., OWENS, R.J. and WOOF, J.M., 1993, Purification and characterization of chimeric human IgA1 and IgA2 expressed in COS and Chinese hamster ovary cells, *J. Immunol.*, **151**, 4743–4752.

MOSIER, D.E., GULIZIA, R.J., BAIRD, S.M. and WILSON, D.B., 1988, Transfer of a functional immune system to mice with severe combined immunodeficiency, *Nature*, **335**, 256–259.

MUELLER, B.M., REISFELD, R.A. and GILLIES, S.D., 1990, Serum half-life and tumor localization of a chimeric antibody deleted of the CH2 domain and directed against the disialoganglioside GD2, *Proc. Natl Acad. Sci. (USA)*, **87**, 5702–5705.

MULLIGAN, R.C. and BERG, P., 1981, Selection for animal cells that express the *E. coli* gene coding for xanthine–guanine phosphoribosyl transferase, *Proc. Natl Acad. Sci. (USA)*, **78**, 2072–2076.

MURRAY, J.L., MACEY, D.J., KASI, L.P., RIEGER, P., CUNNINGHAM, J., BHADKAMKAR, V., ZHANG, H.Z., SCHLOM, J., ROSENBLUM, M.G. and PODOLOFF, D.A., 1994, Phase II radioimmunotherapy trial with ^{131}I-CC49 in colorectal cancer, *Cancer*, **73**, 1057–1066.

NAKATANI, T., NOMURA, N., HORIGOME, K., OHTSUKA, H. and NOGUCHI, H., 1989, Functional expression of human antibody genes directed against Pseudomonal exotoxin A in mouse myeloma cells, *Bio/Technol.*, **7**, 805–810.

NARULA, J., PETROV, A., PAK, C.K.Y., DITLOW, C., CHEN, F. and KHAW, B.A., 1995, Noninvasive localisation of experimental atherosclerotic lesions with Tc-99m labeled chimeric Z2D3 antibody fab′, *J. Nucl. Med.*, **36**, 138P–139P.

NATANSON, C., HOFFMAN, W.D., SUFFREDINI, A.F., EICHAKER, P.Q. and DANNER, R.L., 1994, Selected treatment strategies for septic shock based on proposed mechanisms of pathogenesis, *Ann. Intern. Med.*, **120**, 771–783.

NAVARRO-TEULON, I., PERALDI-ROUX, S., BERNARDI, T., MARIN, M., PIECHACZYK, M., SHIRE, D., PAU, B. and BIARD-PIECHACZYK, M., 1995, Expression in *E. coli* of soluble and M13 phage-displayed forms of a single-chain antibody fragment specific for digoxin: assessment in a novel drug immunoassay, *Immunotechnol.*, **1**, 41–52.

NEBLOCK, D.S., CHANG, C.H., MASCELLI, M.A., FLEEK, M., STUMPO, L., CULLEN, M.M. and DADDONA, P.E., 1992, Conjugation and evaluation of 7E3xP4B6, a chemically cross-linked bispecific $F(ab')_2$ antibody which inhibits platelet aggregation and localises tissue plasminogen activator to the platelet surface, *Bioconj. Chem.*, **3**, 126–131.

NEDELMAN, M.A., SHEALY, D.J., BOULIN, R., BRUNT, E., SEASHOLTZ, J.I., ALLEN, I.E., MCCARTNEY, J.E., WARREN, F.D., OPPERMANN, H., PANG, R.H.L., BERGER, H.J. and WEISMAN, H.F., 1993, Rapid infarct imaging with a technetium-99m-labeled antimyosin recombinant single-chain Fv: evaluation in a canine model of acute myocardial infarction, *J. Nucl. Med.*, **34**, 234–241.

NERI, D., PETRUL, H., WINTER, G., LIGHT, Y., MARAIS, R., BRITTON, K.E. and CREIGHTON, A.M., 1996, Radioactive labeling of antibody fragments by phosphorylation using human casein kinase II and γ-^{32}P-ATP, *Nature Biotechnol.*, **14**, 485–490.

NEUBERGER, M., WILLIAMS, G.T. and FOX, R.O., 1984, Recombinant antibodies possessing novel effector functions, *Nature*, **312**, 604–608.

NEUMAIER, M., SHIVELY, L., CHEN, F.C., GAIDA, F.J., IlGEN, C., PAXTON, R.J., SHIVELY, J.E. and RIGGS, A.D., 1990, Cloning of the genes for T84.66, an antibody that has a high specificity and affinity for carcinoembryonic antigen, and expression of chimeric human/mouse T84.66 genes in myeloma and Chinese hamster ovary cells, *Cancer Res.*, **50**, 2128–2134.

NEVENS, J.R., MALLIA, A.K., WENDT, M.W. and SMITH, P.K., 1992, Affinity chromatographic purification of immunoglobulin M antibodies utilizing immobilized mannan binding protein, *J. Chromatogr.*, **597**, 247–256.

NEWMAN, R., ALBERTS, J., ANDERSON, D., CARNER, K., HEARD, C., NORTON, F., RAAB, R., REFF, M., SHUEY, S. and HANNA, N., 1992, 'Primatization' of recombinant antibodies for immunotherapy of human diseases: a macaque/human chimeric antibody against human CD4, *Bio/Technol.*, **10**, 1455–1460.

NEWTON, D.L., NICHOLLS, P.J., RYBAK, S.M. and YOULE, R.J., 1994, Expression and characterization of recombinant human eosinophil-derived neurotoxin and eosinophil-derived neurotoxin–anti-transferrin receptor sFv, *J. Biol. Chem.*, **269**, 26739–26745.

NEWTON, D.L., XUE, Y., OLSON, K.A., FETT, J.W. and RYBAK, S.M., 1996, Angiogenin single-chain immunofusions: influence of peptide linkers and spacers between fusion protein domains, *Biochem.*, **35**, 545–553.

NICOLAOU, K.C., SMITH, A.L. and YUE, E.W., 1993, Chemistry and biology of natural and designed enediynes, *Proc. Natl Acad. Sci. (USA)*, **90**, 5881–5888.

NIEBA, L., HONEGGER, A., KREBBER, C. and PLUCKTHUN, A., 1997, Disrupting the hydrophobic patches at the antibody variable/constant domain interface: improved *in vivo* folding and physical chacterization of an engineered scFv fragment, *Protein Engin.*, **10**, 435–444.

NIERODA, C.A., MILENIC, D.E., CARRASQUILLO, J.A., SCHLOM, J. and GREINER, J.W., 1995, Improved tumor radioimmunodetection using a single-chain Fv and γ-interferon: potential clinical applications for radioimmunoguided surgery and γ scanning, *Cancer Res.*, **55**, 2858–2865.

NILSON, B.H.K., LOGDBERG, L., KASTERN, W., BJORCK, L. and AKERSTROM, B., 1993, Purification of antibodies using protein L-binding framework structures in the light chain variable domain, *J. Immunol. Methods*, **164**, 33–40.

NISHINAKA, S., SUZUKI, T., MATSUDA, H. and MURATA, M., 1991, A new cell line for the production of chicken monoclonal antibody by hybridoma technology, *J. Immunol. Methods*, **139**, 217–222.

NITTA, T., SATO, K., YAGITA, H., OKUMURA, K. and ISHII, I., 1990, Preliminary trial of specific targeted therapy against malignant glioma, *Lancet*, **335**, 368–371.

NORMAN, T.J., PARKER, D., ROYLE, L., HARRISON, A., ANTONIW, P. and KING, D.J., 1995a, Improved tumour targeting with recombinant antibody–macrocycle conjugates, *J. Chem. Soc. Chem. Commun.*, 1877–1878.

NORMAN, T.J., PARKER, D., SMITH, F.C. and KING, D.J., 1995b, Towards selective DNA targeting: synthesis of an antibody–macrocycle–intercalator conjugate, *J. Chem. Soc. Chem. Commun.*, 1879–1880.

NORRGREN, K., STRAND, S.E., NILSSON, R., LINDGREN, L. and LILLIEHORN, P., 1991, Evaluation of extracorporeal immunoadsorption for reduction of the blood background in diagnostic and therapeutic applications of radiolabelled monoclonal antibodies, *Antibody Immunoconj. Radiopharm.*, **4**, 907–914.

NYYSSONEN, E., PENTTILA, M., HARKKI, A., SALOHEIME, A., KNOWLES, J.K.C. and KERANEN, S., 1993, Efficient production of antibody fragments by the filamentous fungus *Trichoderma reesei*, *Bio/Technol.*, **11**, 591–595.

O'HARE, M., BROWN, A.N., HUSSAIN, K., GEBHARDT, A., WATSON, G., ROBERTS, L.M., VITETTA, E.S., THORPE, P.E. and LORD, J.M., 1990, Cytotoxicity of a recombinant ricin–A chain fusion protein containing a proteolytically cleavable spacer sequence, *FEBS Lett.*, **273**, 200–204.

OHLIN, M., OWMAN, H., MACH, M. and BORREBAECK, C.A.K., 1996, Light chain shuffling of a high affinity antibody results in a drift in epitope recognition, *Mol. Immunol.*, **33**, 47–56.

ONG, G.L., ETTENSON, D., SHARKEY, R.M., MARKS, A., BAUMAL, R., GOLDENBERG, D.M. and MATTES, M.J., 1991, Galactose-conjugated antibodies in cancer therapy: properties and principles of action, *Cancer Res.*, **51**, 1619–1626.

ORLANDI, R., GUSSOW, D., JONES, P.T. and WINTER, G., 1989, Cloning immunoglobulin variable domains for expression by the polymerase chain reaction, *Proc. Natl Acad. Sci. (USA)*, **86**, 3833–3837.

Ornatowska, M. and Glasel, J.A., 1991, Direct production of Fv fragments from a family of monoclonal IgGs by papain digestion, *Mol. Immunol.*, **28**, 383–391.

O'Shanessy, D.J., 1990, Hydrazido-derivatized supports in affinity chromatography, *J. Chromatogr.*, **510**, 13–21.

Oster, Z.H., Srivasta, S.C., Som, P., Meinken, G.E., Scudder, L.E., Yamamoto, K., Atkins, H.L., Brill, A.B. and Coller, B.S., 1985, Thrombus radioimmunoscintigraphy, an approach using monoclonal antiplatelet antibody, *Proc. Natl. Acad. Sci. (USA)*, **82**, 3465–3468.

Ostermeier, C., Iwata, S., Ludwig, B. and Michel, H., 1995, Fv fragment mediated crystallization of the membrane protein bacterial cytochrome c oxidase, *Nature Struct. Biol.*, **2**, 842–846.

Owen, M., Gandecha, A., Cockburn, B. and Whitelam G., 1992, Synthesis of a functional anti-phytochrome single-chain Fv protein in transgenic tobacco, *Bio/Technol.*, **10**, 790–794.

Owens, R.J. and Robinson, M.K., 1995, *Antibodies against E-selectin*, International Patent Application WO95/26403.

Pack, P., Kujau, M., Schroewckh, V., Knupfer, U., Wenderoth, R., Riesenberg, D. and Pluckthun, A., 1993, Improved bivalent miniantibodies with identical avidity as whole antibodies, produced by high cell density fermentation of *Escherichia coli*, *Bio/Technol.*, **11**, 1271–1277.

Pack, P., Muller, K., Zahn, R. and Pluckthun, A., 1995, Tetravalent miniantibodies with high avidity assembling in *Escherichia coli*, *J. Mol. Biol.*, **246**, 28–34.

Pack, P. and Pluckthun, A., 1992, Miniantibodies: use of amphipathic helices to produce functional, flexibly linked dimeric Fv fragments with high avidity in *Escherichia coli*, *Biochem.*, **31**, 1579–1584.

Packard, B., Edidin, M. and Komoriya, A., 1986, Site-directed labeling of a monoclonal antibody: targeting to a disulphide bond, *Biochem.*, **25**, 3548–3552.

Paddock, S., DeVries, P., Buth, E. and Carroll, S., 1994, Production of monoclonal antibodies by genetic immunization, *Biotechniques*, **16**, 616–620.

Padlan, E.A., 1991, A possible procedure for reducing the immunogenicity of antibody variable domains while preserving their ligand binding properties, *Mol. Immunol.*, **28**, 489–498.

Padlan, E.A., 1994, Anatomy of the antibody molecule, *Mol. Immunol.*, **31**, 169–217.

Paganelli, G., Magnani, P., Zito, F., Villa, E., Sudati, F., Lopalco, L., Rossetti, C., Malcovati, M., Chiolerio, F., Seccamani, E., Siccardi, A.G. and Fazio, F., 1991, Three step monoclonal antibody tumor targeting in carcinoembryonic antigen positive patients, *Cancer Res.*, **51**, 5960–5966.

Page, M.J. and Sydenham, M.A., 1991, High level expression of the humanised antibody CAMPATH–1H in chinese hamster ovary cells, *Bio/Technol.*, **9**, 65–68.

Pai, L.H., Wittes, R., Setser, A., Willingham, M.C. and Pastan, I., 1996, Treatment of advanced solid tumors with immunotoxin LMB-1: an antibody linked to *Pseudomonas* exotoxin, *Nature Med.*, **2**, 350–353.

Pak, K.Y., Nedelman, M.A., Kanke, M., Khaw, B.A., Mattis, J.A., Strauss, H.W., Dean, R.T. and Berger, H.J., 1992, An instant kit method for labeling antimyosin Fab′ with technetium-99m: evaluation in an experimental myocardial infarct model, *J. Nucl. Med.*, **33**, 144–149.

Parham, P., 1986, Preparation and purification of active fragments from mouse monoclonal antibodies, *Immunochemistry* (Weir, D.M., ed.), pp.14.1–14.23, Blackwell Scientific Publications, Oxford.

Park, J.W., Hong, K., Carter, P., Asgari, H., Guo, L.Y., Keller, G.A., Wirth, C., Shalaby, R., Kotts, C., Wood, W.I., Papahadjopoulos, D. and Benz, C.C., 1995, Development of anti-p185^{HER2} immunoliposomes for cancer therapy, *Proc. Natl Acad. Sci. (USA)*, **92**, 1327–1331.

Park, L.S., Friend, D., Price, V., Anderson, D., Singer, J., Prickett, K.S. and Urdal, D.L., 1989, Heterogeneity in human interleukin 3 receptors: a subclass that binds human granulocyte/macrophage colony stimulating factor, *J. Biol. Chem.*, **264**, 5420–5427.

Parkinson, S., 1995, Production of monoclonal antibodies in the milk of transgenic animals, *Exploiting Transgenic Technology*, IBC symposium, San Diego, November.

PASTAN, I.H., ARCHER, G.E., MCLENDON, R.E., FRIEDMAN, H.S., FUCHS, H.E., WANG, Q.C., PAI, L.H., HERNDON, J. and BIGNER, D.D., 1995, Intrathecal administration of single-chain immunotoxin, LMB-7 [B3(Fv)-PE38], produces cures of carcinomatous meningitis in a rat model, *Proc. Natl Acad. Sci. (USA)*, **92**, 2765–2769.

PEAKMAN, T.C., WORDEN, J., HARRIS, R.H., COOPER, H., TITE, J., PAGE, M.J., GEWERT, D.R., BARTHOLEMEW, M., CROWE, J.S. and BRETT, S., 1994, Comparison of expression of a humanised monoclonal antibody in mouse NS0 myeloma cells and chinese hamster ovary cells, *Hum. Antibod. Hybridomas*, **5**, 65–74.

PEDLEY, R.B., BODEN, J.A., BODEN, R., BEGENT, R.H.J., TURNER, A., HAINES, A.M.R. and KING, D.J., 1994, The potential for enhanced tumour localization by poly(ethylene glycol) modification of anti-CEA antibody, *Br. J. Cancer*, **70**, 1126–1130.

PEDLEY, R.B., BODEN, J., BODEN, R., DALE, R. and BEGENT, R.H.J., 1993, Comparative radioimmunotherapy using intact or $F(ab')_2$ fragments of ^{131}I anti-CEA antibody in a colonic xenograft model, *Br. J. Cancer*, **68**, 69–73.

PEDLEY, R.B., DALE, R., BODEN, J.A., BEGENT, R.H.J., KEEP, P. and GREEN, A., 1989, The effect of second antibody clearance on the distribution and dosimetry of radiolabelled anti-CEA antibody in a human colonic tumour xenograft model, *Int. J. Cancer*, **43**, 713–718.

PEI, X.Y., HOLLIGER, P., MURZIN, A.G. and WILLIAMS, R.L., 1997, The 2.0 Å resolution crystal structure of a trimeric antibody fragment with noncognate VH–VL domain pairs shows a rearrangement of VH CDR3, *Proc. Natl Acad. Sci. (USA)*, **94**, 9637–9642.

PELTIER, P., CURTET, C., CHATAL, J.F., LEDOUSSAL, J.M., DANIEL, G., AILLET, G., GRUAZ-GUYON, A., BARBET, J. and DELAAGE, M., 1993, Radioimmunodetection of medullary thyroid cancer using a bispecific anti-CEA/anti-indium-DTPA antibody and an indium-111-labeled DTPA dimer, *J. Nucl. Med.*, **34**, 1267–1273.

PERISIC, O., WEBB, P.A., HOLLIGER, P., WINTER, G. and WILLIAMS, R.L., 1994, Crystal structure of a diabody, a bivalent antibody fragment, *Structure*, **2**, 1217–1226.

PETTERSSON, S., COOK, G.P., BRUGGEMANN, M., WILLIAMS, G.T. and NEUBERGER, M.S., 1990, A second B cell-specific enhancer 3′ of the immunoglobulin heavy chain locus, *Nature*, **344**, 165–168.

PHILLIPS, T.M., 1989, High-performance immunoaffinity chromatography, *Adv. Chromatogr.*, **29**, 133–173.

PHILPOTT, G.W., SCHWARZ, S.W., ANDERSON, C.J., DEHDASHTI, F., CONNETT, J.M., ZINN, K.R., MEARES, C.F., CUTLER, P.D., WELCH, M.J. and SIEGAL, B.A., 1995, RadioimmunoPET: detection of colorectal carcinoma with positron-emitting copper-64-labeled monoclonal antibody, *J. Nucl. Med.*, **36**, 1818–1824.

PIETERSZ, G.A., ROWLAND, A., SMYTH, M.J. and MCKENZIE, I.F.C., 1994, Chemoimmunoconjugates for the treatment of cancer, *Adv. Immunol.*, **56**, 301–387.

PILSON, R.S., LEVIN, W., DESAI, B., REIK, L.M., LIN, P., KORKMAZ-DUFFY, E., CAMPBELL, E., TSO, J.Y., KERWIN, J.A. and HAKIMI, J., 1997, Bispecific humanised anti-IL-2 receptor αβ antibodies inhibitory for both IL-2 and IL-15 mediated proliferation, *J. Immunol.*, **159**, 1543–1556.

PINCUS, S.H. and MCCLURE, J., 1993, Soluble CD4 enhances the activity of immunotoxins directed against gp41 of the human immunodeficiency virus, *Proc. Natl Acad. Sci. (USA)*, **90**, 332–336.

POLAK, J.M. and PRIESTLEY, J.V., eds, 1992, *Electron Microscopic Immunocytochemistry*, Oxford University Press, Oxford.

POLJAK, R.J., AMZEL, L.M., CHEN, B.L., PHIZACKERLY, R.P. and SAUL, F., 1973, Three-dimensional structure of the Fab′ fragment of a human immunoglobulin at 2.8 Å resolution, *Proc. Natl Acad. Sci. (USA)*, **71**, 3440–3444.

POLLACK, S.J., JACOBS, J.W. and SCHULZ, P.G., 1986, Selective chemical catalysis by an antibody, *Science*, **234**, 1570–1573.

PONCET, P., PANCZAK, A., GOUPY, C., GUSTAFSSON, K., BLANPIED, C., CHAVANEL, G., HIRSCH, R. and HIRSCH, F., 1996, Antifection, an antibody-mediated method to introduce genes into lymphoid cells in vitro and in vivo, *Gene Therapy*, **3**, 731–738.

POTTER, K.N., LI, Y. and CAPRA, J.D., 1996, Staphylococcal protein A simultaneously interacts with framework region 1, complementarity determining region 2 and framework region 3 on human VH3 encoded Igs, *J. Immunol.*, **157**, 2982–2988.

POU, K., ONG, H., ADAM, A., LAMOTHE, P. and DELAHAUT, P., 1994, Combined immunoextraction approach coupled to a chemiluminescence enzyme immunoassay for the determination of trace levels of salbutamol and clenbuterol in tissue samples, *Analyst*, **119**, 2659–2662.

POUL, M., CERUTTI, M., CHAABIHI, H., TICCHIONI, M., DERAMOUDT, F., BERNARD, A., DEVAUCHELLE, G., KACZOREK, M. and LEFRANC, M.P., 1995, Cassette baculovirus vectors for the production of chimeric, humanized, or human antibodies in insect cells, *Eur. J. Immunol.*, **25**, 2005–2009.

PRESS, O.W., EARY, J.F., APPELBAUM, F.R., MARTIN, P.J., NELP, W.B., GLENN, S., FISHER, D.R., PORTER, B., MATTHEWS, D.C., GOOLEY, T. and BERNSTEIN, I.D., 1995, Phase II trial of ^{131}I-B1 (anti-CD20) antibody therapy with autologous stem cell transplantation for relapsed B cell lymphomas, *Lancet*, **346**, 336–340.

PRESS, O.W., SHAN, D., HOWELL-CLARK, J., EARY, J., APPELBAUM, F.R., MATTHEWS, D., KING, D.J., HAINES, A.M.R., HAMANN, P., HINMAN, L., SHOCHAT, D. and BERNSTEIN, I., 1996, Comparative metabolism and retention of iodine-125, yttrium-90 and indium-111 radioimmunoconjugates by cancer cells, *Cancer Res.*, **56**, 2123–2129.

PRESTA, L.G., LAHR, S.J., SHIELDS, R.L., PORTER, J.P., GORMAN, C.M., FENDLY, B.M. and JARDIEU, P.M., 1993, Humanization of an antibody directed against IgE, *J. Immunol.*, **151**, 2623–2632.

PRESTA, L.G., SHIELDS, R.L., O'CONNELL, L., LAHR, S.J., PORTER, J.P., GORMAN, C.M. and JARDIEU, P.M., 1994, The binding site on human immunoglobulin E for its high affinity receptor, *J. Biol. Chem.*, **269**, 26368–26373.

PRISYAZHNOY, V.S., FUSEK, M. and ALAKHOV, Y., 1988, Synthesis of high-capacity immunoaffinity sorbents with oriented immobilized immunoglobulins or their Fab′ fragments for isolation of proteins, *J. Chromatogr.*, **424**, 243–253.

PRONGAY, A.J., SMITH, T.J., ROSSMAN, M.G., EHRLICH, L.S., CARTER, C.A. and MCCLURE, J., 1990, Preparation and crystallization of a human immunodeficiency virus p24–Fab complex, *Proc. Natl Acad. Sci. (USA)*, **87**, 9980–9984.

PROUDFOOT, K.A., TORRANCE, C., LAWSON, A.D.G. and KING, D.J., 1992, Purification of recombinant chimeric B72.3 Fab′ and F(ab′)2 using Streptococcal protein G, *Protein Expression Purification*, **3**, 368–373.

PULITO, V.L., ROBERTS, V.A., ADAIR, J.R., ROTHERMEL, A.L., COLLINS, A.M., VARGA, S.S., MARTOCELLO, C., BODMER, M., JOLLIFFE, L.K. and ZIVIN, R.A., 1996, Humanization and molecular modelling of the anti-CD4 monoclonal antibody OKT4A, *J. Immunol.*, **156**, 2840–2850.

QUADRI, S.M., LAI, J., MOHAMMADPOR, H., VRIESENDORP, H.M. and WILLIAMS, J.R., 1993, Assessment of radiolabelled stabilized F(ab′)$_2$ fragments of monoclonal antiferritin in nude mouse model, *J. Nucl. Med.*, **34**, 2152–2159.

QUEEN, C., SCHNEIDER, W.P., SELICK, H.E., PAYNE, P.W., LANDOLFI, N.F., DUNCAN, J.F., AVDALOVIC, N.M., LEVITT, M., JUNGHANS, R.P. and WALDMANN, T.A., 1989, A humanized antibody that binds to the interleukin 2 receptor, *Proc. Natl Acad. Sci. (USA)*, **86**, 10029–10033.

QUESNIAUX, V.F., 1991, Monoclonal antibody technology for cyclosporin monitoring, *Clin. Biochem.*, **24**, 37–42.

QUEZADO, Z.M.N., BANKS, S.M. and NATANSON, C., 1995, New strategies for combatting sepsis: the magic bullets missed the mark but the search continues, *Trends Biotechnol.*, **13**, 56–63.

RAAG, R. and WHITLOW, M., 1995, Single-chain Fvs, *FASEB J.*, **9**, 73–80.

RAJEWSKY, K., 1996, Clonal selection and learning in the immune system, *Nature*, **381**, 751–758.

RAMAN-SURI, C., JAIN, P.K. and MISHRA, G.C., 1995, Development of a piezoelectric crystal based microgravimetric immunoassay for determination of insulin concentration, *J. Biotechnol.*, **39**, 27–34.

RANKIN, E.C.C., CHOY, E.H.S., KASSIMOS, D., KINGSLEY, G.H., SOPWITH, A.M., ISENBERG, D.A. and PANAYI, G.S., 1995, The therapeutic effects of an engineered human anti-tumour necrosis factor alpha antibody (CDP571) in rheumatoid arthritis, *Br. J. Rheumatol.*, **34**, 334–342.

REA, D.W. and ULTEE, M.E., 1993, A novel method for controlling the pepsin digestion of antibodies, *J. Immunol. Methods*, **157**, 165–173.

REARDAN, D.T., MEARES, C.F., GOODWIN, D.A., MCTIGUE, M., DAVID, G.S., STONE, M.R., LEUNG, J.P., BARTHOLOMEW, R.M. and FRINCKE, J.M., 1985, Antibodies against metal chelates, *Nature*, **316**, 265–267.

REES, A.R., STAUNTON, D., WEBSTER, D.M., SEARLE, S.J., HENRY, A.H. and PEDERSEN, J.T., 1994, Antibody design: beyond the natural limits, *Trends Biotechnol.*, **12**, 199–206.

REFF, M.E., CARNER, K., CHAMBERS, K.S., CHINN, P.C., LEONARD, J.E., RAAB, R., NEWMAN, R.A., HANNA, N. and ANDERSON, D.R., 1994, Depletion of B cells in vivo by a chimeric mouse human monoclonal antibody to CD20, *Blood*, **83**, 435–445.

REIST, C.J., ARCHER, G.E., KURPAD, S.N., WIKSTRAND, C.J., VAIDANATHAN, G., WILLINGHAM, M.C., MOSCATELLO, D.K., WONG, A.J., BIGNER, D.D. and ZALUTSKY, M.R., 1995, Tumor-specific anti-epidermal growth factor receptor variant III monoclonal antibodies: use of the tyramine–cellobiose radioiodination method enhances cellular retention and uptake in tumour xenografts, *Cancer Res.*, **55**, 4375–4482.

REIST, C.J., GARG, P.K., ALSTON, K.L., BIGNER, D.D. and ZALUTSKY, M.R., 1996, Radioiodination of internalising monoclonal antibodies using *N*-succinimidyl-5-iodo-3-pyridine carbonate, *Cancer Res.*, **56**, 4970–4977.

REITER, Y., BRINKMANN, U., KREITMAN, R.J., JUNG, S.H., LEE, B. and PASTAN, I., 1994, Stabilization of the Fv fragments in recombinant immunotoxins by disulfide bonds engineered into conserved framework regions, *Biochem.*, **33**, 5451–5459.

REITER, Y. and PASTAN, I., 1996, Antibody engineering of recombinant Fv immunotoxins for improved targeting of cancer: disulfide stabilized Fv immunotoxins, *Clin. Cancer Res.*, **2**, 245–252.

REITER, Y., WRIGHT, A.F., TONGE, D.W. and PASTAN, I., 1996, Recombinant single-chain and disulfide-stabilized Fv-immunotoxins that cause complete regression of a human colon cancer xenograft in nude mice, *Int. J. Cancer*, **67**, 113–123.

RENNER, C., JUNG, W., SAHIN, U., DENFIELD, R., POHL, C., TRUMPER, L., HARTMANN, F., DIEHL, V., VAN LIER, R. and PFREUNDSCHUH, M., 1994, Cure of xenografted human tumors by bispecific monoclonal antibodies and human T cells, *Science*, **264**, 833–835.

RENNER, C. and PFREUNDSCHUH, M., 1995, Tumor therapy by immune recruitment with bispecific antibodies, *Immunol. Rev.*, **145**, 179–209.

RETTIG, W.J., GARIN-CHESA, P., HEALEY, J.H., SU, S.L., JAFFE, E.A. and OLD, L.J., 1992, Identification of endosialin, a cell surface glycoprotein of vascular endothelial cells in human cancer, *Proc. Natl Acad. Sci. (USA)*, **89**, 10832–10836.

RHEINNECKER, M., HARDT, C., ILAG, L.L., KUFER, P., GRUBER, R., HOESS, A., LUPAS, A., ROTTENBERGER, C., PLUCKTHUN, A. and PACK, P., 1996, Multivalent antibody fragments with high functional affinity for a tumor-associated antigen, *J. Immunol.*, **157**, 2989–2997.

RHIND, S.K., KING, D.J., BODEN, J., BODEN, R., PEDLEY, R.B., SEARLE, F.A., BEGENT, R.H.J., BAGSHAWE, K.D., ABRAMS, D., YARRANTON, G.T., BODMER, M., SECHER, D.S. and ADAIR, J., 1990, Rapid clearance of aglycosyl recombinant chimeric B72.3(γ4) antibody in athymic mice bearing human colorectal xenografts, *Antibod. Immunoconj. Radiopharm.*, **3**, 55.

RICHARDSON, J.H., SODROSKI, J.G., WALDMANN, T.A. and MARASCO, W.A., 1995, Phenotypic knockout of the high-affinity human interleukin-2 receptor by intracellular single-chain antibodies against the alpha subunit of the receptor, *Proc. Natl Acad. Sci. (USA)*, **92**, 3137–3141.

RIDDER, R., SCHMITZ, R., LEGAY, F. and GRAM, H., 1995, Generation of rabbit monoclonal antibody fragments from a combinatorial phage display library and their production in the yeast *Pichia pastoris*, *Bio/Technol.*, **13**, 255–260.

RIDGEWAY, J.B.B., PRESTA, L.G. and CARTER, P., 1996, Knobs-into-holes engineering of antibody CH3 domains for heavy chain heterodimerization, *Protein Engin.*, **9**, 617–621.

Riechmann, L., Clark, M., Waldmann, H. and Winter, G., 1988a, Reshaping human antibodies for therapy, *Nature*, **332**, 323–327.

Riechmann, L., Foote, J. and Winter, G., 1988b, Expression of an antibody Fv fragment in myeloma cells, *J. Mol. Biol.*, **203**, 825–828.

Riechmann, L., Weill, M. and Cavanagh, J., 1992, Improving the antigen affinity of an antibody Fv-fragment by protein design, *J. Mol. Biol.*, **224**, 913–918.

Riethmuller, G., Schneider-Gadicke, E., Schlimok, G., Schmiegel, W., Raab, R., Hoffken, K., Gruber, R., Pichlmaier, H., Hirche, H., Pichlmayr, R., Buggisch, P. and Witte, J., 1994, Randomised trial of monoclonal antibody for adjuvant therapy of resected Dukes' C colorectal carcinoma, *Lancet*, **343**, 1177–1183.

Rigl, T., Rivera, H.N., Patel, M.T., Ball, R.T., Stults, N.L. and Smith, D.F., 1995, Bioluminescence immunoassays for human endocrine hormones based on aqualite, a calcium activated photoprotein, *Clin. Chem.*, **41**, 1363–1364.

Rini, J.M., Schulze-Gahmen, U. and Wilson, I.A., 1992, Structural evidence for induced fit as a mechanism for antibody–antigen recognition, *Science*, **255**, 959–965.

Roberts, S., Cheetham, J.C. and Rees, A.R., 1987, Generation of an antibody with enhanced affinity and specificity for its antigen by protein engineering, *Nature*, **328**, 731–734.

Rodrigues, M.L., Presta, L.G., Kotts, C.E., Wirth, C., Mordenti, J., Osaka, G., Wong, W.L.T., Nuijens, A., Blackburn, B. and Carter, P., 1995, Development of a humanized disulfide-stabilized anti-p185^{HER2} Fv-β-lactamase fusion protein for activation of a cephalosporin doxorubicin prodrug, *Cancer Res.*, **55**, 63–70.

Rodrigues, M.L., Snedcor, B., Chen, C., Wong, W.L.T., Garg, S., Blank, G.S., Maneval, D. and Carter, P., 1993, Engineering Fab′ fragments for efficient F(ab′)$_2$ formation in *Escherichia coli* and for improved *in vivo* stability, *J. Immunol.*, **151**, 6954–6961.

Rodwell, J., Alvarez, V., Lee, C., Lopes, A., Goers, J., King, H., Powsner, H. and McKearn, T., 1986, Site-specific covalent modification of mAbs: *in vitro* and *in vivo* evaluations, *Proc. Natl Acad. Sci. (USA)*, **83**, 2632–2636.

Roes, J. and Rajewsky, K., 1993, Immunoglobulin D (IgD)-deficient mice reveal an auxiliary receptor function for IgD in antigen-mediated recruitment of B cells, *J. Exp. Med.*, **177**, 45–55.

Roguska, M.A., Pedersen, J.T., Keddy, C.A., Henry, A.H., Searle, S.J., Lambert, J.M., Goldmacher, V.S., Blatter, W.A., Rees, A.R. and Guild, B.C., 1994, Humanization of murine monoclonal antibodies through variable domain resurfacing, *Proc. Natl Acad. Sci. (USA)*, **91**, 969–973.

Rondelli, D., Andrews, R.G., Hansen, J.A., Ryncarz, R., Faerber, M.A. and Anasetti, C., 1996, Alloantigen presenting function of normal human CD34+ hematopoietic cells, *Blood*, **88**, 2619–2675.

Rong, H., Deftos, L.J., Ji, H. and Bucht, E., 1997, Two-site immunofluorometric assay of intact salmon calcitonin with improved sensitivity, *Clin. Chem.*, **43**, 71–75.

Rongen, H.A.H., van der Horst, H.M., Oosterhout, A.J.M., Bult, A. and van Bennekom, W.P., 1997, Application of xanthine oxidase-catalyzed luminol chemiluminescence in a mouse interleukin-5 immunoassay, *J. Immunol. Methods*, **197**, 161–169.

Ross, A.H., Herlyn, D. and Koprowski, H., 1987, Purification of monoclonal antibodies from ascites using ABx liquid chromatography column, *J. Immunol. Methods*, **102**, 227–231.

Rousseaux, J., Rousseaux-Prevost, R. and Bazin, H., 1983, Optimal conditions for the preparation of Fab and F(ab′)$_2$ fragments from monoclonal IgG of different rat IgG subclasses, *J. Immunol. Methods*, **64**, 141–146.

Roux, P., Jeanteur, P. and Piechaczyk, M., 1989, A versatile and potentially general approach to the targeting of specific cell types by retroviruses, *Proc. Natl Acad. Sci. (USA)*, **86**, 9079–9083.

Rubin, R.H., Baltimore, D., Chen, B.K., Wilkinson, R.A. and Fischman, A.J., 1996, *In vivo* tissue distribution of CD4 lymphocytes in mice determined by radioimmunoscintigraphy with an ^{111}In-labeled anti-CD4 monoclonal antibody, *Proc. Natl Acad. Sci. (USA)*, **93**, 7460–7463.

RUBIN, R.H., FISCHMAN, A.J., CALLAHAN, R.J., KHAW, B.A., KEECH, F., AHMAD, M., WILKINSON, R. and STRAUSS, H.W., 1989, ^{111}In labeled nonspecific immunoglobulin scanning in the detection of focal infection, *N. Engl. J. Med.*, **321**, 935–940.

RUNGE, M.S., BODE, C., MATSUEDA, G.R. and HABER, E., 1988, Conjugation to an antifibrin monoclonal antibody enhances the fibrinolytic potency of tissue plasminogen activator *in vitro*, *Biochem.*, **27**, 1153–1157.

RUNGE, M.S., QUERTERMOUS, T., ZAVODNY, P.J., LOVE, T.W., BODE, C., FREITAG, M., SHAW, S.Y., HUANG, P.L., CHOU, C.C., MULLINS, D., SCHNEE, J.M., SAVARD, C.E., ROTHENBERG, M.E., NEWELL, J.B., MATSUEDA, G.R. and HABER, E., 1991, A recombinant chimeric plasminogen activator with high affinity for fibrin has increased thrombolytic potency *in vitro* and *in vivo*, *Proc. Natl Acad. Sci. (USA)*, **88**, 10337–10341.

RYBAK, S.M., HOOGENBOOM, H.R., MEADE, H.M., RAUS, J.C.M., SCHWARTZ, D. and YOULE, R.J., 1992, Humanization of immunotoxins, *Proc. Natl Acad. Sci. (USA)*, **89**, 3165–3169.

SAGARA, H., MATSUDA, H., WADA, N., YAGITA, H., FUKUDA, T., OKUMURA, K., MAKINO, S. and RA, C., 1997, A monoclonal antibody against very late activation antigen-4 inhibits eosinophil accumulation and late asthmatic response in a guinea pig model of asthma, *Int. Arch. Allergy Immunol.*, **112**, 287–294.

SAHAGAN, B.G., DORAI, H., SALTZGABER-MULLER, J., TONEGUZZO, F., GUINDON, C.A., LILLY, S.P., MCDONALD, K.W., MORRISSEY, D.V., STONE, B.A. and DAVIS, G.L., 1986, A genetically engineered murine/chimeric antibody retains specificity for human tumour-associated antigen, *J. Immunol.*, **137**, 1066–1074.

SALEH, M.N., KHAZAELI, M.B., WHEELER, R.H., ALLEN, L., TILDEN, A.B., GRIZZLE, W., REISFELD, R.A., YU, A.L., GILLIES, S.D. and LOBUGLIO, A.F., 1992, Phase I trial of the chimeric anti-GD2 monoclonal antibody ch14.18 in patients with malignant melanoma, *Hum. Antibod. Hybridomas*, **3**, 19–24.

SANDHU, J., SHPITZ, B., GALLINGER, S. and HOZUMI, N., 1994, Human primary immune response in SCID mice engrafted with human peripheral blood lymphocytes, *J. Immunol.*, **152**, 3806–3813.

SANO, T., SMITH, C.L. and CANTOR, C.R., 1992, Immuno-PCR: very sensitive antigen detection by means of specific antibody–DNA conjugates, *Science*, **258**, 120–122.

SARMAY, G., LUND, J., ROZSNYAY, Z., GERGELY, J. and JEFFERIS, R., 1992, Mapping and comparison of the interaction sites on the Fc region of IgG responsible for triggering antibody dependent cellular cytotoxicity (ADCC) through different types of human Fcγ receptor, *Mol. Immunol.*, **29**, 633–639.

SASSO, D.E., GIONFRIDDO, M.A., THRALL, R.S., SYRBU, S.I., SMILOWITZ, H.M. and WEINER, R.E., 1996, Biodistribution of indium-111-labeled antibody directed against intercellular adhesion molecule-1, *J. Nucl. Med.*, **37**, 656–661.

SATO, H., WATANABE, K., AZUMA, J., KIDAKA, T. and HORI, M., 1989, Specific removal of IgE by therapeutic immunoadsorption system, *J. Immunol. Methods*, **118**, 161–168.

SATO, K., TSUCHIYA, M., SALDANHA, J., KOISHIHARA, Y., OHSUGI, Y., KISHIMOTO, T. and BENDIG, M.M., 1994, Humanization of a mouse anti-human interleukin-6 receptor antibody comparing two methods for selecting human framework regions, *Mol. Immunol.*, **31**, 371–381.

SAUER-ERICKSSON, A.E., KLEYWEGT, G.J., UHLEN, M. and JONES, T.A., 1995, Crystal structure of the C2 fragment of Streptococcal protein G in complex with the Fc domain of human IgG, *Structure*, **3**, 265–278.

SAWYER, J.R., SCHLOM, J. and KAHMIRI, S.V.S., 1994, The effects of induction conditions on production of a soluble anti-tumor sFv in *Escherichia coli*, *Protein Engin.*, **7**, 1401–1406.

SCHIABLE, T., DEWOODY, K., WEISMAN, H., LINE, B., KEENAN, A. and ALAVI, A., 1992, Accurate diagnosis of acute deep vein thrombosis with technetium99m antifibrin scintigraphy: final phase III results, *J. Nucl. Med.*, **33** (suppl.), 848.

SCHIER, R., MCCALL, A., ADAMS, G.P., MARSHALL, K.W., MERRITT, H., YIM, M., CRAWFORD, R.S., WEINER, L.M., MARKS, C. and MARKS, J.D., 1996, Isolation of picomolar affinity anti-c-erb-B2 single-chain Fv by molecular evolution of the complementarity determining regions in the center of the antibody binding site, *J. Mol. Biol.*, **263**, 551–567.

SCHMIDT, T.G.M. and SKERRA, A., 1994, One-step affinity purification of bacterially produced proteins by means of the strep-tag and immobilized recombinant core streptavidin, *J. Chromatogr.*, **676**, 337–343.

SCHOTT, M.E., FRAZIER, K.A., POLLOCK, D.K. and VERBANAC, K.M., 1993, Preparation, characterization, and *in vivo* biodistribution properties of synthetically cross-linked multivalent antitumour antibody fragments, *Bioconjugate Chem.*, **4**, 153–165.

SCHOTT, M.E., MILENIC, D.E., YOKOTA, T., WHITLOW, M., WOOD, J.F., FORDYCE, W.A., CHENG, R.C. and SCHLOM, J., 1992, Differential metabolic patterns of iodinated versus radiometal chelated anticarcinoma single-chain Fv molecules, *Cancer Res.*, **52**, 6413–6417.

SCHRAPPE, M., BUMOL, T.F., APELGREN, L.D., BRIGGS, S.L., KOPPEL, G.A., MARKOWITZ, D.D., MUELLER, B.M. and REISFELD, R.A., 1992, Long-term growth suppression of human glioma xenografts by chemoimmunoconjugates of 4-desacetylvinblastine-3-carboxyhydrazide and monoclonal antibody 9.2.27, *Cancer Res.*, **52**, 3838–3844.

SCHRIER, D.M., STEMMER, S.M., JOHNSON, T., KASLIWAL, R., LEAR, J., MATTHES, S., TAFFS, S., DUFTON, C., GLENN, S.D., BUTCHKO, G., CERIANI, R.L., ROVIRA, D., BUNN, P., SHPALL, E.J., BEARMAN, S.I., PURDY, M., CAGNONI, P. and JONES, R.B., 1995, High-dose ^{90}Y Mx-DTPA-BrE3 and autologous stem cell support (AHSCS) for the treatment of advanced breast cancer: a phase 1 trial, *Cancer Res.*, **55**, 5921s–5924s.

SCHULTZ, P.G. and LERNER, R.A., 1995, From molecular diversity to catalysis: lessons from the immune system, *Science*, **269**, 1835–1842.

SCHUMACHER, J., KLIVENYI, G., MATYS, R., KIRCHGEBNER, H., HAUSER, H., MAIER-BORST, W. and MATZKU, S., 1990, Uptake of indium-111 in the liver of mice following administration of indium-111 DTPA-labeled monoclonal antibodies, *J. Nucl. Med.*, **31**, 1084–1093.

SCHUSTER, J.M., GARG, P.K., BIGNER, D.D. and ZALUTSKY, M.R., 1991, Improved therapeutic efficacy of a monoclonal antibody radioiodinated using *N*-succinimidyl-3-(tri-n-butylstannyl) benzoate, *Cancer Res.*, **51**, 4164–4169.

SEARLE, F., BIER, C., BUCKLEY, R.G., NEWMAN, S., PEDLEY, R.B., BAGSHAWE, K.D., MELTON, R.G., ALWAN, S.M. and SHERWOOD, R.F., 1986, The potential of carboxypeptidase G2–antibody conjugates as anti-tumour agents, *Br. J. Cancer*, **53**, 377–384.

SEIDMAN, J.G. and LEDER, P., 1978, The arrangement and rearrangement of antibody genes, *Nature*, **276**, 790–795.

SELF, C.H., DESSI, J.L. and WINGER, L.A., 1994, High-performance assays of small molecules: enhanced sensitivity, rapidity and convenience demonstrated with a non-competitive immunometric anti-immune complex assay system for digoxin, *Clin. Chem.*, **40**, 2035–2041.

SEON, B.K., MATSUNO, F., HARUTA, Y., KONDO, M. and BARCOS, M., 1997, Long-lasting complete inhibition of human solid tumors in SCID mice by targeting endothelial cells of tumor vasculature with anti-human endoglin immunotoxin, *Clin. Cancer Res.*, **3**, 1031–1044.

SHARKEY, R.M., JUWEID, M., SHEVITZ, J., BEHR, T., DUNN, R., SWAYNE, L.C., WONG, G.Y., BLUMENTHAL, R.D., GRIFFITHS, G.L., SIEGAL, J.A., LEUNG, S., HANSEN, H.J. and GOLDENBERG, D.M., 1995, Evaluation of a complementarity-determining region-grafted (humanized) anti-carcinoembryonic antigen monoclonal antibody in preclinical and clinical studies, *Cancer Res.*, Suppl. **55**, 5935s–5945s.

SHARMA, S.K., BAGSHAWE, K.D., BURKE, P.J., BODEN, R.W. and ROGERS, G.T., 1990, Inactivation and clearance of an anti-CEA carboxypeptidase G2 conjugate in blood after localisation in a xenograft model, *Br. J. Cancer*, **61**, 659–662.

SHARON, J. and GIVOL, D., 1976, Preparation of Fv fragment from the mouse myeloma XRPC-25 immunoglobulin possessing anti-dinitrophenyl activity, *Biochem.*, **15**, 1591–1594.

SHAW, D.R., KHAZAELI, M.B. and LOBUGLIO, A.F., 1988, Mouse/human chimeric antibodies to a tumour-associated antigen: biologic activity of the four human IgG subclasses, *J. Natl Cancer Inst.*, **80**, 1553–1559.

SHIH, L.B., GOLDENBERG, D.M., XUAN, H., LU, H., SHARKEY, R.M. and HALL, T.C., 1991, Anthracycline immunoconjugates prepared by a site-specific linkage via an amino-dextran intermediate carrier, *Cancer Res.*, **51**, 4192–4198.

SHIN, S.U., FRIDEN, P., MORAN, M. and MORRISON, S.L., 1994, Functional properties of antibody insulin-like growth factor fusion proteins, *J. Biol. Chem.*, **269**, 4979–4985.

SHIN, S.U., FRIDEN, P., MORAN, M., OLSON, T., KANG, Y., PARDRIDGE, W.M. and MORRISON, S.L., 1995, Transferrin–antibody fusion proteins are effective in brain targeting, *Proc. Natl Acad. Sci. (USA)*, **92**, 2820–2824.

SHIN, S.U., WEI, C.F., AMIN, A.R., THORBECKE, G.J. and MORRISON, S.L., 1992, Structural and functional properties of mouse–human chimeric IgD, *Human Antibod. Hybrid.*, **3**, 65–74.

SHIN, S.U., WU, D., RAMANTHAN, R., PARDRIDGE, W.M. and MORRISON, S.L., 1997, Functional and pharmacokinetic properties of antibody–avidin fusion proteins, *J. Immunol.*, **158**, 4797–4804.

SHOPES, B., 1992, A genetically engineered human IgG mutant with enhanced cytolytic activity, *J. Immunol.*, **148**, 2918–2922.

SIEGALL, C.B., LIGGITT, D., CHACE, D., MIXAN, B., SUGAI, J., DAVIDSON, T. and STEINITZ, M., 1997, Characterization of vascular leak syndrome induced by the toxin component of *Pseudomonas* exotoxin-based immunotoxins and its potential inhibition with non-steroidal anti-inflammatory drugs, *Clin. Cancer Res.*, **3**, 339–345.

SIMMONS, L.C. and YANSURA, D.G., 1996, Translational level is a critical factor for the secretion of heterologous proteins in *Escherichia coli*, *Nature Biotech.*, **14**, 629–634.

SIMON, T. and RAJEWSKY, K., 1992, A functional antibody mutant with an insertion in the framework region 3 loop of the VH domain: implications for antibody engineering, *Protein Engin.*, **5**, 229–234.

SIMS, M.J., HASSAL, D.G., BRETT, S., ROWAN, W., LOCKYER, M.J., ANGEL, A., LEWIS, A.P., HALE, G., WALDMANN, H. and CROWE, J.S., 1993, A humanized CD18 antibody can block function without cell destruction, *J. Immunol.*, **151**, 2296–2308.

SINGER, I.I., KAWKA, D.W., DEMARTINO, J.A., DAUGHERTY, B.L., ELLISTON, K.O., ALVES, K., BUSH, B.L., CAMERON, P.M., CUCA, G.C., DAVIES, P., FORREST, M.J., KAZAZIS, D.M., LAW, M.F., LENNY, A.B., MACINTYRE, D.E., MEURER, R., PADLAN, E.A., PANDYA, S., SCHMIDT, J.A., SEAMANS, T.C., SCOTT, S., SILBERKLANG, M., WILLIAMSON, A.R. and MARK, G.E., 1993, Optimal humanization of 1B4, an anti-CD18 murine monoclonal antibody is achieved by correct choice of human V-region framework sequences, *J. Immunol.*, **150**, 2844–2857.

SISSON, T.H. and CASTOR, C.W., 1990, An improved method for immobilising IgG antibodies to protein A agarose, *J. Immunol. Methods*, **127**, 215–220.

SIVOLAPENKO, G.B., DOULI, V., PECTASIDES, D., SKARLOS, D., SIRMALIS, G., HUSSAIN, R., COOK, J., COURTNEY-LUCK, N.S., MERKOURI, E., KONSTANTINIDES, K. and EPENETOS, A.A., 1995, Breast cancer imaging with radiolabelled peptide from complementarity-determining region of antitmour antibody, *Lancet*, **346**, 1662–1666.

SKERRA, A., PFITZINGER, I. and PLUCKTHUN, A., 1991, The functional expression of antibody Fv fragments in *Escherichia coli*: improved vectors and a generally applicable purification technique, *Bio/Technol.*, **9**, 273–278.

SKERRA, A. and PLUCKTHUN, A., 1988, Assembly of a functional immunoglobulin Fv fragment in *Escherichia coli*, *Science*, **240**, 1038–1041.

SKERRA, A. and PLUCKTHUN, A., 1991, Secretion and *in vivo* folding of the Fab fragment of the antibody McPC603 in *Escherichia coli*: influence of disulphides and cis-prolines, *Protein Engin.*, **4**, 971–979.

SLAMON, D.J., CLARK, G.M., WONG, S.G., LEVIN, W.J., ULLRICH, A. and MCQUIRE, W.L., 1987, Human breast cancer: correlation of relapse and survival with amplification of the HER-2/neu oncogene, *Science*, **235**, 177–182.

SMITH, G.P., 1985, Filamentous fusion phage: novel expression vectors that display cloned antigens on the virion surface, *Science*, **228**, 1315–1317.

SMITH, R.I.F., COLOMA, M.J. and MORRISON, S.L., 1995, Addition of a μ-tailpiece to IgG results in polymeric antibodies with enhanced effector functions including complement mediated cytolysis by IgG4, *J. Immunol.*, **154**, 2226–2236.

SMITH, R.I.F. and MORRISON, S.L., 1994, Recombinant polymeric IgG: an approach to engineering more potent antibodies, *Bio/Technol.*, **12**, 683–688.

SONGSIVILAI, S., CLISSOLD, P.M. and LACHMAN, P.J., 1989, A novel strategy for producing chimeric bispecific antibodies by gene transfection, *Biochem. Biophys. Res. Commun.*, **164**, 271–276.

SOUTHERN, P.J. and BERG, P., 1982, Transformation of mammalian cells to antibiotic resistance with a bacterial gene under control of the SV40 early region promoter, *J. Mol. Appl. Genet.*, **1**, 327–341.

SPADA, S. and PLUCKTHUN, A., 1997, Selectively infective phage (SIP) technology: a novel method for *in vivo* selection of interacting protein–ligand pairs, *Nature Med.*, **3**, 694–696.

SPEIKER-POLET, H., SETHUPATHI, P., YAM, P.C. and KNIGHT, K.L., 1995, Rabbit monoclonal antibodies: generating a fusion partner to produce rabbit–rabbit hybridomas, *Proc. Natl Acad. Sci. (USA)*, **92**, 9348–9352.

SPITZNAGEL, T.M. and CLARK, D.S., 1993, Surface-density and orientation effects on immobilized antibodies and antibody fragments, *Bio/Technol.*, **11**, 825–829.

SPRINGER, T.A., 1994, Traffic signals for lymphocyte recirculation and leukocyte emigration: the multistep paradigm, *Cell*, **76**, 301–314.

STACK, W.A., MANN, S.D., ROY, A.J., HEATH, P., SOPWITH, M., FREEMAN, J., HOLMES, G., LONG, R., FORBES, A., KAMM, M.A. and HAWKEY, C.J., 1997, Randomised controlled trial of CDP571 antibody to tumour necrosis factor-α in Crohn's disease, *Lancet*, **349**, 521–524.

STANCOVSKI, I., SCHINDLER, D.G., WAKS, T., YARDEN, Y., SELA, M. and ESHHAR, Z., 1993, Targeting of T-lymphocytes to neu/HER2-expressing cells using chimeric single chain Fv receptors, *J. Immunol.*, **151**, 6577–6582.

STANKER, L.H., VANDERLAAN, M. and JUAREZ-SALINAS, H., 1985, One-step purification of mouse monoclonal antibodies from ascites fluid by hydroxylapatite chromatography, *J. Immunol. Methods*, **79**, 157–169.

STANLEY, C.J., JOHANNSSON, A. and SELF, C.H., 1985, Enzyme amplification can enhance both the speed and sensitivity of immunoassays, *J. Immunol. Methods*, **83**, 89–95.

STEIN, R., GOLDENBERG, D.M., THORPE, S.R. and MATTES, M.J., 1997, Advantage of a residualising iodine radiolabel for radioimmunotherapy of xenografts of human non-small cell carcinoma of the lung, *J. Nucl. Med.*, **38**, 391–395.

STEISS, R.G., CARRASQUILLO, J.A., MCCABE, R., BOOKMAN, M.A., REYNOLDS, J.C., LARSON, S.M., SMITH, J.W., CLARK, J.W., DAILEY, V., DEL VECCHIO, S., SHUKE, N., PINSKY, C.M., URBA, W.J., HASPEL, M., PERENTESIS, P., PARIS, B., LONGO, D.L. and HANNA, M.G., JR, 1990, Toxicity, immunogenicity and tumor radioimmunodetecting ability of two human monoclonal antibodies in patients with metastatic colorectal carcinoma, *J. Clin. Oncol.*, **8**, 476–490.

STEPHENS, P. and COCKETT, M., 1989, A versatile set of mammalian expression vectors, *Nucl. Acids Res.*, **17**, 7110.

STEPHENS, S., EMTAGE, S., VETTERLEIN, O., CHAPLIN, L., BEBBINGTON, C., NESBITT, A., SOPWITH, M., ATHWAL, D., NOVAK, C. and BODMER, M., 1995, Comprehensive pharmacokinetics of a humanized antibody and analysis of residual anti-idiotypic responses, *Immunol.*, **85**, 668–674.

STEWART, J.S.W., HIRD, V., SNOOK, D., DHOKIA, B., SIVOLAPENKO, G.B., HOOKER, G., TAYLOR-PAPADIMITRIOU, J., SULLIVAN, M., LAMBERT, H.E., COULTER, C., MASON, W.P., SOUTTER, W.P. and EPENETOS, A.A., 1990, Intraperitoneal yttrium-90 labeled monoclonal antibody in ovarian cancer, *J. Clin. Oncol.*, **8**, 1941–1950.

STEWART, J.S.W., HIRD, V., SNOOK, D., SULLIVAN, M., HOOKER, G., COURTENAY-LUCK, N., SIVOLAPENKO, G., GRIFFITHS, G., MYERS, M.J., LAMBERT, H.E. and EPENETOS, A.A., 1989, Intraperitoneal radioimmunotherapy for ovarian cancer: pharmacokinetics, toxicity and efficacy of I-131 labeled monoclonal antibodies, *Int. J. Radiat. Oncol. Biol. Phys.*, **16**, 405–413.

STRATTON, J.R., CERQUEIRA, M.D., DEWHURST, T.A. and KOHLER, T.R., 1994, Imaging arterial thrombosis: comparison of technetium-99m labeled monoclonal antifibrin antibody and indium-111 labeled platelets, *J. Nucl. Med.*, **35**, 1731–1737.

Su, X., Prestwood, A.K. and McGraw, R.A., 1992, Production of recombinant porcine tumor necrosis factor alpha in a novel *E. coli* expression system, *Biotechniques*, **13**, 756–762.

Suitters, A.J., Foulkes, R., Opal, S.M., Palardy, J.E., Emtage, J.S., Rolfe, M., Stephens, S., Morgan, A., Holt, A.R., Chaplin, L.C., Shaw, N.E., Nesbitt, A.M. and Bodmer, M.W., 1994, Differential effect of isotype on efficacy of anti-tumor necrosis factor α chimeric antibodies in experimental septic shock, *J. Exp. Med.*, **179**, 849–856.

Sun, L.K., Fung, M.S.C., Sun, W.N.C., Sun, C.R.Y., Chang, W. and Chang, T.W., 1995, Human IgA monoclonal antibodies specific for a major ragweed pollen antigen, *Bio/Technol.*, **13**, 779–786.

Sutton, B.J. and Gould, H.J., 1993, The human IgE network, *Nature*, **366**, 421–428.

Tai, M.S., Mudgett-Hunter, M., Levinson, D., Wu, G.M., Haber, E., Oppermann, H. and Huston, J.S., 1990, A bifunctional fusion protein containing Fc binding fragment B of Staphylococcal protein A amino terminal to antidigoxin single-chain Fv, *Biochem.*, **29**, 8024–8030.

Takahashi, H., Odaka, A., Kawaminami, S., Matsunaga, C., Kato, K., Shimada, I. and Arata, Y., 1991, Multinuclear NMR study of the structure of the Fv fragment of anti-dansyl mouse IgG2a antibody, *Biochem.*, **30**, 6611–6619.

Tao, M.H. and Morrison, S.L., 1989, Studies of aglycosylated chimeric mouse–human IgG, *J. Immunol.*, **143**, 2595–2601.

Tao, M.H., Smith, R.I.F. and Morrison, S.L., 1993, Structural features of human immunoglobulin G that determine isotype-specific differences in complement activation, *J. Exp. Med.*, **178**, 661–667.

Tavladoraki, P., Benvenuto, E., Trinca, S., DeMartinis, D., Cattaneo, A. and Galeffi, P., 1993, Transgenic plants expressing a functional single-chain Fv are specifically protected from virus attack, *Nature*, **366**, 469–472.

Tempest, P.R., Bremner, P., Lambert, M., Taylor, M., Furze, L.M., Carr, F.J. and Harris, W.J., 1991, Reshaping a human monoclonal antibody to inhibit human respiratory syncytial virus infection in vivo, *Bio/Technol.*, **9**, 266–271.

Terskikh, A., Couty, S., Pelegrin, A., Hardman, N., Hunziker, W. and Mach, J.P., 1994, Dimeric recombinant IgA directed against carcino-embryonic antigen, a novel tool for carcinoma localization, *Mol. Immunol.*, **31**, 1313–1319.

Thakur, M.L., DeFulvio, J., Park, C.H., Damjanov, A., Yaghsezian, H., Jungkind, D., Epstein, A. and McAfee, J.G., 1991, Technetium labeled proteins for imaging inflammatory foci, *Nucl. Med. Biol.*, **18**, 605–612.

Thakur, M.L., Marcus, C.S., Henneman, P., Butler, J., Sinow, R., Diggles, L., Minami, C., Mason, G., Klein, S. and Rhodes, B., 1996, Imaging inflammatory diseases with neutrophil-specific technetium-99m-labeled monclonal antibody anti-SSEA-1, *J. Nucl. Med.*, **37**, 1789–1795.

Thommes, J., Halfar, M., Lenz, S. and Kula, M.R., 1995, Purification of monoclonal antibodies from whole hybridoma fermentation broth by fluidized bed adsorption, *Biotech. Bioengin.*, **45**, 205–211.

Thorpe, G.H.G. and Kricka, L.J., 1986, Enhanced chemiluminescence reactions catalysed by horseradish peroxidase, *Methods Enzymol.*, **133**, 331–354.

Thorpe, P., Wallace, P., Knowles, P., Relf, M., Brown, N., Watson, G., Blakey, D. and Newell, D., 1988, Improved antitumour effects of immunotoxins prepared with deglycosylated ricin A-chain and hindered disulphide linkages, *Cancer Res.*, **48**, 6396–6403.

Tinubu, S.A., Hakimi, J., Kondas, J.A., Bailon, P., Familletti, P.C., Spence, C., Crittenden, M.D., Parenteau, G.L., Dirbas, F.M., Tsudo, M., Bacher, J.D., Kasten-Sportes, C., Martinucci, J.L., Goldman, C.K., Clark, R.E. and Waldmann, T.A., 1994, Humanized antibody directed to the IL-2 receptor b-chain prolongs cardiac allograft survival, *J. Immunol.*, **153**, 4330–4338.

Tomasi, T.B., 1992, The discovery of secretory IgA and the mucosal immune system, *Immunol. Today*, **13**, 416–418.

TOMIZUKA, K., YOSHIDA, H., UEJIMA, H., KUGOH, H., SATO, K., OHGUMA, A., HAYASAKA, M., HANAOKA, K., OSHIMURA, M. and ISHIDA, I., 1997, Functional expression and germline transmission of a human chromosome fragment in chimaeric mice, *Nature Genet.*, **16**, 133–143.

TOM-MOY, M., BAER, R.L., SPIRA-SOLOMON, D. and DOHERTY, T.P., 1995, Atrazine measurements using surface transverse wave devices, *Anal. Chem.*, **67**, 1510–1516.

TORCHILIN, V.P., 1994, Immunoliposomes and PEGylated immunoliposomes: possible use for targeted delivery of imaging agents, *Immunomethods*, **4**, 244–258.

TORNOE, I., TITLESTAD, I.L., KEJLING, K., ERB, K., DITZEL, H.J. and JENSENIUS, J.C., 1997, Pilot scale purification of human monoclonal IgM (COU-1) for clinical trials, *J. Immunol. Methods*, **205**, 11–17.

TRAIL, P.A., WILLNER, D., LASCH, S.J., HENDERSON, A.J., HOFSTEAD, S., CASAZZA, A.M., FIRESTONE, R.A., HELLSTROM, I. and HELLSTROM, K.E., 1993, Cure of xenografted human carcinomas by BR96–doxorubicin immunoconjugates, *Science*, **261**, 212–215.

TRAMONTANO, A., JANDA, K.D. and LERNER, R.A., 1986, Catalytic antibodies, *Science*, **234**, 1566–1570.

TRELOAR, P.H., NKOHKWO, A.T., KANE, J.W., BARBER, D. and VADGAMA, P.M., 1994, Electrochemical immunoassay: simple kinetic determination of alkaline phosphatase enzyme labels in limited and excess reagent systems, *Electroanalysis*, **6**, 561–566.

TRILL, J.J., SHATZMAN, A.R. and GANGULY, S., 1995, Production of monoclonal antibodies in COS and CHO cells, *Curr. Opin. Biotechnol.*, **6**, 553–560.

TSAI, P.K., BURKE, C.J., IRWIN, J.W., BRUNER, M.W., TUNG, J.S., HOLLIS, G.F., MARK, G.E., KESSLER, J.A., BOOTS, L.J., CONLEY, A.J. and MIDDAUGH, C.R., 1995, Enhancing the avidity of a human recombinant anti-HIV-1 monoclonal antibody through oligomerization, *J. Pharm. Sci.*, **84**, 866–870.

TSCHMELITSCH, J., BARENDSWAARD, E., WILLIAMS, C., YAO, T.J., COHEN, A.M., OLD, L.J. and WELT, S., 1997, Enhanced antitumour activity of combination radioimmunotherapy (131I-labeled monoclonal antibody A33) with chemotherapy (fluorouracil), *Cancer Res.*, **57**, 2181–2186.

TSUI, P., TORNETTA, M.A., AMES, R.S., BANKOSKY, B.C., GRIEGO, S., SILVERMAN, C., PORTER, T., MOORE, G. and SWEET, R.A., 1996, Isolation of a neutralising human RSV antibody from a dominant, non-neutralising immune repertoire by epitope-blocked panning, *J. Immunol.*, **157**, 772–780.

TURNER, A., KING, D.J., FARNSWORTH, A.P.H., RHIND, S.K., PEDLEY, R.B., BODEN, J., BODEN, R., MILLICAN, T.A., MILLAR, K., BOYCE, B., BEELEY, N.R.A., EATON, M.A.W. and PARKER, D., 1994, Comparative biodistributions of indium-111-labelled macrocycle chimeric B72.3 antibody conjugates in tumour-bearing mice, *Br. J. Cancer*, **70**, 35–41.

TUTT, A., GREENMAN, J., STEVENSON, G.T. and GLENNIE, M.J., 1991a, Bispecific $F(ab'\gamma)_3$ antibody derivatives for redirecting unprimed cytotoxic T cells, *Eur. J. Immunol.*, **21**, 1351–1358.

TUTT, A., STEVENSON, G.T. and GLENNIE, M.J., 1991b, Trispecific $F(ab')_3$ derivatives that use cooperative signaling via the TCR/CD3 complex and CD2 to activate and redirect resting cytotoxic T cells, *J. Immunol.*, **147**, 60–69.

UEDA, H., TSUMOTO, K., KUBOTA, K., SUZUKI, E., NAGAMUNE, T., NISHIMURA, H., SCHEULER, P.A., WINTER, G., KUMAGAI, I. and MAHONEY, W.C., 1996, Open sandwich ELISA: a novel immunoassay based on the interchain interaction of antibody variable region, *Nature Biotechnol.*, **14**, 1714–1718.

ULTEE, M.E., BRIDGER, G.J., ABRAMS, M.J., LONGLEY, C.B., BURTON, C.A., LARSEN, S.K., HENSON, G.W., PADMANABHAN, S., GAUL, F.E. and SCHWARTZ, D.A., 1997, Tumor imaging with technetium-99m-labeled hydrazinonicotinamide–Fab′ conjugates, *J. Nucl. Med.*, **38**, 133–138.

UTTENREUTHER-FISCHER, M.M., HUANG, C.S. and YU, A.L., 1995, Pharmacokinetics of human–mouse chimeric anti-GD2 mAb ch14.18 in a phase I trial in neuroblastoma patients, *Cancer Immunol. Immunother.*, **41**, 331–338.

VALLERA, D.A., PANOSKALTIS-MORTARI, A., JOST, C., RAMAKRISHNAN, S., EIDE, C.R., KREITMAN, R.J., NICHOLLS, P.J., PENNELL, C. and BLAZAR, B.R., 1996, Anti-graft versus host disease effect of DT390-anti-CD3 sFv, a single chain Fv fusion immunotoxin specifically targeting the CD3 epsilon moiety of the T cell receptor, *Blood*, **88**, 2342–2353.

VALLERA, D.A., TAYLOR, P.A., PANOSKALTSIS-MORTARI, A. and BLAZAR, B.R., 1995, Therapy of ongoing graft versus host disease induced across the major or minor histocompatibility barrier in mice with anti-CD3 F(ab′)2-ricin toxin A chain immunotoxin, *Blood*, **86**, 4367–4375.

VAN DER LUBBE, P.A., DIJKMANS, B.A., MARKUSSE, H.M., NASSANDER, U. and BREEDVELD, F.C., 1995, A randomised double-blind placebo controlled study of CD4 monoclonal antibody therapy in early rheumatoid arthritis, *Arthritis Rheum.*, **38**, 1097–1106.

VAN DER WEIDE, J., HOMAN, H.C., COZIJNSEN-VAN RHEENAN, E., VIVIE-KIPP, Y., POORTMAN, J. and KRAAIJENHAGEN, R.J., 1992, Nonisotopic binding assay for measuring vitamin B12 and folate in serum, *Clin. Chem.*, **38**, 766–768.

VAN HOF, A.C., MOLTHOFF, C.F.M., DAVIES, Q., PERKINS, A.C., VERHEIJEN, R.H.M., KENEMANS, P., DEN HOLLANDER, W., WILHELM, A.J., BAKER, T.S., SOPWITH, M., FRIER, M., SYMONDS, E.M. and ROOS, J.C., 1996, Biodistribution of 111indium labeled engineered human antibody CTM01 in ovarian cancer patients: influence of protein dose, *Cancer Res.*, **56**, 5179–5185.

VAUGHAN, T.J., WILLIAMS, A.J., PRITCHARD, K., OSBOURN, J.K., POPE, A.R., EARNSHAW, J.C., MCCAFFERTY, J., HODITS, R.A., WILTON, J. and JOHNSON, K.S., 1996, Human antibodies with sub-nanomolar affinities isolated from a large non-immunized phage display library, *Nature Biotechnol.*, **14**, 309–314.

VERHAAR, M.J., KEEP, P.A., HAWKINS, R.E., ROBSON, L., CASEY, J.L., PEDLEY, B., BODEN, J.A., BEGENT, R.H.J. and CHESTER, K.A., 1996, Technetium-99m radiolabelling using a phage derived single-chain Fv with a C-terminal cysteine, *J. Nucl. Med.*, **37**, 868–872.

VERHOEYEN, M., MILSTEIN, C. and WINTER, G., 1988, Reshaping human antibodies: grafting an anti-lysozyme activity, *Science*, **239**, 1534–1536.

VITETTA, E.S. and UHR, J.W., 1994, Monoclonal antibodies as agonists: an expanded role for their use in cancer therapy, *Cancer Res.*, **54**, 5301–5309.

VOGEL, C.A., GALMICHIE, M.C. and BUCHEGGER, F., 1997, Radioimmunotherapy and fractionated radiotherapy of human colon cancer liver metastases in nude mice, *Cancer Res.*, **57**, 447–453.

VOGL, M., PARISCH, M., KLEIN, C. and MULLER, M.M., 1996, Evaluation of the Syva Emit^R 2000 digoxin assay on a Hitachi 717 analyzer, *Clin. Biochem.*, **29**, 389–391.

VOLA, R., LOMBARDI, A., TARDITI, L., BJORCK, L. and MARIANI, M., 1995, Recombinant proteins L and LG: efficient tools for purification of murine immunoglobulin G fragments, *J. Chromatogr.*, **668**, 209–218.

VOLLMERS, H.P., WOZNIAK, E., STEPIEN-BOTSCH, E., ZIMMERMAN, U. and MULLER-HERMELINK, H.K., 1996, A rapid method for purification of monoclonal human IgM from mass culture, *Human Antibod. Hybridomas*, **7**, 37–41.

VOSS, A., NIERSBACH, M., HAIN, R., HIRSCH, H.J., LIAO, Y.C., KREUZALER, F. and FISCHER, R., 1995, Reduced virus infectivity in *N. tabacum* secreting a TMV-specific full-size antibody, *Mol. Breed.*, **1**, 39–50.

VUILLEZ, J.P., MORO, D., BRICHON, P.Y., ROUVIER, E., BRAMBILLA, E., BARBET, J., PELTIER, P., MEYER, P., SARRAZIN, R. and BRAMBILLA, C., 1997, Two-step immunoscintigraphy for non-small-cell lung cancer staging using a bispecific anti-CEA/anti-indium-DTPA antibody and an indium-111-labeled DTPA dimer, *J. Nucl. Med.*, **38**, 507–511.

WALDMANN, T.A., WHITE, J.D., CARRASQUILLO, J.A., REYNOLDS, J.C., PAIK, C.H., GANSOW, O.A., BRECHBIEL, M.W., JAFFE, E.S., FLEISHER, T.A., GOLDMAN, C.K., TOP, L.E., BAMFORD, R., ZAKNOEN, S., ROESSLER, E., KSTEN-SPORTES, C., ENGLAND, R., LITOU, H., JOHNSON, J.A., JACKSON-WHITE, T., MANNS, A., HANCHARD, B., JUNGHANS, R.P. and NELSON, D., 1995, Radioimmunotherapy of interleukin-2Rα-expressing adult T-cell leukemia with yttrium-90-labelled anti-Tac, *Blood*, **86**, 4063–4075.

WALKER, W. and GALLAGHER, G., 1994, The *in vivo* production of specific human antibodies by vaccination of human-PBL-SCID mice, *Immunol.*, **83**, 163–170.

WANG, Y., HU, Q., MADRI, J.A., ROLLINS, S.A., CHODERA, A. and MATIS, L.A., 1996, Amelioration of lupus-like autoimmune disease in NZB/WF1 mice after treatment with a blocking monoclonal antibody specific for complement component C5, *Proc. Natl Acad. Sci. (USA)*, **93**, 8563–8568.

WANG, Y., ROLLINS, S.A., MADRI, J.A. and MATIS, L.A., 1995, Anti-C5 monoclonal antibody therapy prevents collagen-induced arthritis and ameliorates established disease, *Proc. Natl Acad. Sci. (USA)*, **92**, 8955–8959.

WARD, E.S., GUSSOW, D., GRIFFITHS, A.D., JONES, P.T. and WINTER, G., 1989, Binding activities of a repertoire of single immunoglobulin variable domains secreted from *Escherichia coli*, *Nature*, **341**, 544–546.

WASHBURN, L.C., SUN, T.T.H., LEE, Y.C.C., BYRD, B.L., HOLLOWAY, E.C., CROOK, J.E., STUBBS, J.B., STABIN, M.G., BRECHBIEL, M.W., GANSOW, O.A. and STEPLEWSKI, Z., 1991, Comparison of five bifunctional chelate techniques for ^{90}Y-labeled monoclonal antibody CO17-1A, *Nucl. Med. Biol.*, **18**, 313–321.

WAWRZYNCZAK, E.J., CUMBER, A.J., PARNELL, G.D., JONES, P.T. and WINTER, G., 1992, Blood clearance in the rat of a recombinant mouse monoclonal antibody lacking the N-linked oligosaccharide side-chains of the CH2 domains, *Mol. Immunol.*, **29**, 213–220.

WEBBER, K.O., REITER, Y., BRINKMANN, U., KREITMAN, R. and PASTAN, I., 1995, Preparation and characterization of a disulfide-stabilized Fv fragment of the anti-Tac antibody, comparison with its single-chain analog, *Mol. Immunol.*, **32**, 249–258.

WEBER, R.W., BOUTIN, R.H., NEDELMAN, M.A., LISTER-JAMES, J., and DEAN, R.T., 1990, Enhanced kidney clearance with an ester linked 99m-Tc-radiolabelled antibody Fab′-chelator conjugate, *Bioconjug. Chem.*, **1**, 431–437.

WEEKS, I., BEHESTI, I., MCCAPRA, F., CAMPELL, A.K. and WOODHEAD, J.S., 1983, Acridinium esters as high specific activity labels in immunoassay, *Clin. Chem.*, **29**, 1474–1479.

WEIDEN, P.L., BREITZ, H.B., SEILER, C.A., BJORN, M.J., RATLIFF, B.A., MALLETT, R., BEAUMIER, P.L., APPELBAUM, J.W., FRITZBERG, A.R. and SALK, D., 1993, Rhenium-186-labeled chimeric antibody NR-LU-13: pharmacokinetics, biodistribution and immunogenicity relative to murine analog NR-LU-10, *J. Nucl. Med.*, **34**, 2111–2119.

WEIJTENS, M.E.M., WILLEMSEN, R.A., VALERIO, D., STAM, K. and BOLHUIS, R.L.H., 1996, Single chain Ig/g gene-redirected human T-lymphocytes produce cytokines, specifically lyse tumour cells, and recycle lytic capacity, *J. Immunol.*, **157**, 836–843.

WEINER, L.M., CLARK, J.I., DAVEY, M., LI, W.S., DE PALAZZO, I.G., RING, D.B. and ALPAUGH, R.K., 1995, Phase I trial of 2B1, a bispecific monoclonal antibody targeting c-erb-b2 and FcγRIII, *Cancer Res.*, **55**, 4586–4593.

WEISS, E., CHATALLIER, J. and ORFANOUDAKIS, G., 1994, *In vivo* biotinylated recombinant antibodies: construction, characterization and application of a bifunctional Fab-BCCP fusion protein produced in *E. coli*, *Protein Expression Purification*, **5**, 509–517.

WELLING, G.W., VAN GORKUM, J., DAMHOF, R.A. and DRIJFHOUT, J.W., 1991, A ten residue fragment of an antibody (mini-antibody) directed against lysozyme as a ligand in immunoaffinity chromatography, *J. Chromatogr.*, **548**, 235–242.

WELS, W., HARWERTH, I.M., ZWICKL, M., HARDMAN, N., GRONER, B. and HYNES, N.E., 1992, Construction, bacterial expression and characterization of a bifunctional single-chain antibody–phosphatase fusion protein targeted to the human erbb-2 receptor, *Bio/Technol.*, **10**, 1128–1132.

WELT, S., SCOTT, A.M., DIVGI, C.R., KEMENY, N.E., FINN, R.D., DAGHIGHIAN, F., ST GERMAIN, J., CARSWELL-RICHARDS, E., LARSON, S.M. and OLD, L.J., 1996, Phase I/II study of iodine-125-labelled monoclonal antibody A33 in patients with advanced colon cancer, *J. Clin. Oncol.*, **14**, 1787–1797.

WENDLING, D., RACADOT, E., WIDENES, J. and the FRENCH INVESTIGATORS GROUP, 1996, Randomized, double-blind, placebo controlled multicenter trial of murine anti-CD4 therapy in rheumatoid arthritis, *Arthritis Rheum.*, **39**, S245, abstract 1303.

WENTWORTH, P., DATTA, A., BLAKEY, D., BOYLE, T., PARTRIDGE, L.J. and BLACKBURN, G.M., 1996, Towards antibody-directed ‘abzyme’ prodrug therapy, ADAPT: carbamate prodrug activation by a catalytic antibody and its *in vitro* application to human tumor cell killing, *Proc. Natl Acad. Sci. (USA)*, **93**, 799–803.

WERLEN, R.C., LANKINEN, M., OFFORD, R.E., SCHUBIGER, P.A., SMITH, A. and ROSE, K., 1996, Preparation of a trivalent antigen binding construct using polyoxime chemistry: improved biodistribution and potential for therapeutic application, *Cancer Res.*, **56**, 809–815.

WERLEN, R.C., LANKINEN, M., ROSE, K., BLAKEY, D., SHUTTLEWORTH, H., MELTON, R. and OFFORD, R.E., 1994, Site-specific conjugation of an enzyme and an antibody fragment, *Bioconj. Chem.*, **5**, 411–417.

WERTHER, W.A., GONZALEZ, T.N., O'CONNOR, S.J., MCCABE, S., CHAN, B., HOTALING, T., CHAMPE, M., FOX, J.A., JARDIEU, P.M., BERMAN, P.W. and PRESTA, L.G., 1996, Humanization of an anti-lymphocyte function associated antigen LFA-1 monoclonal antibody and reengineering of the humanised antibody for binding to rhesus LFA-1, *J. Immunol.*, **157**, 4986–4995.

WHITLOW, M., BELL, B.A., FENG, S.L., FILPULA, D., HARDMAN, K.D., HUBERT, S.L., ROLLENCE, M.L., WOOD, J.F., SCHOTT, M.E., MILENIC, D.E., YOKOTA, T. and SCHLOM, J., 1993, An improved linker for single-chain Fv with reduced aggregation and enhanced proteolytic stability, *Protein Engin.*, **6**, 989–995.

WHITLOW, M. and FILPULA, D., 1991, Single-chain Fv proteins and their fusion proteins, *Methods*, **2**, 97–105.

WHITLOW, M., FILPULA, D., ROLLENCE, M.L., FENG, S.L. and WOOD, J.F., 1994, Multivalent Fv's: characterization of single-chain Fv oligomers and preparation of a bispecific Fv, *Protein Engin.*, **7**, 1017–1026.

WHITTLE, N., ADAIR, J., LLOYD, C., JENKINS, L., DEVINE, J., SCHLOM, J., RAUBITSHEK, A., COLCHER, D. and BODMER, M., 1987, Expression in COS cells of a mouse/human chimaeric B72.3 antibody, *Protein Engin.*, **1**, 499–505.

WILKINSON, I., JACKSON, C.A., LANG, G.M., HOLFORD-STREVENS, V. and SEHON, A., 1987, Tolerance induction in mice by conjugates of monoclonal immunoglobulins and monomethoxypolyethylene glycol, *J. Immunol.*, **139**, 326–331.

WILLIAMS, R.O., FELDMANN, M. and MAINI, R.N., 1992, Anti-tumour necrosis factor ameliorates joint disease in murine collagen induced arthritis, *Proc. Natl Acad. Sci. (USA)*, **89**, 9784–9788.

WILLIAMS, W.V., KIEBER-EMMONS, T., VONFELDT, J., GREENE, M.I. and WEINER, D.B., 1991, Design of bioactive peptides based on antibody hypervariable region structures, *J. Biol. Chem.*, **266**, 5182–5190.

WILLINS, J.D. and SGOUROS, G., 1995, Modelling analysis of platinum-195m for targeting individual blood-borne cells in adjuvant radioimmunotherapy, *J. Nucl. Med.*, **36**, 315–319.

WINTER, G., GRIFFITHS, A.D., HAWKINS, R.E. and HOOGENBOOM, H.R., 1994, Making antibodies by phage display technology, *Annu. Rev. Immunol.*, **12**, 433–455.

WOOD, C.R., BOSS, M.A., KENTEN, J.H., CALVERT, J.E., ROBERTS, N.A. and EMTAGE, J.S., 1985, The synthesis and *in vivo* assembly of functional antibodies in yeast, *Nature*, **314**, 446–449.

WOODHEAD, J.S., ADDISON, G.M. and HALES, C.N., 1974, *Br. Med. Bull.*, **30**, 44.

WRIGHT, A. and MORRISON, S.L., 1994, Effect of CH2-associated carbohydrate structure on the functional properties and *in vivo* fate of chimeric mouse–human immunoglobulin G, *J. Exp. Med.*, **180**, 1087–1096.

WU, X.C., NG, S.C., NEAR, R.I. and WONG, S.L., 1993, Efficient production of a functional single-chain antidigoxin antibody via an engineered *Bacillus subtilis* expression–secretion system, *Bio/Technol.*, **11**, 71–76.

YALLOW, R.S. and BERSON, S.A., 1960, Immunoassay of endogenous plasma insulin in man, *J. Clin. Invest.*, **39**, 1157–1175.

YAMAGUCHI, Y., KIM, H., KATO, K., MASUDA, K., SHIMADA, I. and ARATA, Y., 1995, Proteolytic fragmentation with high specificity of mouse immunoglobulin G, *J. Immunol. Methods*, **181**, 259–267.

YAMAIZUMI, M., MEKADA, E., UCHIDA, T. and OKADA, Y., 1978, One molecule of diptheria toxin fragment A introduced into a cell can kill the cell, *Cell*, **15**, 245–250.

YANG, W.P., GREEN, K., PINZ-SWEENEY, S., BRIONES, A.T., BURTON, D.R. and BARBAS, C.F., III, 1995, CDR walking mutagenesis for the affinity maturation of a potent human anti-HIV-1 antibody into the picomolar range, *J. Mol. Biol.*, **254**, 392–403.

YARNOLD, S. and FELL, H.P., 1994, Chimerization of antitumour antibodies via homologous recombination conversion vectors, *Cancer Res.*, **54**, 506–512.

YARRANTON, G.T. and MOUNTAIN, A., 1992, Expression of proteins in prokaryotic systems – principles and case studies, *Protein Engineering, a Practical Approach* (Rees, A.R., ed.), pp. 303–325, Oxford University Press, Oxford.

YEDNOCK, T.A., CANNON, C., FRITZ, L.C., SANCHEZ-MADRID, F., STEINMAN, L. and KARIN, N., 1992, Prevention of experimental autoimmune encephalomyelitis by antibodies against α4β1 integrin, *Nature*, **356**, 63–66.

YEUNG, D., GILL A., MAULLE, C. and DAVIES, R.J., 1995, Detection and quantitation of biomolecular interactions with optical biosensors, *Trends Anal. Chem.*, **14**, 49–56.

YOKOTA, T., MILENIC, D.E., WHITLOW, M. and SCHLOM, J., 1992, Rapid tumor penetration of a single-chain Fv and comparison with other immunoglobulin forms, *Cancer Res.*, **52**, 3402–3408.

YOKOYAMA, K., IKEBUKURO, K., TAMIYA, E., KARUBE, I., ICHIKI, N. and ARIKAWA, Y., 1995, Highly sensitive quartz crystal immunosensors for multisample detection of herbicides, *Anal. Chim. Acta*, **304**, 139–145.

YORKE, E.D., BEAUMIER, P.L., WESSELS, B.W., FRITZBERG, A.R. and MORGAN, A.C., JR, 1991, Optimal antibody–radionuclide combinations for clinical radioimmunotherapy: a predictive model based on mouse pharmacokinetics, *Nucl. Med. Biol.*, **18**, 827–835.

YOUNG, W.W., TAMURA, Y., WOLOCK, D.M. and FOX, J.W., 1984, Staphylococcal protein A binding to the Fab fragments of mouse monoclonal antibodies, *J. Immunol.*, **133**, 3163–3166.

YUAN, Q., STRAUCH, K.L., LOBB, R.R. and HEMLER, M.E., 1996, Intracellular single-chain antibody inhibits integrin VLA-4 maturation and function, *Biochem. J.*, **318**, 591–596.

ZALIPSKY, S. and LEE, C., 1992, Use of functionalized poly(ethylene glycol)s for modification of polypeptides, *Poly(Ethylene Glycol) Chemistry* (Milton Harris, J., ed.), pp. 347–371, Plenum, New York.

ZALUTSKY, M.R., MCLENDON, R.E., GARG, P.K., ARCHER, G.E., SCHUSTER, J.M. and BIGNER, D.D., 1994, Radioimmunotherapy of neoplastic meningitis in rats using an α-particle-emitting immunoconjugate, *Cancer Res.*, **54**, 4719–4725.

ZHU, Z., LEWIS, G.D. and CARTER, P., 1995, Engineering high affinity humanized Anti-p185HER2/anti-CD3 bispecific $F(ab')_2$ for efficient lysis of p185HER2 overexpressing tumor cells, *Int. J. Cancer*, **62**, 319–324.

ZHU, Z., PRESTA, L.G., ZAPATA, G. and CARTER, P., 1997, Remodeling domain interfaces to enhance heterodimer formation, *Protein Sci.*, **6**, 781–788.

ZHU, Z., ZAPATA, G., SHALABY, R., SNEDECOR, B., CHEN, H. and CARTER, P., 1996, High level secretion of a humanized diabody from *Escherichia coli*, *Bio/Technol.*, **14**, 192–196.

ZOU, Y.R., MULLER, W., GU, H. and RAJEWSKY, K., 1994, Cre-loxP-mediated gene replacement: a mouse strain producing humanized antibodies, *Current Biol.*, **4**, 1099–1103.

ZU PULITZ, J., KUBASEK, W.L., DUCHENE, M., MARGET, M., VON SPECHT, B. and DOMDEY, H., 1990, Antibody production in baculovirus-infected insect cells, *Bio/Technol.*, **8**, 651–654.

Index

abrin 139
abzymes 189–90
acute myeloid leukemia 36, 37, 70, 108, 137, 144
ADEPT (antibody-directed enzyme prodrug therapy) 139, 144–6
adhesion 151, 152–3, 154–5, 159
adjuvants 15–17
adventitious agents 161, 176
affinity 2, 6–7, 13–14, 96, 102, 190, 191–2
  antibody engineering 27, 29, 35, 42, 45, 48, 55
  cancer therapy 126, 137, 140
  generation of MAbs 15, 19–21, 23–6
  immunoassays 78, 79, 88, 92
  infectious diseases 148
  inflammatory diseases 153, 154, 156, 157, 158
  pharmacokinetics 67, 73, 74
  production of MAbs 172, 175, 177, 179, 180, 182
  RAID 108, 114, 115
affinity chromatography 176, 177, 179–84
affinity maturation 2, 13, 25–6
aggregation 46, 47, 149–50
  effector functions 55, 59, 63, 65
  production of MAbs 169, 170, 171
aglycosyl 73
AIDS therapy 75
alcohol abuse 101
alkaline phosphatase 83, 85–7, 98, 100–1
allergies 11, 18, 29, 99, 105, 151, 158–9
allotypes 13
Alzheimer's disease 118
amines 59–60, 63
amino acids 4, 6, 34, 68, 155, 191–2
  antibody fragments 45, 46, 47, 48
  effector functions 57, 58–9
  immunopurification 103, 104
  multiple specificities 53
ammonium sulphate 176
amphipathic helix 48, 49
anaphylatic shock 18, 29
anion exchange 179–83
antibiotics 148, 163, 168
antibody-dependent cellular cytotoxicity (ADCC) 6–7, 18, 119–20, 157, 173
  cancer therapy 121–5
  effector functions 54, 55, 57, 58
anti-carcinoembryonic antigen (anti-CEA) 40, 72
antigen presenting cells (APC) 152
anti-idiotype response 31, 120, 123–4, 191
  overcoming immunogenicity 35, 37
anti-Tac antibody 36, 37
apoptosis 121, 122, 146
arabinose 171
arthritis 35–7, 100–1, 118, 153–4
  therapy 151, 153–8
artichoke mottled crinkle virus 189
asthma 11, 99, 158–9
auger emitters 128, 132, 135, 136
autoimmune diseases 151, 153–4, 155, 156
avidin 74, 83, 88, 89, 115, 137
avidity 5, 9, 14, 27, 67, 79, 104
  antibody fragments 42, 48, 49, 50
  effector functions 55, 57, 63, 65
  generation of MAbs 21, 24
  multiple specificities 52, 53
  RAID 108, 109
  therapeutic applications 130, 148

B cells 2–3, 11–13, 32, 70
  generation of MAbs 14–20, 22, 23
  lymphoma 32, 70, 122, 123
  therapeutic applications 122, 123, 157
Bacillus subtilis 172
background binding 78, 85
bacteria 1, 23, 93, 147, 148–9
  cancer therapy 127, 132, 139, 145
  production of MAbs 163, 167–70, 174, 176, 177
bacteriophage 21, 31
basophils 158
BIAcore 95–6, 102
biodistribution 67–75, 113–14, 165
  cancer therapy 130, 131, 133–5, 137

biosensors 190
biotin 24, 67, 74, 115, 137, 184
  immunoassays 83, 88–9
biotin-carboxyl carrier protein (BCCP) 88–9
bispecificity 17, 27, 50–4, 74, 92, 171, 179
  antibody fragments 43, 44, 46
  cancer therapy 125–7, 141, 146
  cardiovascular disease 151
  effector functions 58
  inflammatory diseases 157
  RAID 114, 115
blocking 119–20, 121, 153–5, 156
Bolton-Hunter reagent 82, 83
bone marrow 99, 105, 157
  cancer therapy 128, 133, 134, 136–7, 138, 142
bovine serum albumin 15, 89
brain acetylcholinesterase 91
breast cancer 36, 70, 91, 99, 187, 191
  RAID 108, 109
  therapy 123, 126, 134, 138, 141–3
bromelain 41, 42

C1q 6–7, 57, 58
camels 48, 191
CAMPATH-1H 35–7, 122, 156, 164
cancer 40, 120–47, 148
  chemotherapy causing heart problems 116
  imaging agents 107
  immunoassays 91
  immunocytochemistry 99
  immunosensors 95
  overcoming immunogenicity 32
  pharmacokinetics 70, 75, 126, 129, 130
  prospects 189, 190
  RAID 108, 109, 114–15, 136
  unmodified antibodies 121–3
  see also tumours
carbohydrates 1, 5, 7, 9, 14, 15, 73
  cancer therapy 141, 143
  effector functions 57, 63, 64, 65
  immunoassays 85, 89
  immunopurification 103
  production of MAbs 162, 166, 173–5
carcinoembryonic antigen (CEA) 36, 40, 70, 72
  cancer therapy 138, 144
  RAID 109, 110, 111
carcinomas *see* cancer
carcinomatous meningitis 141
cardiovascular diseases 149–51, 154–5
  see also heart diseases
catabolic degradation 69
catalytic antibodies 189–90
cation exchange 179–83
cell culture 175–6
cell cycle arrest 121, 122
cellulose 79
centrifugation 176–7, 180, 182, 183
chain shuffling 26, 192
chemical conjugates 58–63, 65, 85
chemical cross-linking 51–3, 72, 73, 104
  antibody fragments 43, 44, 48–50
  cancer therapy 122, 126, 129, 130, 131, 134
  effector functions 57, 61
chemical modification 38–40, 75
chemiluminescence 83, 87–8, 89, 90, 95
  western blotting 100, 101
chemotherapy 122, 123, 138–9, 142–3
  heart disease 116
chickens 15
chimeric antibodies 29–37, 119–20, 189
  cancer therapy 121, 122, 127, 130, 136
  cardiovascular diseases 149, 150
  effector functions 54, 55, 56, 57
  inflammatory diseases 153, 156
  overcoming immunogenicity 29–37, 38
  pharmacokinetics 69–73
  production 164–7, 171, 173, 175, 184
chloramine 80, 82
Chinese hamster ovary (CHO) cells 29, 35, 73, 163–7
chromosomes 11, 52, 164, 165, 171
  generating MAbs 17, 19, 21
class switching 13, 15
clearance 74–5, 136–7, 185
  cancer therapy 128–30, 135, 136–7, 144, 145
  pharmacokinetics 68, 69, 71, 72, 73, 74–5
  RAID 110, 114–17
  RIGS 116
cloned enzyme donor immunoassay (CEDIA) 90
clones and cloning 28–9, 30, 52, 78, 99
  generation of MAbs 16, 18, 20, 24
  production of MAbs 161, 162, 164, 168, 172
codon 24
colloidal gold 98, 100
colorectal cancer 32, 70, 107
  RAID 108–11, 114–15
  RIGS 116
  therapy 120, 123–4, 135–6, 138, 141–2, 144
colorimetric reactions 87
complement activation 6–9, 18, 162
  effector functions 54–5, 57, 58
  therapeutic applications 119, 120, 122, 149
complement-mediated lysis 121, 122
complementarity determining regions (CDRs) 6, 27, 103
  antibody fragments 46, 48
  effector functions 58
  overcoming immunogenicity 30, 31, 33, 34, 35
  phage display 23, 25, 26
  production of MAbs 162, 170, 182
  prospects 190, 191
  RAID 117
  therapeutic applications 122, 148, 150, 155
constant domains 4–7, 9–12, 103
  antibody fragments 42, 49
  cancer therapy 121, 122
  cardiovascular diseases 150
  effector functions 55, 57, 65, 66–7
  generation of MAbs 20, 25
  inflammatory diseases 154–5, 156
  multiple specificities 53
  overcoming immunogenicity 29–31, 35, 37, 38
  pharmacokinetics 68, 73, 74, 75
  production of MAbs 162, 166, 170, 177, 182
COS cells 29, 35, 162–3
costs 28, 42, 105, 111, 192, 193
  production of MAbs 161, 173, 174, 176, 179
cows 175
Cre-*loxP* 31

cyanuric chloride 38, 39
cyclosporin A 40
cysteines 111, 167
antibody fragments 41, 43–6, 48, 49
cancer therapy 129, 131, 141
effector functions 57, 59, 63, 65
multiple specificities 53
cytokines 20, 28, 71, 99, 120
cancer therapy 125, 126, 127
effector functions 65, 67
infectious diseases 148, 149
inflammatory diseases 151–3, 155–8
cytomegalovirus infection 99, 162–3, 164
cytoplasm 187–8, 189
cytosol 187–8
cytotoxic drugs 120, 122, 125, 139–42, 144, 145

dermatitis and eczema 158
dextran 38, 40, 54, 143
diabetes 156
diabodies 46–9, 58, 74, 127, 171
multiple specificities 51, 54
digoxin 91, 92
dihydrofolate reductase (DHFR) 164–5
dimers and dimerisation 48, 49, 53, 67, 72, 92
antibody fragments 46–50
cancer therapy 122
effector functions 57
multiple specificities 53, 54
pharmacokinetics 73
diptheria toxin (DT) 139, 140, 158
disulphide bonds 4–7, 9, 11, 72
antibody fragments 40, 41, 43, 45, 46, 48, 49
cancer therapy 129, 140, 141, 143, 145
effector functions 55, 56, 59, 63, 65
multiple specificities 53
production of MAbs 169, 170, 171
RAID 111
disulphide-linked Fvs (dsFvs) 45, 46, 66, 169
cancer therapy 141
dot blots 18
drug abuse 77, 90–2, 93, 104
drug conjugates 142–4
drugs 28, 51, 71, 119, 120
cytotoxic 120, 122, 125, 139–42, 144, 145
effector functions 54, 58

effector cell targeting 125–7
effector functions 4–8, 28, 29, 40, 54–67, 119–20, 193
cancer therapy 121, 122
infectious diseases 149
inflammatory diseases 154, 155
pharmacokinetics 66, 75
production of MAbs 162, 166, 175
elution 178–81, 183–5
embryonic stem cells 21, 31
endothelium 151–4
endosome 69
endotoxins 148–9, 185
environmental and food monitoring 91, 93–4, 104, 106
enzyme linked immunoadsorbent assay (ELISA) 18, 83–7, 89, 92, 101, 102
enzyme multiplied immunoassay technique (EMIT) 90
enzymes 120, 187, 189, 191
antibody fragments 41–2, 46
cancer therapy 139, 142, 144, 145
effector functions 54, 58, 62, 63, 65, 67
generation of MAbs 16, 17–18
immunoassays 83–90, 92
immunocytochemistry 98
immunosensors 94–5
production of MAbs 163, 164–5, 171, 173, 182
western blotting 100
eosinophils 125, 145, 159
epidermal growth factor (EGF) 121
epithelial cells 11, 57
epithelial membrane antigen (EMA) 124
epitopes 2, 26, 31, 188
cancer therapy 123
generation of MAbs 15, 19, 25–6
immunoassays 78, 79, 81, 91, 92
immunocytochemistry 97
immunopurification 104
immunosensors 95
infectious diseases 147–8
inflammatory diseases 153, 156
western blotting 100
Epstein-Barr virus 19, 20
Escherichia coli 167–72
antibody fragments 43, 45, 50
cancer therapy 127, 139, 141, 145
effector functions 67
immunoassays 85, 88, 89
multiple specificities 53, 54
overcoming immunogenicity 35
phage display 22, 24
production of MAbs 163, 167–72, 181–4
western blotting 101, 102
ethylene glycol 178
eukaryotics 172–3
exons 11–13
expanded bed chromatography 180–1, 183
experimental autoimmune encephalomyelitis (EAE) 155
expression systems 25, 28–9, 43, 161–7
cell culture 175–6
E. coli 167–72
insect cells 175
mammalian cells 161–7
plants 173–4
stable 162, 163–7
transgenic animals 174–5
transient 162–3
extracorporeal immunoadsorption 74

Fab (fragment antigen binding) 5–6, 40–2, 43–5, 50, 187
cancer therapy 122, 130
cardiovascular diseases 149–50
fusion proteins 65, 66, 67
immunoassays 88–9
immunocytochemistry 97
infectious diseases 147–8
overcoming immunogenicity 35, 37, 38
phage display 23, 24, 25
pharmacokinetics 68, 72

production of MAbs 167–71, 173, 174, 181–2
prospects 191
RAID 110, 114, 116, 117
structural biology 106
western blotting 101
Fab′ 42, 43–5, 48, 50
cancer therapy 126–7, 129, 131, 134–5, 141
cardiovascular diseases 151
effector functions 65, 66
immunoassays 88
immunopurification 104
multiple specificities 51, 53
overcoming immunogenicity 35
pharmacokinetics 73
production of MAbs 163, 167, 169–72, 174, 183–4
prospects 188
RAID 109–11, 117, 118
western blotting 101
$F(ab')_2$ 5, 40, 41, 42, 43, 50
cancer therapy 129, 130, 135, 136, 138, 144, 145
effector functions 59, 65, 66, 67
immunoassays 85
immunocytochemistry 97
inflammatory diseases 157, 158
multiple specificities 53, 54
overcoming immunogenicity 35, 37
pharmacokinetics 68, 71–4
production of MAbs 167, 181, 182
RAID 109
western blotting 101
Fc (fragment crystaline) receptors 4–8, 75, 120, 189
antibody fragments 40, 41
cancer therapy 122, 125, 130
cardiovascular diseases 150
effector functions 54–7, 63, 65, 66
immunoassays 85
immunocytochemistry 97
immunopurification 103
infectious diseases 148
inflammatory diseases 154–6, 158, 159
pharmacokinetics 68, 69, 71, 73, 75
production of MAbs 162, 176, 177, 182
FcRn (neonatal Fc receptor) 69, 73, 74
Fd 43, 45, 167
Fd′ 43, 53, 167
fertility testing 90–1
fibrinogen 149, 150
ficin 41, 42
field-effect transistor devices (immunoFET) 94
FLAG system 104, 184
flow cytometry 99
fluorescein 59, 87
fluorescence 18, 59, 63
flow cytometry 99
immunoassays 83, 87–8, 89, 90–2
immunocytochemistry 98
immunosensors 95
fluorescence activated cell sorting (FACS) 99
fluoroimmunoassay 89, 91
formazan 86, 87
fractionation on protein A chromatography 52
Freund adjuvant 16–17
fungi 1, 173
fusion 29, 85
antibody fragments 48, 49, 50
effector functions 58, 65–7
generating MAbs 15–17, 19, 23–4
multiple specificities 51, 53, 54
production of MAbs 163
fusion proteins 65–7, 104, 188
cancer therapy 127, 141, 142
cardiovascular diseases 150–1
immunoassays 85, 87, 88–9, 92
production of MAbs 167–9, 173
western blotting 101–2
Fv 40, 42–3, 45–8, 49–50, 106
cancer therapy 129, 130
immunocytochemistry 97
immunopurification 105
immunosensors 95
overcoming immunogenicity 35, 37
pharmacokinetics 72, 73
production of MAbs 167–70, 183, 184
RAID 109
see also scFv

galactose 75, 166–7
galactosylation 75
gel filtration 176–7, 180–3, 185
gelonin 66, 67
genes and genetic engineering 1, 2, 11–13, 18, 27–75
amplification 164–5
antibody fragments 40–50
cancer therapy 123
effector functions 54–67
isolation in variable regions 28–9
multiple specificities 50–4
overcoming immunogenicity 29–40
phage display 21–6
pharmacokinetics and biodistribution 67–75
production of MAbs 162–5, 167–8, 170–1, 174–5, 183
prospects 187–93
rescue 20
transgenic animals 20–1, 174–5
glutamine 24, 164–6
glycoproteins 130, 135
glycosylation 9, 11, 35, 73, 101
cancer therapy 121, 141
effector functions 63, 65
production of MAbs 162, 166–8, 173–5
goats 175
graft versus host disease 36, 37, 70, 157–8
granulocytes 145, 153
growth factors 121, 146
guinea pigs 155, 159, 182

half-life 38, 75, 128
pharmacokinetics 68–75
RAID 108, 110
haloacetyl 59, 60
hamsters 15
HAT selection 16, 18
heart disease 107, 116, 120, 149–51, 154–5
heavy chains 4–7, 9–12, 28
antibody fragments 42, 43, 49
effector functions 55–7, 65, 67

generation of MAbs 17, 21, 25, 26
immunoassays 85
inflammatory diseases 155
multiple specificities 51–4
overcoming immunogenicity 31, 33, 34
pharmacokinetics 72, 74
production of MAbs 162–4, 168, 172–5, 182–3
prospects 190, 191
RAID 116
western blotting 101
hepatitis (B and C) 85, 147, 191
herbicides and pesticides 91, 93–4, 106
herpes simplex viruses 147
heterobifunctional reagents 59, 61, 63
heterodimers 53, 54
heterohybridomas 19, 20
heteromyelomas 19
hinge regions 4–5, 7, 9–11
antibody fragments 40–1, 43, 48, 49
cancer therapy 129, 131
effector functions 55, 56, 59, 63, 67
immunopurification 104, 105
multiple specificities 53
pharmacokinetics 68, 72
production of MAbs 167
RAID 109, 111
histamine 158
histidines 59, 65, 184
Hodgkin lymphoma 126
homobifunctional reagents 59, 61, 63
homodimers 53
homogeneous assays 89
hormones 88, 89, 92, 93, 95
horseradish peroxidase 83, 85, 94, 98, 100–1
human anti-mouse antibody (HAMA) 18, 69, 138
overcoming immunogenicity 29, 31, 37, 40
human chorionic gonadotrophin (hCG) 90–1, 94
human cytomegalovirus (hCMV) 162–3, 164
human immunodeficiency virus (HIV) 26, 101, 106, 147–8
non-tumour RAID 118
overcoming immunogenicity 34
pharmacokinetics 75
prospects 188, 190
humanised antibodies 119, 120
cancer therapy 121–3, 126, 130, 134, 136, 142, 144–5
effector functions 54, 55, 56, 57
infectious diseases 147
inflammatory diseases 153, 155, 156, 157–9
multiple specificities 52
overcoming immunogenicity 29–37
pharmacokinetics 69–72
production of MAbs 164, 167, 170–2, 175, 182
hybrid hybridomas 14–18
cancer therapy 126
cardiovascular disease 151
generation of MAbs 14–18, 19–21, 23, 26
inflammatory disease 157
multiple specificities 51, 52
overcoming immunogenicity 31
production of MAbs 161–4, 175–6, 181
variable region genes 28, 29
hydrophobic interaction chromatography 176, 180, 181, 183, 192
hypermutation 2, 13
hypervariable regions 5–6
hypogammaglobulinemia 68

imidoesters 59–60
immobilisation
cancer therapy 136
immunoassays 78–9, 81, 85–8, 89–90
immunopurification 103–6
immunosensors 94, 95–6
production of MAbs 176, 178–9, 181–4
immortalisation 16, 19, 23
immune response 1, 2, 119, 120
cancer therapy 122–4, 134, 142, 145
generation of MAbs 16–18
inflammatory disease 156, 158
overcoming immunogenicity 29, 33, 37, 38, 40
production of MAbs 167
immunisation 119, 123
generation of MAbs 15–20, 22, 23, 26
immunoadhesins 75
immunoaffinity chromatography 24, 102–5, 136
immunoassays 77–92
immunoaffinity chromatography 104
western blotting 102
immunoblotting 100–2
immunochromatographics 90
immunoconjugates 128, 132, 133, 134
immunocytochemistry 87, 97–9
immunofluorescence 18
immunogenicity 28, 29–40, 193
antibody fragments 37–8, 49
cancer therapy 32, 121, 123, 126, 142, 145
cardiovascular diseases 150
chemical modification 38–40
chimeric and humanised antibodies 29–37, 38
effector functions 55
inflammatory diseases 156
multiple specificities 53
production of MAbs 174
RAID 108, 116
immunoglobulin (Ig) 2–15, 27, 28, 31, 98, 183, 189
antibody fragments 46, 50
effector functions 54–8
generation of MAbs 15, 19, 20–1
IgA 4, 6, 9–11, 13, 42, 57
production of MAbs 173–5, 182
IgA1 9–10, 56
IgA2 9–10, 56, 57
IgD 4, 6, 9–11, 42, 56
IgE 4, 6, 9–11, 13, 42, 105, 158–9
IgG 4–15, 21, 85, 96, 177, 187
antibody fragments 40–2, 49, 50
cancer therapy 122, 127, 129–31, 141, 144
cardiovascular diseases 150
effector functions 54–9, 62, 63, 65, 66
immunopurification 103, 104, 105
infectious diseases 147–8
inflammatory diseases 153, 158
multiple specificities 51, 52–4
overcoming immunogenicity 30, 35, 37, 38, 40
pharmacokinetics 68–75
production of MAbs 162–4, 166–9, 173–5

purification 176, 177, 178–82
RAID 109–11, 117, 118
western blotting 101, 102
IgG1 6–9, 12–13, 41, 42, 153
cancer therapy 121, 122, 129
effector functions 54–5, 56
pharmacokinetics 68, 71, 73
purification 177–9
IgG2 6–9, 13, 154, 177
cancer therapy 129
effector functions 54, 55, 56
pharmacokinetics 68
IgG2a 6–8, 42, 55, 165, 177, 181
cancer therapy 121, 122, 124
IgG2b 6–8, 42, 55, 177
cancer therapy 121, 122
pharmacokinetics 73
IgG2c 177
IgG3 6–9, 11, 13, 42, 177
cancer therapy 121
effector functions 54–5, 56
pharmacokinetics 68, 73
IgG4 6–9, 42, 101, 166, 177
cancer therapy 129
effector functions 54, 55, 56
inflammatory diseases 153–6
pharmacokinetics 68, 71, 73
IgH 11
IgM 4, 6, 9–11, 13–15, 19–21, 149, 181
antibody fragments 42
anti-idiotype response 37
cancer therapy 122
effector functions 56, 57, 58
pharmacokinetics 68
production of MAbs 166, 172
purification 181, 182
immunohistochemistry 18, 50
immunoliposomes 143
immunopurification 102–6
see also purification
immunoradiometric assay (IRMA) 79–83, 85, 91
immunosensors 92–7
electrochemical 93, 94–5
mass detecting 93–4
optical 93, 95–7
immunosuppression 29, 35, 40
immunotherapy 122, 174
immunotoxins 139–42, 148, 157–8, 168
cancer therapy 132, 139–42, 143–4, 147
inclusion bodies 167, 169
infectious diseases 107, 116, 117–18, 147–9
inflammatory bowel disease 151, 153–4
inflammation and inflammatory diseases 122, 147–9, 151–9
imaging agents 107
RAID 116, 117–18
insects 175
integrins 188
intercellular adhesion molecule-1 (ICAM-1) 55, 118
interleukins (IL) 34, 99, 104, 148, 167, 187
cancer therapy 121, 126, 127
fusion proteins 66, 67
inflammatory diseases 151, 153, 156–7, 159
iodogen 80

J chain 9–11
jellyfish 88

keyhole limpet haemocyanin 15
kidneys 72, 120, 130, 135, 136, 189
RAID 114

lactose 168, 171
lanthanide chelates 87–8
leucine zippers 48, 49, 51, 53
leukemias 120, 135, 136, 137, 157
leukocytes 7, 117–18, 151–5
ligands 96, 153
cancer therapy 121, 133–7, 140
RAID 111, 112, 113, 115
light chains 4–13, 28, 29, 85, 190, 191
antibody fragments 42, 43, 45, 46, 48
effector functions 58, 63, 65, 67
generation of MAbs 17, 21, 25, 26
multiple specificities 51, 52, 53
overcoming immunogenicity 31, 33, 34
production of MAbs 162–4, 167–8, 172–5, 182, 184
linkers 30, 54, 102
antibody fragments 44–9
cancer therapy 130, 131, 136, 141, 142
effector functions 58–60, 63, 65
lipids 89, 179
liposomes 89, 143, 188
liquid tumour types 138–9
liver 18, 74, 75, 136, 144
cancer therapy 136, 141, 144
RAID 109, 113, 114
lung cancer 107, 135, 138, 143
RAID 108, 109, 111, 115
lymphocytes 2, 19, 20, 118, 189
therapeutic applications 122, 148, 151, 153, 156
see also B cells; T cells
lymphomas 108, 120, 134, 137, 141–2
Hodgkin 126
non-Hodgkin 35, 36, 122, 137, 156
lysine 38, 42, 59, 60, 102
effector functions 59, 60, 63, 65
immunoassays 79, 83
lysis and lysing 2, 6, 89, 121–2, 169, 171, 189
cancer therapy 126–7
effector functions 54, 55, 57, 58
thrombi 150
lysozymes 46, 58, 69, 123, 190, 191

macrophages 4, 7, 125, 145, 151, 153
major histocompatibility complex (MHC) class II molecules 2, 15, 127, 156
maleimide 38, 39, 53, 104, 113
antibody fragment 43, 44, 50
cancer therapy 129, 130, 134
effector functions 59–60
mannose 181
mast cells 158–9
mediate signal transduction 119
melanoma 32, 70, 108, 127
memory cells 2, 3, 13, 15, 157
meningococcal meningitis 148
metastatic carcinoma 99
methotraxate 164

mice 11, 41–2, 51, 119
cancer therapy 120–3, 126, 130–1, 135–7, 143–4, 147
cardiovascular diseases 150
effector functions 54, 55, 56
generation of MAbs 14–15, 17–21, 25–6
immunocytochemistry 97, 98
immunoglobulins 6–8
infectious diseases 148, 149
inflammatory diseases 153, 155, 156
overcoming immunogenicity 29–37, 38, 40
pharmacokinetics 69, 71–4, 75
production of MAbs 161, 165–6, 172–3, 175–82
transgenic 20–1, 31
microbial hosts 172–3
milk 174
monkeys 31, 35, 72, 155, 156, 157, 162
monocytes 7, 122, 125, 145
monomers 46, 47, 58
mononuclear phagocytic cells 126
multiple sclerosis 151, 155, 156
multivalent antibody fragments 48–50
mutagenesis 6, 26, 148, 155, 190, 192
overcoming immunogenicity 30, 34
pharmacokinetics 68, 73
mycobacteria 17
mycoplasma 161
mycotoxins 91
myeloid leukemia 36, 37, 70, 108, 137, 144
myelomas 15–20, 42, 68, 122, 162–7

natural killer (NK) cells 7, 20, 125, 126, 145
nematodes 189
neoplastic meningitis 135
neuroblastoma 32, 108, 127
non-Hodgkin lymphoma 35, 36, 122, 137, 156
non-isotopic immunoassays 83–5
nonlymphoid cells 162
NSO cells 165–6, 167, 178

oligonucleotides 23, 28, 30, 34, 89
ovarian cancer 32, 70, 75, 107, 108–9, 187
therapy 135, 138, 143

pancarcinoma 108
papain 40–2, 182
pepsin 41–2, 182
peptides 2, 104, 151, 175, 184, 190, 191
antibody fragments 45, 46, 48, 49
cancer therapy 127, 141
effector functions 54, 67
generation of MAbs 15, 23
multiple specificities 53
RAID 117
Peptostreptococcus magnus protein L 182–3
periodate oxidation 63, 64, 65
peroxidase 50
pH 41, 59, 94, 102
pharmacokinetics 69, 74
production of MAbs 169, 176, 178–80, 184
phage display 21–6, 28, 54, 106, 158, 192
cancer therapy 121, 146
immunoassays 92
immunosensors 95
infectious diseases 147, 148
pharmacokinetics 73
RAID 109
phagocytosis 4, 7, 8, 119, 120, 148, 153
effector functions 54, 58
pharmacokinetics 28, 40, 67–75, 165, 166, 193
cancer therapy 70, 75, 126, 129, 130
effector functions 66, 75
RAID 108
phenotypes 189
phosphate starvation 171
plants 67, 173–4, 189, 192
toxins 132, 139, 141
plasmids 24, 188
production of MAbs 162–3, 168, 171, 172, 174
plasminogen 150–1
polyacrylamide gel electrophoresis (PAGE) 100
SDS-Page 50, 56, 178, 183, 184
polycations 188
polyclonal antisera 77–9, 97, 100, 119, 147
poly(ethylene glycol)(PEG) 15, 17, 59, 73
cancer therapy 142, 143
overcoming immunogenicity 29, 38, 39, 40
polylysine 188
polymerase 34
chain reaction (PCR) 22, 23, 29, 30, 34–5, 89
polymers and polymerisation 9, 11, 100
cancer therapy 122, 143
effector functions 57, 58
overcoming immunogenicity 29, 38
polymorphonuclear neutrophils (PMN) 125
polypeptides 4, 9, 11, 45, 48, 53, 100
positron emission tomography (PET) 110, 114, 136
pregnancy testing 90–1
primary recovery 176, 180
primatised antibody 31, 156
primer annealing 23, 29
prodrugs 75, 120, 190
ADEPT 139, 144–6
promoters/enhancers 162–5, 168, 171, 173
prostate cancer 91, 95, 107, 108
proteases 172, 173
protein disulphide isomerase (PDI) 171
proteins 1, 2, 4–5, 10, 12, 65–7, 106, 187–8
antibody fragments 42, 43, 45, 46, 49, 50
cancer therapy 127, 130, 137, 139–43, 145–6
cardiovascular diseases 150–1
effector functions 56, 58, 59, 61, 63, 65–7
generation of MAbs 14–15, 17, 23–4
immunoassays 83, 85, 87–9, 92
immunopurification 102–5
immunosensors 94
infectious diseases 147–8
inflammatory diseases 153
multiple specificities 52, 53
overcoming immunogenicity 30, 33, 34, 38, 40
pharmacokinetics 67, 68, 71, 73, 75
production of MAbs 162–3, 167–85
purification 102–5, 176, 177–85
RAID 111
western blotting 100, 101–2
see also fusion proteins
proteolysis 5, 11, 15, 38, 40–2, 129
antibody fragments 40–2, 43, 50
effector functions 55, 63, 65

  multiple specificities 53
  overcoming immunogenicity 37
  pharmacokinetics 73
  reverse 63, 65, 67
Pseudomonas exotoxin (PE) 20, 139–42
psoriasis 151, 154, 156
pulmonary embolism 92
purification 42, 51, 102–6, 176–85
  production of MAbs 161, 169, 175, 176–85
  prospects 190, 192
  therapeutic use 185
pyridyl disulphides 59, 60

quadromas 51

rabbits 15, 38, 151, 173, 182
radioimmunoassay (RIA) 18, 77, 78–9, 91–2
radioimmunoconjugates 126
radioimmunodetection (RAID) 106, 107–16
  cancer therapy 108, 109, 114–15, 136
  non-tumour 116–18
radioimmunoguided surgery (RIGS) 106, 116
radioimmunoscintigraphy (RIS) 106, 114, 116
radioimmunotherapy (RIT) 105, 113, 128–39
radioiodination 75, 82, 110–11, 133
radioisotopes 58, 120, 132–6
  cancer therapy 123, 128, 130, 131, 132–6, 138
radiolabelling
  antibody fragments 44, 45, 48
  cancer therapy 122, 126, 128, 131, 133–8
  effector functions 59, 63, 65, 67
  immunoassays 78, 79, 83
  immunopurification 105
  in vivo diagnostics 106
  pharmacokinetics 72, 74
  RAID 107, 110–15, 117
  RIGS 116
radionuclides 28, 67, 71
  cancer therapy 128, 135, 136
  RAID 110, 111, 114
random coupling 102
rats 15, 17, 51, 55, 119
  cancer therapy 122, 135, 141
  production of MAbs 161, 176, 177
recombinant antibodies 21, 29, 40, 42–50, 119–20, 192
  cancer therapy 122, 126, 129, 140–2, 144
  effector functions 54–5, 56–7
  immunoassays 88, 90
  infectious diseases 148
  inflammatory diseases 153, 158
  multiple specificities 52, 53
  overcoming immunogenicity 31, 37
  pharmacokinetics 72
  production of MAbs 161–5, 174–5, 177, 181, 183
  RAID 111
  western blotting 101, 102, 104, 105
redox 169
respiratory syncytial virus (RSV) 147–8
retroviruses 189
rheumatoid arthritis 35–7, 100–1, 118, 153–4
  therapy 151, 153–8
rhinitis 158
ribosomes 140, 168–9
ricin 139–43, 147, 148, 157–8

salmon calcitonin 91
saporin 141
sarcoidosis 99
SDS-Page 50, 56, 178, 183, 184
selectins 151–4
selective precipitation 181
selectively infective phage (SIP) 24
self antigens 15, 26, 152
sensitivity 85–9
  flow cytometry 99
  immunoassays 85–9, 90, 91
  immunosensors 93–4
  RAID 115, 117
  RIGS 116
  western blotting 100
sepharose 79
sequential affinity purification 51
serine 56, 57
severe combined immune deficiency (SCID) mice 20
sheep 175
shock 148, 149, 153
side-effects 142, 150, 155, 157
signal transduction 121, 123
single chain Fv (scFv) 45–6, 47–50, 187–9
  cancer therapy 127, 129, 141, 142, 144, 145
  cardiovascular diseases 150
  fusion proteins 66, 67
  immunoassays 85, 88, 89
  immunopurification 104–6
  inflammatory diseases 154
  multiple specificities 51, 53, 54
  phage display 23, 24
  pharmacokinetics 72, 73
  production of MAbs 167–75, 182–4
  RAID 109–11, 117
  RIGS 116
  western blotting 101
single domain antibodies (dAbs) 46
single photon emission computerised tomography (SPECT) 106
site-specific attachment 63–5
size of antibodies 15, 27–8, 54, 66, 72, 176
  RAID 108
  western blotting 100
sodium dododecyl sulphate (SDS) 100, 101
  SDS-Page 50, 56, 178, 183, 184
sodium iodide 79, 82
somatic mutation 21
specificity 1, 11, 27, 28, 42, 58
  cancer therapy 123, 126, 127
  cardiovascular diseases 151
  generation of MAbs 14–15, 21, 24, 26
  immunoassays 78, 84, 85, 91, 92
  infectious diseases 147
  inflammatory diseases 157
  multiple 50–4
  overcoming immunogenicity 35
  prospects 187–9, 192
  RAID 107, 108
  structure of antibodies 2–4, 6
stable expression system 162, 163–7
staphylococcal nuclease 65, 66
staphylococcal protein A 68
  production of MAbs 167, 176–83, 185

streptavidin 24, 49–50, 74, 98, 137, 184
immunoassays 83, 88–9
RAID 114, 115, 116
Streptococcal protein G 176, 177, 179–83
superantigens 127
surface plasmon resonance (SPR) 95–6
systemic lupus erythematosus 154, 156
systemic vasculitis 36, 37

T cells 2, 148, 155–8, 189
cancer therapy 123, 125–7, 145
generation of MAbs 15, 20, 21
inflammatory diseases 151–3, 155
Tac-receptors 36, 37, 157
targeted therapy 109, 110, 114–16, 125–7, 146–7
tetrameric fragments 49–50
thioesters 59
thiols 38, 53
antibody fragments 41, 43, 44
cancer therapy 129, 135
effector functions 59, 60, 63
immunopurification 104, 105
RAID 111, 112
thiophilic adsorption 180
thrombi 117, 150–1, 191
thyroid cancer 132
TNF 37, 148, 149, 151, 153–4
tobacco mosaic virus 189
toxins 28, 51, 54, 91, 119, 120
diptheria 139, 140, 158
pharmacokinetics 71, 75
plants 132, 139, 141
see also endotoxins; immunotoxins
transgenic organisms 20–1, 31, 174–5, 192, 193
transient expression systems 162–3
transplant rejection 99, 120, 137, 151, 155, 156–7
tresyl chloride 38, 39
triabodies 47, 48
trimers 47, 48, 50
triomas 19, 51
trispecificity 27, 53
trypsin 42
tryptophan 59, 65
tumour-associated antigens 107–9, 123, 128, 146
tumours 37, 46, 50, 119, 120–47, 187–9, 191
effector functions 54, 57
generation of MAbs 15, 17, 19, 26
immunocytochemistry 97, 99
immunopurification 104, 105
overcoming immunogenicity 37, 40
pharmacokinetics 71, 72, 74, 75
production of MAbs 176
RAID 107–16
RIGS 116
see also cancer
tyrosine 59, 79–80, 187

unmodified antibodies 121–3

vaccination and vaccines 120, 123, 147
valency 27, 28, 66
variable domains 4–6, 12, 28–9, 105, 188, 191–2
antibody fragments 45–8
cardiovascular diseases 150
generation of MAbs 20–1, 23, 25–6
multiple specificities 54
overcoming immunogenicity 30, 31, 33, 34–5, 38
purification 182, 183
vascular leak syndrome 142
vascular targeting 146–7
vectors 24, 188
production of MAbs 165, 167, 169, 175
veneering 34
vesicular stomatitis virus (VSV) 148
vinylsulphone 59
viruses 1, 17, 19, 93, 127, 147–8
production of MAbs 161, 162–3, 175, 185
prospects 188–9, 191

western blots 18, 100–2

xanthine 163

yeasts 172–3
yield of MAbs 19, 161–71, 176, 179–80, 183, 193